VOLUME ONE HUNDRED AND FIFTEEN

# Advances in
# CANCER RESEARCH

VOLUME ONE HUNDRED AND FIFTEEN

# ADVANCES IN CANCER RESEARCH

## Applications of Viruses for Cancer Therapy

Series Editors

**KENNETH D. TEW**

*Cell and Molecular Pharmacology, John C. West Chair of Cancer Research, Department of Cell and Molecular Pharmacology & Experimental Therapeutics at MUSCCharleston, South Carolina, USA*

**PAUL B. FISHER**

*Professor and Chair, Department of Human & Molecular Genetics,*
*Director, VCU Institute of Molecular Medicine,*
*Thelma Newmeyer Corman Chair in Cancer Research, VCU Massey Cancer Center,*
*Virginia Commonwealth University, School of Medicine, Richmond, Virginia, USA*

Edited by

**DAVID T. CURIEL**

*Director, Cancer Biology Division, Director, Biologic Therapeutics Center*
*Department of Radiation Oncology, Washington University School of Medicine*
*St. Louis, MO, USA*

**PAUL B. FISHER**

*Professor and Chair, Department of Human & Molecular Genetics,*
*Director, VCU Institute of Molecular Medicine,*
*Thelma Newmeyer Corman Chair in Cancer Research, VCU Massey Cancer Center,*
*Virginia Commonwealth University, School of Medicine, Richmond, Virginia, USA*

ELSEVIER

AMSTERDAM • BOSTON • HEIDELBERG • LONDON
NEW YORK • OXFORD • PARIS • SAN DIEGO
SAN FRANCISCO • SINGAPORE • SYDNEY • TOKYO
Academic Press is an imprint of Elsevier

Academic Press is an imprint of Elsevier
525 B Street, Suite 1900, San Diego, CA 92101-4495, USA
225 Wyman Street, Waltham, MA 02451, USA
32 Jamestown Road, London, NW1 7BY, UK
The Boulevard, Langford Lane, Kidlington, Oxford, OX51GB, UK
Radarweg 29, PO Box 211, 1000 AE Amsterdam, The Netherlands

First edition 2012



ISBN: 978-0-12-398342-8
ISSN: 0065-230X

For information on all Academic Press publications
visit our website at store.elsevier.com

Printed and bound by CPI Group (UK) Ltd, Croydon, CR0 4YY

Transferred to digital print 2012

# CONTENTS

# CONTRIBUTORS


**Ramon Alemany**
Translational Research Laboratory, Institut Català d'Oncologia-IDIBELL, L'Hospitalet de Llobregat, Barcelona, Spain

**Matthew S. Beatty**
Division of Cancer Biology, Department of Radiation Oncology, Washington University School of Medicine in St. Louis, St. Louis, Missouri, and Division of Molecular and Cellular Pathology, Department of Pathology, University of Alabama at Birmingham, Birmingham, Alabama, USA

**Donald J. Buchsbaum**
Department of Radiation Oncology, University of Alabama at Birmingham, Birmingham, Alabama, USA

**Vincenzo Cerullo**
Laboratory of Immunovirotherapy, Division of Biopharmaceutics and Pharmacokinetics, Faculty of Pharmacy, and Cancer Gene Therapy Group, Molecular Cancer Biology Program & Transplantation Laboratory & Haartman Institute, University of Helsinki, Helsinki, Finland

**David T. Curiel**
Division of Cancer Biology, Department of Radiation Oncology, Washington University School of Medicine in St. Louis, St. Louis, Missouri, USA

**Swadesh K. Das**
Department of Human and Molecular Genetics, and VCU Institute of Molecular Medicine, School of Medicine, Virginia Commonwealth University, Richmond, VA, USA

**Rupesh Dash**
Department of Human and Molecular Genetics, School of Medicine, Virginia Commonwealth University, Richmond, VA, USA

**Tanja D. de Gruijl**
Department of Medical Oncology, VU University Medical Center/Cancer Center Amsterdam, Amsterdam, The Netherlands

**Paul Dent**
VCU Institute of Molecular Medicine; VCU Massey Cancer Center, and Department of Neurosurgery, School of Medicine, Virginia Commonwealth University, Richmond, VA, USA

**Paul B. Fisher**
Department of Human and Molecular Genetics; VCU Institute of Molecular Medicine, and VCU Massey Cancer Center, School of Medicine, Virginia Commonwealth University, Richmond, VA, USA

**Akseli Hemminki**
Cancer Gene Therapy Group, Molecular Cancer Biology Program & Transplantation Laboratory & Haartman Institute, University of Helsinki, Helsinki, Finland

**Ziyue Karen Jiang**
Department of Molecular and Medical Pharmacology, and Institute of Molecular Medicine, University of California Los Angeles, Los Angeles, California, USA

**Sergey A. Kaliberov**
Department of Radiation Oncology, Washington University in St. Louis, St. Louis, Missouri, USA

**Anniina Koski**
Cancer Gene Therapy Group, Molecular Cancer Biology Program & Transplantation Laboratory & Haartman Institute, University of Helsinki, Helsinki, Finland

**Devanand Sarkar**
Department of Human and Molecular Genetics; VCU Institute of Molecular Medicine, and VCU Massey Cancer Center, School of Medicine, Virginia Commonwealth University, Richmond, VA, USA

**Siddik Sarkar**
Department of Human and Molecular Genetics, School of Medicine, Virginia Commonwealth University, Richmond, VA, USA

**Makoto Sato**
Department of Molecular and Medical Pharmacology, and Institute of Molecular Medicine, University of California Los Angeles, Los Angeles, California, USA

**Karoly Toth**
Department of Molecular Microbiology and Immunology, Saint Louis University School of Medicine, St. Louis, Missouri, USA

**Markus Vähä-Koskela**
Cancer Gene Therapy Group, Molecular Cancer Biology Program & Transplantation Laboratory & Haartman Institute, University of Helsinki, Helsinki, Finland

**Rieneke van de Ven**
Department of Medical Oncology, VU University Medical Center/Cancer Center Amsterdam, Amsterdam, The Netherlands, and Laboratory of Molecular and Tumor Immunology, Robert W. Franz Cancer Research Center at the Earle A. Chiles Research Institute, Providence Cancer Center, Portland, Oregon, USA

**Xiang-Yang Wang**
Department of Human and Molecular Genetics; VCU Institute of Molecular Medicine, and VCU Massey Cancer Center, School of Medicine, Virginia Commonwealth University, Richmond, VA, USA

**William S.M. Wold**
Department of Molecular Microbiology and Immunology, Saint Louis University School of Medicine, St. Louis, Missouri, USA

**Lily Wu**
Department of Molecular and Medical Pharmacology; Institute of Molecular Medicine, and Department of Urology, David Geffen School of Medicine, University of California Los Angeles, Los Angeles, California, USA

CHAPTER ONE

# Cancer Terminator Viruses and Approaches for Enhancing Therapeutic Outcomes

**Swadesh K. Das**[*,†], **Siddik Sarkar**[*], **Rupesh Dash**[*], **Paul Dent**[†,‡,§], **Xiang-Yang Wang**[*,†,‡], **Devanand Sarkar**[*,†,‡], **Paul B. Fisher**[*,†,‡,1]

[*]Department of Human and Molecular Genetics, School of Medicine, Virginia Commonwealth University, Richmond, VA, USA
[†]VCU Institute of Molecular Medicine, School of Medicine, Virginia Commonwealth University, Richmond, VA, USA
[‡]VCU Massey Cancer Center, School of Medicine, Virginia Commonwealth University, Richmond, VA, USA
[§]Department of Neurosurgery, School of Medicine, Virginia Commonwealth University, Richmond, VA, USA
[1]Corresponding author: e-mail address: pbfisher@vcu.edu

## Contents



## Abstract

No single or combinatorial therapeutic approach has proven effective in decreasing morbidity or engendering a cure of metastatic cancer. In principle, conditionally replication-competent adenoviruses that induce tumor oncolysis through cancer-specific replication hold promise for cancer therapy. However, a single-agent approach may not be adequate to completely eradicate cancer in a patient because most cancers arise from abnormalities in multiple genetic and signal transduction pathways and

*Advances in Cancer Research*, Volume 115
ISSN 0065-230X
http://dx.doi.org/10.1016/B978-0-12-398342-8.00001-X

targeting disseminated metastases is difficult to achieve. Based on these considerations, a novel class of cancer destroying adenoviruses have been produced, cancer terminator viruses (CTVs), in which cancer-specific replication is controlled by the *progression-elevated gene-3* promoter and replicating viruses produce a second transgene encoding an apoptosis-inducing and immunomodulatory cytokine, either *melanoma differentiation-associated gene-7*/interleukin-24 (*mda-7*/IL-24) or *interferon-γ*. This review focuses on these viruses and ways to improve their delivery systemically and enhance their therapeutic efficacy.

## 1. INTRODUCTION

Cancer exacts an enormous global health burden, affecting every geographical region and socioeconomic level and accounting for one in eight deaths worldwide, more than HIV/AIDS, tuberculosis, and malaria combined (Cancer Figurs & Facts, 2011). Cancer is characterized by uncontrolled cell growth with frequent spread of abnormal cells from a primary tumor site to distant sites in the body, often culminating in death. The etiology of this devastating disease includes both exogenous (tobacco, infectious organisms, chemicals, radiation) and endogenous factors (inherited mutations, epigenetic changes, hormones). Although, more than 60% of all cancer deaths occur in low- and middle-income countries, developed countries are also equally affected. According to recent statistics, in 2011, diagnosis of approximately 1.5 million new cancer cases was expected in the United States and about 15 thousand affected individuals were anticipated to die every day (Cancer Figurs & Facts, 2011).

A key challenge facing oncology researchers/clinicians in developing effective anticancer therapies is the molecular heterogeneity of tumors, which frequently increase during neoplastic progression. Major advances in genomics, transcriptomics, and proteomics continue to shed light on the underlying pathophysiology of tumors resulting in a shift from traditional cancer treatment, classically based on cytotoxics or antihormone therapies, and launching the concept and implementation of targeted gene therapy. In 1964, a trio of Nobel laureates (Edward Tatum, Joshua Lederberg, and Arthur Kornberg) suggested that it would be possible to cure often-fatal genetic disorders, like cystic fibrosis, muscular dystrophy, and multiple sclerosis, by replacing the defective gene with a functional one. In the past decades, tremendous efforts have been made to translate gene therapy of cancer into the clinical arena. The efficient delivery of therapeutic genes with appropriate targeted gene expression are crucial factors for promoting clinically effective gene therapies. Two broad approaches have been used to deliver DNA to cells, namely, viral vectors and nonviral vectors. Although the nonviral methods for gene therapy provide certain advantages including simple large-scale production and low-host immunogenicity, viral vectors are naturally evolved vehicles, thereby efficiently transferring their genes into host cells. This property of viruses makes them desirable agents for engineering viral vector-based systems to deliver therapeutic genes.

## 2. VIRAL VECTORS FOR CANCER GENE THERAPY

Viruses are obligate intracellular parasites and highly evolutionary conserved biological machines that very resourcefully gain access, transfer their DNA, and exploit the host cell machinery to ensure their replication (Thomas, Ehrhardt, & Kay, 2003). Use of the viral infection pathway, while circumventing the subsequent expression of viral genes that leads to replication and toxicity through deleting all or part of the coding regions of the viral genome, embodies virus-based vectors with necessary properties for efficient gene transfer. Although a number of viruses have been developed as potential gene transfer vectors, based on the packing capacity of transgenes, host range, tropism, and inflammatory potential (Wu & Zhou, 2011), interest has been focused on a few select viruses including adenoviruses (Ads), adeno-associated viruses, retroviruses, and herpes simplex virus type-1. A brief description of various virus vectors with their advantages/disadvantages is presented in Table 1.1. Several additional virus vectors, derived from vaccinia virus, human cytomegalovirus, Epstein Barr virus, poxviruses, and foamy virus, have also been tested for gene therapy approaches in both basic and clinical research. In this review, we have integrated recent information related to Ad vector biology and methods to enhance their therapeutic potential.

## 3. ADENOVIRUSES AS VECTORS FOR CANCER GENE THERAPY

Ads, a nonenveloped DNA virus, were first isolated from adenoid tissues in 1953 (Enders et al., 1956). This virus is associated with mild upper respiratory tract infections in humans. Based on hemagglutinin properties, 51 serotypes of Ads have been identified which are classified into six subgroups (A–F). Because of its association with mild disease and relatively easy

**Table. 1.1** General characteristic of few commonly used viral vectors in cancer gene therapy

| Virus | Family | Genome/insert size/ particle size/envelope | Advantages | Disadvantages |
|---|---|---|---|---|
| Adenovirus | Adenoviridae | dsDNA, 36 kb/7.5 kb/ 90 nm/No | Infects dividing and nondividing cells; nontoxic to infecting cells; efficient gene transfer; high-viral titers achievable | Small insert size; postinfection of immune response leading to decreased infectivity to desired cells |
| Herpes simplex virus | Herpesviridae | dsDNA, 152 kb/ 40–50 kb/ 120–200 nm/Yes | Infects dividing and nondividing cells; prolonged gene expression achievable; large insert size for transgene; high-viral titers achievable; sensitive to acyclovir/ ganciclovir | Postinfection of immune response leading to decreased infectivity to desired cells; possibility of herpes encephalitis |
| Retrovirus | Retroviridae | ssRNA, 7–11 kb/8 kb/ 10 nm/Yes | High-transduction efficiency; easy to design; chronic infection; integration into host genome resulting into prolonged gene expression; viral proteins not expressed in the host, thereby limiting the chances of reconstitution of replication-competent virus | Infects dividing cells only; low titers; random site integration; possibility of latent diseases, like malignancy or immunodeficiency |
| Adeno-associated virus | Paroviridae | ssDNA, 5 kb/5 kb/ 20 nm/No | Infects dividing and nondividing cells; lack of pathogenicity; nontoxic to host cells; stable integration into the host cells at a specific site in the human chromosome 19 | Low insert size; provoke humoral immunity on infection; large-scale production is labor intensive |

genome to manipulate compared with other Ad serotypes, human serotype 5 (Ad.5) of species C has gained common use as a vector for gene therapy.

As a gene transfer vector, Ads have discrete advantages including high-transfection efficiency irrespective of the growth status of cells, inability to integrate into the host genome, easy capsid/genome modifications for retargeting tropism to different tissues or tumors, and efficient production of high virus yields. As of March 2011, $\sim$25% of the gene therapy clinical trials included Ads as vectors (http://wiley.co.uk/genetherapy/clinica). Gendicine is the first commercialized gene therapy medicine for the treatment of cancer that uses a shuttling system based on Ad to carry a p53 gene to limit cancer growth (Zhaohui, 2005). However, the long-term clinical efficacy of this vector has been disappointing, despite over 300 clinical trials that have shown it to be well tolerated and efficient in gene transfer.

## 3.1. Biology of adenoviruses

The Ad genome is composed of linear double-stranded DNA roughly 36 kb long of which up to 30 kb can be exchanged with foreign DNA (Smith, 1995; Tooze, 1980). Gene expression is divided into four early (*E1* through *E4*) transcriptional units, functioning in part as master transcriptional regulators, and late (*L1* through *L5*) transcripts which code for structural proteins (Fig. 1.1). As a minimum, three regions of the viral genome can accept insertions or substitutions of DNA to generate helper-independent viruses: a region in *E1* (including *E1A* and *E1B*), a region in *E3*, and a short region between *E4* and the end of genome. Since the *E1A* gene is indispensable for virus replication, in the first generation of Ads, the modification of the Ad genome was based on deletion of the entire *E1A* and/or partial deletions of *E1B* and *E3* genes to produce replication-incompetent vectors providing appropriate space to insert transgenes (Fig. 1.1). However, even in the absence of *E1* gene products, there was low-level viral replication of these vectors inducing $CD4^+$- and $CD8^+$-dependent immune responses leading to a reduced duration of gene expression in *in vivo* systems (Yang, Su, & Wilson, 1996; Yang & Wilson, 1995). To decrease toxicity and prolong gene expression, newer second-generation Ad vectors have been created which lack *E2A* and contain mutations and/or deletions of the viral *E4* gene (Dedieu et al., 1997; Yang et al., 1994). In another approach, "gutted" or "gutless" Ad vectors were formed by removing the complete viral coding regions leaving only the inverted terminal repeats (ITR).

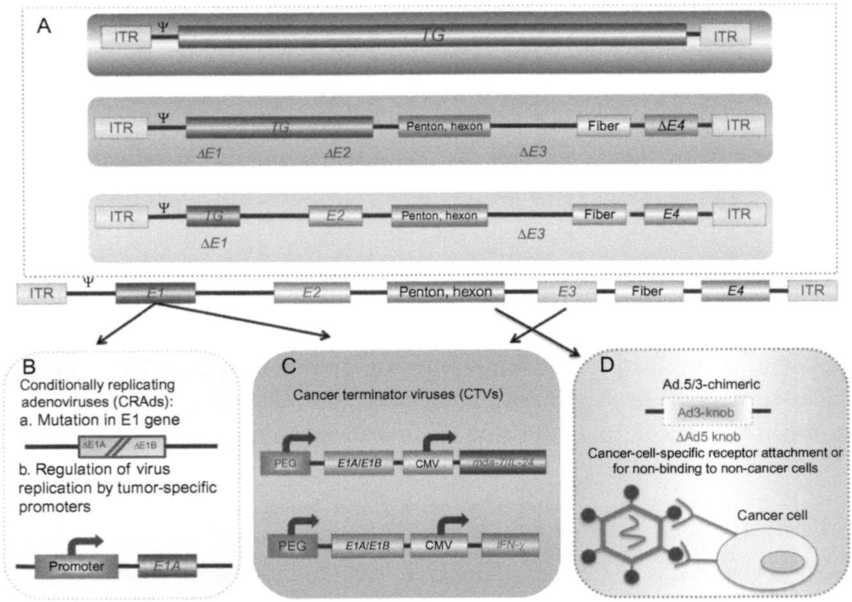


**Figure 1.1** Genetic modifications of the adenoviral genome result in oncolytic and cancer therapeutic viruses. Serotype 5 of adenovirus (Ad.5) is the most commonly used vector for gene therapy. In an effort to decrease the immune response, when delivered inside the human body, and increase cancer cell infectivity, a number of modifications in the Ad.5 genome are resulting in unique genetically modified Ad vectors (Box A). Further modifications, including mutations in the E1 region of Ad to permit cancer cell-specific replication or tumor-specific promoters driving the *E1A* gene, result in the development of conditionally replicating Ads (CRAds; Box B). Recombinant therapeutic Ads are being constructed in which nonessential genes of the Ad genome are deleted (e.g., E3 region) and replaced with therapeutic transgenes (e.g., *mda-7*/IL-24, IFN-γ). This type of "armed" oncolytic adenovirus is referred to as a cancer terminator virus (CTV; Box C). Modifications in the fiber domain of the Ad.5 genome have also been used to generate chimeric Ad.5/3 viruses (Box D). These viruses retain high infectivity in cancer cells that use the Coxsackie and adenovirus receptor (CAR) for entry of Ad.5 and also provide entry into cancer cells in a CAR-independent manner, thereby enhancing efficacy of these viruses for cancer gene therapy. (See Page 1 in Color Section at the back of book.)

Therapeutic genes transmitted by gutless virus have been successfully transported to the liver of mouse models for various human diseases, such as hemophilia A and B, obesity, familial hypercholesterolemia, diabetes, and chronic viral hepatitis, with encouraging consequences in the context of long-term expression with reduced and transient cellular immune responses (reviewed by Alba, Bosch, & Chillon, 2005). However, in most cases, stimulating a noticeable and durable antitumor response necessitates multiple administrations of replication-incompetent Ads, which can stimulate an immune response, promoting viral clearance (Sarkar, Su, & Fisher, 2006). In these contexts, conditionally replication-competent adenoviruses (CRCAs) are presently being evaluated because of their effectiveness in killing cancer cells by viral replication, thus requiring a reduced number of administrations (Fig. 1.1).

## 3.2. Conditionally replication-competent adenoviruses for cancer therapy

Considering their therapeutic potential, CRCAs exemplify a new class of anticancer agents. Ironically, although advances in genetic engineering have allowed the development of replication-defective Ad vectors and the long history of their use as simple delivery vehicles for therapeutic genes (Gomez-Manzano et al., 1996; Roth & Cristiano, 1997), replication-competent wild-type strains were actually the first to be delivered safely to patients (Smith, Huebner, Rowe, Schatten, & Thomas, 1956). Now more than half-century later, replication-competent Ads are once again in the limelight and are being analyzed as therapeutic agents (Heise, Williams, Xue, Propst, & Kirn, 1999; Kirn & McCormick, 1996). Moreover, through advanced genetic engineering CRCAs are now being used with the intention of maximizing tumor-selective replication.

The most important consideration when using a CRCA is to guarantee cancer cell-specific replication. To achieve this objective, a number of distinctive strategies have been considered and are currently being tested. One of the tactics for optimizing tumor selectivity is to delete gene functions that are critical for efficient viral replication in normal cells, but not in tumor cells. For example, the functions of the *p53* gene product are compromised both in tumors and in Ad-infected cells. Since wild-type Ads contain the *E1B-55kd* gene that is responsible for p53 binding and inactivation, it was presumed that the mutant Ad with deletion of *E1B-55kd* gene would be unable to inactivate p53 in normal cells and would thus be ineffective in replicating proficiently. In contrast, cancer cells missing functional p53 would be anticipated to be sensitive to viral replication and cytolysis. Based on this assumption, *E1B-55K*d-deficient Ads were generated (ONYX-015 or *dl1520*) that were predicted to replicate in p53 mutant cells, a common feature in many cancer cells, but not in p53 wild-type (normal) cells. Accordingly, it was predicted that p53 mutant tumor cells would be eliminated in a replication-dependent manner both *in vitro* and *in vivo*

(Bischoff et al., 1996). However, the p53-dependent specificity became controversial and several studies confirmed that numerous tumor cell lines with normal p53 gene sequences were also somewhat sensitive to the effects of *dl1520* (Hall, Dix, O'Carroll, & Braithwaite, 1998; Heise et al., 1997; Turnell, Grand, & Gallimore, 1999). Recent data confirmed that it is the loss of *E1B-55K*d-mediated late viral RNA export, rather than p53-inactivation that restricts ONYX-015 replication in primary cells (O'Shea et al., 2004). Cancer cells that support replication of ONYX-015 provide the RNA export function of the E1B-55Kd protein. In an alternate approach, exploiting the retinoblastoma protein pathway in normal cells, Fueyo et al. engineered an Ad mutant by removing the Rb-binding CR2 region of E1A (*dl922/947*) (Fueyo et al., 2000) that selectively permits Ads to propagate in human cancer cells deficient in the Rb pathway. A number of additional modifications in *dl922/947* have been created and assessed for cancer-selective activity (Johnson et al., 2002; Kim, Cho, Kim, Jung, & Yun, 2002; Lee et al., 2002).

The second strategy to develop CRAds is to limit the expression of *E1A* gene product to tumor tissues through the use of tumor- or tissue-specific promoters. Ideally, when a specific promoter controls critical regulators of viral replication, replication would be limited to tissues where this particular promoter is active. *E1A*, as the key regulator of Ad replication by regulating S-phase entry, offers the natural choice to control virus replication via a specific promoter (Whyte, Ruley, & Harlow, 1988). Other Ad genes, such as *E1B*, *E2*, and *E4*, under a distinct promoter can also be used to develop conditionally replicating viruses (Nettelbeck, 2003). Tissue- or tumor-specific promoters can replace endogenous viral sequences in order to direct viral replication to a particular target tissue. For example, the prostate-specific antigen (PSA) promoter/enhancer element has been inserted upstream of the *E1A* gene, thereby creating a prostate-specific oncolytic virus (Rodriguez et al., 1997) and a Phase I clinical trial by intraprostatic injection following radiotherapy has established safety and viral dose-dependent reduction in PSA levels (DeWeese et al., 2001). However, no objective clinical response in terms of tumor regression was observed. To increase the specificity, a second prostate-specific enhancer sequence was subsequently inserted upstream of the *E1B* gene causing improved selectivity over the first-generation virus (Heise & Kirn, 2000). Human telomerase is very active in more than 85% of primary cancers regardless of their anatomic origin and its activity correlates closely with human telomerase reverse transcriptase (hTERT) expression (Fujiwara, Urata, & Tanaka, 2007). Since the hTERT promoter is preferentially activated in human cancer cells, recombinant viruses were produced in which *E1A* expression was under regulatory control of the hTERT promoter. These Ads show restricted replication to telomerase positive tumor cells and efficiently lyse the target tumor cells (Fujiwara, Shirakawa, & Kagawa, 2011; Huang, Savontaus, Shinozaki, Sauter, & Woo, 2003; Irving et al., 2004; Kim et al., 2003; Lanson, Friedlander, Schwarzenberger, Kolls, & Wang, 2003; Wirth et al., 2003; Zou et al., 2004). Survivin is often overexpressed in brain tumors and its promoter has been employed to drive *E1A* to treat malignant gliomas (Van Houdt et al., 2006). β-Catenin-responsive promoters have also been used to drive Ad early genes in cancers with an activated Wnt-signaling pathway, such as colorectal and liver cancers (Fuerer & Iggo, 2002). To target malignant melanoma, Nettelbeck et al., developed a novel Ad (Ad. TyrΔ2Δ24) by replacing the Ad *E1A* promoter with the promoter/enhancer of the melanocyte and melanoma-specific tyrosinase gene and by establishing mutations in the *E1A* gene to prevent the mutant protein from interacting with and inactivating pRb and p300, thus prohibiting viral replication in normal cells (Nettelbeck, Rivera, Balague, Alemany, & Curiel, 2002). Ad.TyrΔ2Δ24 has been further improved by incorporating the tyrosinase promoter to drive the *E4* gene (Ad.2Xtyr) and the efficacy of this approach was documented in *in vitro* organotypic raft cultures (Banerjee et al., 2004). In a similar respect, the cyclooxygenase-2 promoter was used to control the expression of *E1A* of Ad and the oncolytic activity of such viruses was verified in various tumors, including cervical, ovarian, and pancreatic cancers (Hoffmann & Wildner, 2006). MUC1/DF3 is a transmembrane mucin protein normally expressed on the apical borders of secretory epithelial cells and the expression of this promoter is specific to multiple tumors, including breast, myeloma, and pancreatic cancers (Fukazawa et al., 2010). This promoter has been used to drive *lacZ* or human *somatostatin receptor subtype 2* (*hSSTR2*) for targeting breast and pancreatic cancers, respectively (Chen et al., 1995). A similar approach has been followed using various cancer-selective promoters such as carcinoembryonic antigen (Li et al., 2003), alpha-fetoprotein (AFP) (Hallenbeck et al., 1999), and their preclinical applications have been tested and found to be effective in diverse animal models (Li et al., 2003; Rodriguez et al., 1997).

Although oncolytic Ads have been found to be safe in clinical trials, as a single agent they engender very restricted clinical responses and it is only in combination with radiation therapy or chemotherapy that they show noticeable effects. In these contexts, arming these viruses with additional agents offers potential to augment their oncolytic properties. Ad.TK$^{RC}$, an *E1B-55K*d-deleted Ad (ONYX-015) in which HSV-Tk (herpes simplex virus thymidine kinase) was connected to *E1A* by an internal ribosome entry site, showed enhanced growth inhibition of colon cancer xenografts in nude mice in the presence of gancyclovir (Wildner, Blaese, & Morris, 1999). Kanai and colleagues engineered a recombinant Ad expressing the HSV-Tk gene under the human α-fetoprotein enhancer and promoter domain to target hepatocellular cancers (Kanai et al., 1996). In another study, tBid, a truncated BH3-interacting domain death agonist, driven by the AFP promoter was packaged in an Ad and cancer-specific antitumoral activity was established both *in vitro* and *in vivo* (Miao et al., 2006). The bacterial nitroreductase gene was inserted at the site of the *E1B-55K*d deletion to generate CRCA-NTR (PS1217H6), showing increased efficacy in the presence of the prodrug CD1954 (Chen et al., 2004). Ad.5-CD/TKrep that delivers a cytosine deaminase/herpes simplex virus-1 thymidine kinase fusion gene in an ONYX-015 backbone efficiently decreased serum PSA in a Phase I clinical trial in prostate cancer patients (Freytag et al., 2002). However, these clinical responses were further augmented when this viral approach was combined with radiotherapy (Freytag et al., 2003). In addition to suicide genes, immunomodulatory genes, such as granulocyte-macrophage colony-stimulating factor and the cofactor B7-1, have been combined in a replicating Ad backbone documenting antitumor activity and conferring long-lasting immunity against a tumor rechallenge in a syngeneic mouse melanoma model (Choi et al., 2006). Oncolytic Ads, in which replication is driven by the hTERT promoter, have been engineered to express interferon-γ (IFN-γ) causing regression of liver cancer xenografts in both immunocompetent and immunodeficient animals (Su et al., 2006). Apoptosis-inducing genes have also been incorporated in a replicating Ad backbone such as Ad.HS4.AFP.E1a/TNF-related apoptosis-inducing ligand (TRAIL), in which an HS4 insulator-containing α-fetoprotein promoter driving *E1A* expression is coupled with TRAIL expression, which displayed enhanced oncolysis of hepatocellular carcinomas (Ye et al., 2005).

Multiple approaches are currently being evaluated to increase stringent cancer-specific replication, and various structural permutations and combinations with diverse agents with potent cancer-inhibitory and -destroying actions are being evaluated. To enable cancer-specific replication, we employed a unique cancer-specific promoter, effective in all cancers studied so far, to regulate expression of the *E1A* gene, and to enhance the therapeutic potential of these viruses, we also engineered these Ads to produce apoptosis-inducing and immune-modulating genes and named these viruses "cancer terminator viruses" (CTVs) (Ye et al., 2005). In the remainder of this review, we focus on the recent advances in our comprehension of the mode of action of the CTVs and their potential applications as part of unique and promising cytokine-based gene therapy strategy for effectively treating human cancers.

## 4. CANCER TERMINATOR VIRUSES: EFFICACIOUS REAGENTS FOR CANCER GENE THERAPY

We postulated that inducing tumor cell lysis (through viral replication) combined with induction of apoptosis by means of a secreted cytokine (such as *melanoma differentiation-associated gene-7* (*mda-7*)/interleukin-24 (IL-24) or IFN-γ) would provide a potent tumor- and metastasis-destroying agent when combined into one therapeutic reagent or platform. The CTV is a first-generation single therapeutic type 5 adenovirus reagent employing a cancer-specific promoter derived from the rodent *progression-elevated gene-3* promoter (PEG-Prom) that can specifically initiate viral replication specifically in tumor cells, but not in normal cells. To further improve the efficacy of this CTV, we engineered the CRCA to produce a tumor-specific killing molecule, either the novel therapeutic IL-10 family member cytokine gene *mda-7*/IL-24 or the immune-modulating cytokine *IFN-γ*. Attractively, both the newly generated virus and its encoded therapeutic molecules enter the circulation and "'seek and destroy" other tumor cells, whether in the local tumor environment or disseminated throughout the body (metastatic), without harming normal healthy cells. In principle, these unique features of the CTV would make it particularly useful for patients whose cancer has stopped responding to other treatments and has metastasized.

### 4.1. Development of the CTV

#### *4.1.1 mda-7/IL-24: A novel cancer therapeutic cytokine gene*

Dedifferentiation is one of the hallmarks of cancer (Sachs, 1987) and induction of differentiation, causing cancer cells to revert back to normal phenotypes, is considered as a potential anticancer therapeutic strategy (Borden, Lotan, Levens, Young, & Waxman, 1993; Fisher, Prignoli, Hermo, Weinstein, & Pestka, 1985; Jiang, Lin, Su, Goldstein, & Fisher, 1995; Jiang et al., 1996; Sachs, 1978, 1987, 1990). When human melanoma cells are treated with a combination of fibroblast interferon (IFN-β) and the protein kinase C activator mezerein terminal differentiation results which is characterized by acquisition of morphological, biochemical, and molecular attributes of normal melanocytes along with an irreversible loss in proliferative capacity and tumorigenic activity (Fisher et al., 1986, 1985; Staudt, Depass, Sarkar, & Fisher, 2009; Staudt & Dittmer, 2007). To detect the molecules potentially regulating terminal differentiation, a subtraction hybridization approach was performed between proliferating and terminally differentiated human melanoma cells resulting in the cloning of a spectrum of genes called *melanoma differentiation-associated* (*mda*) genes (Jiang et al., 1995; Jiang, Shah, & Hilt, 1993; Staudt et al., 2009).

*mda*-7, a novel gene at the time of its initial cloning by Jiang and Fisher in 1993 (Fisher et al., 1986, 1985; Staudt et al., 2009; Staudt & Dittmer, 2007), belongs to the IL-10 family of cytokines and is also designated as IL-24 based on the presence of an IL-10 signature motif, chromosomal localization in a cytokine cluster at 1q32–33, and functional properties (Caudell et al., 2002; Fisher et al., 2003; Sauane et al., 2003). The *mda*-7/IL-24 cDNA encodes a protein of 206 amino acids with a predicted size of ~24 kDa that contains a 49-amino acid signal peptide. The protein has three glycosylation sites at amino acids 95, 109, and 126, and the secreted product is heavily glycosylated showing multiple higher molecular weight bands in Western blots (Emdad et al., 2009). The observation that MDA-7/IL-24 protein is expressed in normal human melanocytes as well as in cells of hematopoietic lineages, while its expression is progressively lost during melanoma progression, suggests a potential tumor suppressor function of *mda*-7/IL-24 (Huang et al., 2001). Indeed, a series of studies, employing a replication-incompetent adenovirus expressing *mda*-7/IL-24 (Ad.*mda*-7) or recombinant MDA-7/IL-24 protein, confirmed that *mda*-7/IL-24 induces apoptosis in a diverse group of tumors including melanoma, glioblastoma, leukemia, and carcinomas of breast, kidney, cervix, colorectum, liver, lung, ovary, and prostate without causing any detrimental effects to normal cells, including melanocytes, astrocytes, fibroblasts, and mesothelial and epithelial cells (Chada et al., 2004; Dash, Bhutia, et al., 2010; Emdad et al., 2009; Fisher, 2005; Fisher et al., 2003; Gopalkrishnan, Sauane, & Fisher, 2004; Gupta et al., 2006; Inoue et al., 2006; Lebedeva et al., 2007; Lebedeva, Su, et al., 2003; Liang et al., 2011; Sauane et al., 2003). In addition to antitumor activity, Yang et al. recently demonstrated that MDA-7/IL-24 is also involved in regulation of human myeloid leukemic cell differentiation (Yang et al., 2011). The molecular mechanism underlying cancer-specific apoptosis induction by *mda*-7/IL-24 involves accumulation of MDA-7/IL-24 protein in the endoplasmic reticulum (ER) where it interacts with the chaperone protein BiP/GRP78 resulting in the activation of the ER stress or unfolded protein response (Sauane et al., 2004). Additionally, treatment with Ad.*mda*-7 also results in generation of ceramides and reactive oxygen species in mitochondria and these events culminate into activation of p38 MAP kinase (MAPK) pathway and induction of growth arrest and DNA damage (GADD)-inducible genes leading to a predominantly intrinsic pathway of apoptosis induction (Sarkar et al., 2002; Su et al., 2003). Recently, in prostate cancer MDA-7/IL-24 was identified to interact with clusterin and modulate the relative levels of soluble and nuclear isoforms promoting growth arrest at the G2/M phase and apoptosis (Bhutia et al., 2011). Depending upon the cell type, *mda*-7/IL-24 can either induce toxic autophagy (Park et al., 2008) or initial protective autophagy (Bhutia et al., 2010) that switches to apoptosis (reviewed by Dash, Bhutia, et al., 2010; Emdad et al., 2009). In addition to cancer-specific apoptosis induction, *mda*-7/IL-24 embodies several attributes that make it an ideal gene therapeutic for cancer. Secreted MDA-7/IL-24 protein binds to its cognate IL-20R1/IL-20R2 or IL-20R1/IL-22R1 receptors on the cell surface and induces expression of endogenous *mda*-7/IL-24 by stabilizing *mda*-7/IL-24 mRNA (Sauane et al., 2008). This phenomenon is observed both in normal and in cancer cells, and since normal cells are resistant to the toxic effects of MDA-7/IL-24, they might serve as a continuous source of MDA-7/IL-24, induced by exogenous MDA-7/IL-24, to kill neighboring and distant cancer cells therefore significantly magnifying the anticancer effect. Additionally, MDA-7/IL-24 inhibits tumor angiogenesis, synergizes with radiation, chemotherapy, and monoclonal antibody therapies and stimulates an antitumor immune response and together these events mount a potent "bystander" response so that MDA-7/IL-24 can eradicate both primary tumors to which it is delivered as well as distant tumors resembling metastases (Chada et al., 2006; Cunningham et al., 2005; Fisher, 2005; Gupta et al., 2006; Ramesh et al., 2003; Sarkar et al., 2007; Sauane et al., 2008; Su, Emdad, Sarkar, et al., 2005; Tong et al., 2005).

The findings gleaned from *in vitro* studies and mouse models have been confirmed in patients with advanced cancers (both carcinomas and melanomas) in a Phase I clinical trial using a replication-incompetent serotype 5 adenovirus expressing MDA-7/IL-24, Ad.*mda*-7 (INGN-241) (Cunningham et al., 2005; Eager, Harle, & Nemunaitis, 2008; Fisher et al., 2003; Lebedeva et al., 2007; Tong et al., 2005). This study documented that MDA-7/IL-24 was well tolerated and demonstrated evidence of significant (~44%) clinical activity. Similar to the preclinical studies, potent "*bystander*" antitumor activity was also apparent in this clinical study. Infection of a small percentage (10–30%) of tumor cells with Ad.*mda*-7 resulted in detectable MDA-7/IL-24 protein levels and increased tumor cell apoptosis many centimeters from the site of original injection. A majority of the patients showed a marked increase in $CD3^{+}$ and $CD8^{+}$ T cells as well as transient increases in circulating cytokines, such as IL-6 and TNF-α, indicating activation of an antitumor immune response (Tong et al., 2005). These provocative results provide direct support for using *mda*-7/IL-24 in developing a potentially effective gene-based therapy for cancer.

#### 4.1.2 Progression-elevated gene-3 promoter, a unique cancer-specific gene promoter for controlling therapeutic gene expression

While *mda*-7/IL-24 is a tumor suppressor, *progression-elevated gene-3* (*PEG-3*) is an oncogene that was cloned by subtraction hybridization in our laboratory employing E11, a mutant adenovirus type 5 (H5ts125)-transformed rat embryo cell clone that forms small, slow-growing, compact tumors, and E11-NMT, a clone of E11 passaged through nude mice (Su, Gopalkrishnan, Narayan, Dent, & Fisher, 2002) that forms rapidly growing, highly aggressive tumors (Babiss, Zimmer, & Fisher, 1985; Fisher, Bozzone, & Weinstein, 1979; Su, Shi, & Fisher, 1997). Rodent PEG-3 is a C-terminal truncated mutant form of the growth arrest and DNA damage inducible gene-34 (GADD-34), a frequent event in rodent tumorigenesis, and functions as a dominant negative inhibitor of the apoptosis-inducing function of GADD-34 (Hollander, Poola-Kella, & Fornace, 2003; Su, Emdad, Sauane, et al., 2005; Su et al., 1997). *PEG-3* is overexpressed in E11-NMT cells versus E11 cells and also in normal cloned rat embryo fibroblasts (CREF) displaying a transformed/tumorigenic phenotype upon expression of diverse oncogenes, including Ha-*ras*, v-*src*, human papilloma virus type-18-transforming genes, and a specific mutant of adenovirus 5 (H5hr1), relative to parental CREF cells (Su et al., 1997). Overexpression of *PEG-3*, by a replication-incompetent adenovirus (Ad.*PEG-3*), induces genomic instability which is associated with modulation of expression of genes regulating centrosomal duplication (Emdad et al., 2005), and an aggressive phenotype characterized by increased invasion, angiogenesis, and metastasis and associated with elevated metalloproteinases, MMP2 and MMP9, and vascular endothelial growth factor, in human tumor cells, but not in normal cells

(Su, Shi, & Fisher, 2000; Su et al., 2001). In association with the oncogenic properties of *PEG-3*, the promoter region of the *PEG-3* gene (PEG-Prom) showed 8- to 10-fold more activity in CREF transformed with either Ha-*ras* or v-*raf* than the parental CREF cells, and a minimum region of the *PEG-Prom* that extends from −118 to +194 (when the transcription initiation site is regarded as +1) was shown to be sufficient for the increased activity associated with transformation and cancer progression (Su et al., 2000, 2001). This region contains a binding site for polyoma enhancer activator-3 (PEA-3) at −104 and for activator protein-1 (AP-1) at +8, and sequence-specific mutational analysis revealed that both of these transcription factors are important for regulating the basal and oncogene-induced activity of the *PEG-Prom* (Su et al., 2000, 2001). Transgenes, such as luciferase, green fluorescence protein (GFP), or herpes thymidine kinase-1 (HSV-Tk), driven under PEG-Prom, are expressed uniquely in cancer cells and not in normal cells confirming the cancer-specific activity of the *PEG-Prom* (Su, Emdad, Sauane, et al., 2005). A recent study using metastatic breast carcinoma or melanoma models has demonstrated that, upon intravenous administration, *PEG-Prom* facilitates transgene expression in all the cancer cells dispersed throughout the mouse body with no expression detected in normal cells, thus facilitating detection of tumor metastasis by *in vivo* imaging (Bhang, Gabrielson, Laterra, Fisher, & Pomper, 2011).

Based on the above considerations, exploiting the cancer-specific expression property of the PEG-Prom, CTVs were generated in which viral replication is controlled by a minimal active region of this promoter. As indicated above, to enhance the therapeutic potential of these CTVs, we further modified/engineered these Ads to produce either *mda*-7/IL-24 (Ad.PEG-*E1A*-*mda*-7 or Ad.*CTV-m7*) or IFN-γ (Ad.PEG-*E1A*-IFN-γ or Ad.*CTV*-γ) under the control of a cytomegalovirus (CMV) promoter.

### 4.1.3 CTV construction

To generate CTVs, two shuttle vectors, pE1.2 and pE3.1, were used (Sarkar, Su, Vozhilla, Park, Randolph, et al., 2005) with the *PEG-Prom* driving the *E1A* gene (PEG-Prom-*E1A*) and the CMV promoter controlling the therapeutic gene (*mda*-7/IL-24 or IFN-γ) were inserted into the multiple cloning sites of the vectors pE1.2 and pE3.1, respectively (Fig. 1.2). These transgene cassettes were inserted into the Ad backbone in a four-fragment ligation exploiting the presence of unique restriction enzyme sites in the shuttle vectors and in the Ad plasmid. The ligation product was transformed into *Escherichia coli* and clones were selected for resistance to ampicillin

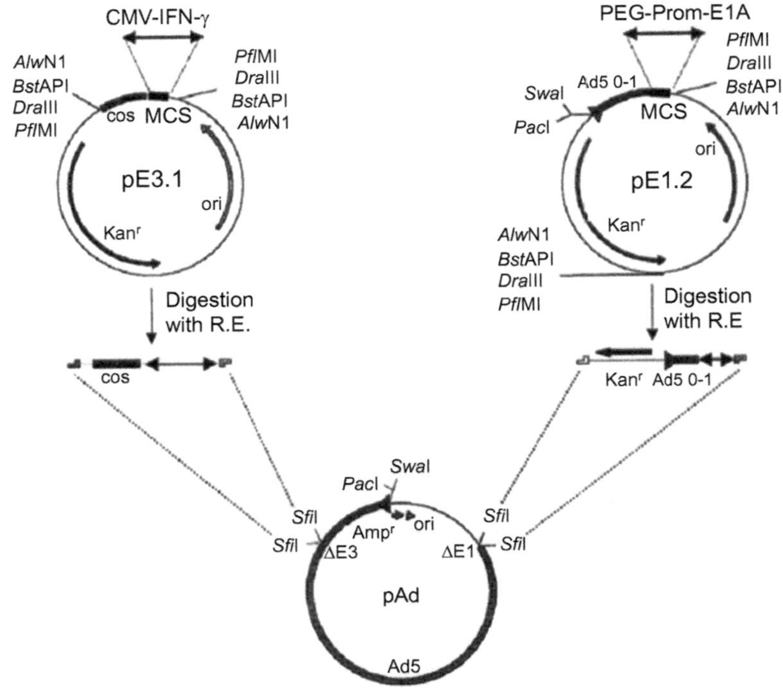


**Figure 1.2** Constructing a CTV using recombinant DNA techniques. The cancer-specific PEG-Prom-driven *E1A* gene (PEG-Prom-E1A) and the CMV promoter-driven *mda-7*/IL-24 or IFN-γ gene (CMV-*mda-7* or CMV-IFN-γ) were cloned into the multiple cloning site of the shuttle vectors pE1.2 and pE3.1, respectively. The cassettes containing the promoter and the transgene were further digested by different restriction enzymes as shown in the figure and ligated into *Sfi*I-digested adenoviral vector. *Taken from Sarkar, Su, Vozhilla, Park, Randolph, et al. (2005).*

(ampicillin-resistance gene provided by adenoviral plasmid) and kanamycin (kanamycin-resistance gene provided by the fragment from the shuttle vector). The cosmid DNA was amplified by standard large-scale preparation using a CsCl gradient, digested with *Pac*I restriction enzyme, and transfected into human embryonic kidney-293 cells for *in vivo* recombination. The Ad was purified and viral titer was determined by measuring absorbance at 260 nm and using BD AdenoX rapid titer kit (BD Biosciences).

## 4.2. Preclinical studies with CTVs

The relative therapeutic efficiency of the CTVs, both Ad.PEG-*E1A*-*mda*-7 (Ad.*CTV*-*m*7) and/or Ad.PEG-*E1A*-*IFN*-*γ* (Ad.*CTV*-*γ*), was evaluated in various cancer models (Sarkar et al., 2006, 2008, 2005; Sarkar, Su, Vozhilla,

Park, Randolph, et al., 2005). As a control Ad.CMV-*E1A-mda-7* or Ad. CMV-*E1A-IFN-γ*, in which viral replication is controlled by the *CMV* promoter and which also expresses *mda-7*/IL-24 or IFN-γ, respectively; Ad. CMV-*E1A* and Ad.PEG-*E1A*, in which viral replication is controlled by the CMV promoter or the PEG-Prom, respectively; and Ad.CMV- *mda-7* and Ad.PEG-*mda-7* or Ad.CMV-*IFN-γ* and Ad.PEG-*IFN-γ*, replication-incompetent Ad in which the *CMV* or the *PEG-Prom* drives *mda-7*/IL-24 or *IFN-γ* expression, respectively, were used. In this particular section, we focus on the results obtained from these preclinical studies.

#### 4.2.1 Ad.CTV-m7 cures primary and distant cancers

Breast cancer is a common female cancer accounting for 32% of all cancers in women (Jemal, Siegel, Xu, & Ward, 2010) and to compare the therapeutic efficacy of Ad.*CTV-m7*, experiments were performed in both immortal mammary epithelial cells and aggressive breast cancer cell lines (Sarkar, Su, Vozhilla, Park, Gupta, et al., 2005). In *in vitro* experiments, infection with Ad.*CTV-m7*-induced viral replication and generated significant MDA-7/ IL-24 production only in cancer cell lines, but not in normal immortal mammary epithelial cells, firmly establishing the cancer cell-selective replication of Ad and *mda-7*/IL-24 expression. All the Ads, except for Ad.*vec*, resulted in significant apoptosis in the breast cancer cell lines. Infection with the replication-competent Ads resulted in predominantly necrosis as evidenced by an increase in late apoptotic cells, whereas infection with Ad.CMV-*mda-7* and Ad.PEG-*mda-7* resulted in predominantly apoptosis as demonstrated by an increase in early apoptotic cells. Substantiating the *in vitro* observations, *in vivo* assays also established that Ad.*CTV-m7* is more efficacious in eradicating not only the primary tumor but also the distant tumors. Experimentally, T47D tumor xenografts were established in both flanks of nude mice and once palpable tumors of $\sim$75 mm$^3$ developed, Ad.*CTV-m7* or other Ad were injected intratumorally only in the left flank tumor. No injections were given to the right-sided tumors. Although Ad.CMV-*E1A* or Ad.PEG-*E1A* inhibited the growth of injected tumors on the left flank, they had little inhibitory effect on tumors on the right side, which was not statistically significant. Ad.CMV-*mda-7* or Ad.PEG-*mda-7* eradicated tumors on the left flanks and significantly inhibited tumor growth on the right flanks. Interestingly, injection of Ad.*CTV-m7* resulted in complete eradication of not only the injected left-sided tumors but also the uninjected right-sided tumors (comparable with a metastasis), providing confidence that this strategy could prove amenable for successfully treating aggressive cancers.

Successful application of this dual cancer-specific targeting strategy in breast cancer provided an impetus for testing the CTVs in other human neoplasms to evaluate potential for eradicating additional primary tumors and metastatic disease. Prostate cancer is one the most important health problems in industrialized countries (Di Lorenzo & De Placido, 2006; Sternberg, 2002) and is refractory to conventional anticancer treatments because of frequent overexpression of antiapoptotic proteins, including Bcl-2 and/or Bcl-$x_L$ (Colombel et al., 1993; Krajewska et al., 1996; Krajewski et al., 1994). Although parental prostate cancer cells are highly susceptible to Ad.*mda*-7-induced apoptosis, stable overexpression of Bcl-2 and Bcl-$x_L$ renders prostate cancer cells resistant to the apoptotic effects of Ad.*mda*-7 (Lebedeva, Sarkar, et al., 2003). To explore the efficacy of the CRCA-based cancer gene therapy approach for eradicating resistant prostate cancer cells, Ad.*CTV-m7* was injected into xenografts derived from DU-145-Bcl-$x_L$ cells in athymic nude mice. Similar to the breast cancer study, tumors were established in both flanks of the mice and therapeutic Ads were injected only into the left flank tumors. Ad.*CTV-m7*-mediated tumor regression was compared with other Ads, including Ad.CMV-*E1A-mda-7*, Ad.CMV-*E1A*, or Ad.PEG-*E1A*, over a period of 6 weeks. This set of experiments revealed that intratumoral injection of Ad.*CTV-m7* could completely eradicate primary and distant tumors (comparable to a metastasis), whereas Ad.CMV-*mda*-7 or Ad.PEG-*mda*-7 only displayed marginal effects on the growth of both left- and right-sided tumors. These results correlate with the resistance profile of DU-145-*Bcl-$x_L$* cells to *mda*-7/IL-24.

The efficacy of Ad.*CTV-m7* was also determined in melanoma (Sarkar et al., 2008), an important public health issue because of rising prevalence in Caucasian populations (Jemal et al., 2010), using a series of melanoma cells including normal immortal human melanocytes. The functionality of all Ads, both replicating and nonreplicating, in cancer/noncancer cells was confirmed by the expression of both E1A and MDA-7/IL-24 protein. *In vitro* cell viability and apoptosis assays established the cancer-specific activity of *mda*-7/IL-24. *In vivo* assays were performed using nude mice containing established human melanoma subcutaneous xenografts on both right and left flanks of the animal. After palpable tumors developed, different Ads were administered to the tumors on the left flank only. No injections were given to the right-sided tumors. The experiment was terminated after 8 weeks since control- and Ad.*vec*-injected animals reached maximum allowable tumor volumes. Similar to both breast and prostate cancer, injecting Ad.

*CTV-m7* completely eliminated not only primary treated tumors but also distant nontreated tumors (established in the opposite flank), thereby implementing a cure in these immune incompetent animals.

Cancer is a progressive disorder involving multiple genetic abnormalities. As such, combinatorial approaches are mandatory to effectively eradicate this disease and combining cancer-specific Ad replication with the multipronged antitumor effects of *mda*-7/IL-24 represents an appealing strategy (Sarkar et al., 2008). This assumption is supported by the observations that, while Ad.*mda*-7 could effectively suppress growth of noninjected distant xenografts over time, these tumors started to regrow, and only Ad. *CTV-m7* could completely eliminate the distant tumors eliciting a "cure." However, the precise mechanism(s) underlying these observations are yet to be defined. Hypothetically, the Ad.*CTV-m7*-induced "cure" in animals might be the summation of the direct apoptosis-inducing properties of MDA-7/IL-24, the indirect antitumor actions of MDA-7/IL-24 (inhibiting angiogenesis and stimulating an antitumor immune response), and *in situ* Ad replication (directly promoting cytolysis) in tumors.

### 4.2.2 Eradication of therapy-resistant pancreatic cancer by Ad.PEG-*E1A-IFN-γ* (Ad.*CTV-γ*)

Based on difficulty in early diagnosis of pancreatic ductal adenocarcinoma (PDAC), this disease is often noticeable after metastasis has occurred and this is one of the reasons it has the shortest average survival time in contrast to all other cancer types (Bardeesy & DePinho, 2002; Jaffee, Hruban, Canto, & Kern, 2002; Russo, Butler, Ove, & Blackstock, 2007). Surgical resection remains the only viable choice for achieving a cure, but only 15–20% of patients, newly diagnosed with pancreatic cancer, are candidates for surgical resection and, regrettably, the majority of resections result in recurrent disease (Donahue & Reber, 2010). Although intensively examined, there are no effective or even palliative therapies for PDAC, and thus developing effective therapeutics remains a top priority. The cancer-specific activity and therapeutic effectiveness of the Ad.*CTV-γ* were evaluated in four pancreatic cancer cell lines, MIA Paca-2, PANC-1, AsPC-1, and BxPC-3, and two normal immortal cell lines, FM-516-SV, normal human melanocytes immortalized by SV40 TAg, and IM-PHFA, primary human fetal astrocytes immortalized by the catalytic subunit of human telomerase (Sarkar, Su, Vozhilla, Park, Randolph, et al., 2005). Monitoring IFN-γ production by ELISA and E1A protein levels by Western blotting after Ad infection confirmed the functionality of these constructs. This study also established that PEG-Prom allows restricted Ad replication in cancer cells, protecting normal cells from growth inhibition because of Ad replication. Mechanistically, Ad.*CTV*-$\gamma$ induced predominantly necrosis as evidenced by an increase in late apoptotic cells, whereas infection with Ad.CMV-*IFN*-$\gamma$ and Ad.PEG-*IFN*-$\gamma$ resulted predominantly in apoptosis as evidenced by an increase in early apoptotic cells. To expand on these *in vitro* studies, *in vivo* assays were done using nude mice containing established AsPC-1 subcutaneous xenografts on both their right and left flanks. Intratumoral injections with Ad.*CTV*-$\gamma$ on the left flank tumors produced marked regression within three injections and completely eradicated tumors on both flanks following seven injections, whereas Ad.CMV-*E1A* or Ad.PEG-*E1A* inhibited the growth of tumors on the left flank, but had little inhibitory effect on the tumors on the right side, which was not statistically significant. Ad.CMV-*IFN*-$\gamma$ or Ad. PEG-*IFN*-$\gamma$ appreciably inhibited tumor growth on both flanks, emphasizing the importance of immune stimulation, but these treatment regimens did not completely destroy the tumors. However, intratumoral injection of Ad.*CTV*-$\gamma$ completely eradicated the primary tumor and dramatically inhibited or eliminated (four out of five animals) the distant tumor (comparable with a metastasis) providing confidence that this strategy may prove amenable for successfully treating aggressive cancers. Further experiments confirmed that Ad.*CTV*-$\gamma$ enters the circulation and replicates in the tumors of the right side and/or the secreted IFN-$\gamma$ induces an antitumor immune response by stimulating natural killer cells that are present in the nude mice. These findings indicate that the replication-competent Ad have the ability to migrate and replicate at distant sites in the animals. In this context, the ability of the PEG-Prom in Ad.*CTV*-$\gamma$ to drive replication is extremely important for ensuring Ad replication in cancer cells, while sparing harmful effects in normal cells.

In addition to the significant direct and indirect antitumor effects of therapeutic genes, this particular strategy has an additional advantage over other combinatorial approaches because of the employment of the PEG-Prom. Most current strategies ensuring cancer selectivity of CRCAs exploit genetic abnormalities in cancer cells such as their p53, RB, or PKR status (Cascallo, Capella, Mazo, & Alemany, 2003; Heise et al., 2000; Johnson et al., 2002). This considerably limits the general application of such strategies in diverse cancer indications, which may lack these alterations, and clinical application of these CRCAs alone often fails to generate any discernable clinical response. In contrast, the minimal region of the PEG-Prom, which was used to construct the CTVs, is controlled by two transcription factors, AP-1 and PEA-3 (Su, Sarkar, Emdad, et al., 2005); the presence of either of these sites is sufficient to ensure cancer selectivity of the PEG-Prom. Since either AP-1 or PEA-3 is overexpressed in a high percentage of cancers, the universal utility of the PEG-Prom to assure cancer selectivity and, therefore, limit nonspecific cytotoxicity is comparable only to that of the promoter of the human telomerase (hTERT) gene (Huang et al., 2003).

Recently, a number of scientific reports have described oncolytic Ads armed with the *mda*-7/IL-24 gene, for example, SG600-IL24, an oncolytic Ad harboring *mda*-7/IL-24 selectively induces apoptosis in different hepatocarcinoma cell lines without affecting a normal liver cell line (Xue et al., 2010). In another approach, an *E1B-55K*d-deleted oncolytic Ad carrying *mda*-7/IL-24 was developed (ZD-55-IL-24) that exhibited better therapeutic efficacy compared with ONXY-015 against hepatic cancer (Xiao et al., 2010). This Ad was more potent in combination with dichloroacetate (Xiao et al., 2010) or cisplatin (Wu et al., 2009). To target liver, Cao et al. developed an α-fetoprotein-controlled oncolytic Ad carrying *mda*-7/IL-24 and demonstrated improved antitumor effects on hepatic cancer (Cao et al., 2011).

## 5. APPROACHES ENHANCING THERAPEUTIC OUTCOMES

Compared to currently available vector approaches, Ad vectors possess the greatest capacity to achieve *in vivo* infection of tumors (Mathis, Stewart, Zhu, & Curiel, 2006). Despite this capacity, broad clinical efficacy of Ad-based treatment regimens remains imperfect because of suboptimal infectivity. The discrepancy between preclinical and clinical studies using type Ad.5 systems can frequently be attributed to differences in the expression of the primary Ad.5 receptor, the Coxsackie and adenovirus receptor (CAR), in primary tumors relative to their cell line counterparts (Kim, Zinn et al., 2002). In addition, other hurdles limiting successful cancer gene/protein therapy include appropriate methods for effectively delivering anticancer therapeutics intravenously without nonspecific trapping in the liver, clearance/neutralization by the immune system, or eliciting immune responses. In this section, we focus on several approaches that can significantly improve therapeutic outcomes.

## 5.1. Improvement of adenoviral vectors to enhance transduction

One approach to circumvent the low efficiency of Ad infection of tumor cells exploits "tropism modification" in which the virus capsid proteins that normally interact with CAR are modified, permitting CAR-independent infectivity of tumor cells. Understanding structural homology between Ad.5 and other Ad serotypes and the cellular entry pathways for Ads has facilitated the development of chimeric vectors in which whole fibers or even the distal knob domains from various Ad serotypes, whose native tropism is associated with receptors other than CAR, are engineered into Ad.5 (Fig. 1.3). In this respect, based on differences in receptor use (Krasnykh, Mikheeva, Douglas, & Curiel, 1996), replacement of the Ad.5 fiber knob domains with the Ad serotype 3 (Ad.3) fibers resulted in enhanced cytopathogenicity of an Ad virotherapy agent toward primary melanoma cells, which was at least four orders of magnitude greater than wild-type Ad.5 (Rivera et al., 2004; Volk et al., 2003). This transductional enhancement was similar for melanoma, ovarian cancer, renal cancer, squamous carcinoma, prostate, and colorectal cancer (Haviv et al., 2002; Kanerva et al., 2002; Kawakami et al., 2003; Rein et al., 2005). Intriguingly, shortening of the Ad.5 fiber shaft in an Ad.5/3 chimera also significantly reduced liver tropism and facilitated enhanced replicative ability compared with its Ad.5 counterpart (Yamamoto & Curiel, 2010). Importantly, Ad.5/3 also infects high-CAR-expressing tumor cells with equal efficacy or even improved activity when compared with Ad.5, thereby providing an expanded scope of utility for Ad.5/3, in both low- and high-CAR-expressing tumor cells. To enhance delivery of *mda*-7/IL-24, we have developed a tropism-modified chimeric adenovirus expressing *mda*-7/IL-24 (Ad.5/3-*mda*-7) and explored its efficacy in the context of prostate cancer (Dash, Dmitriev, et al., 2010). Experiments were performed to directly compare the infectivity of Ad.5/3 chimeric viruses (Ad.5/3-*mda*-7) versus Ad.5 (Ad.5-*mda*-7) viruses in the context of prostate cancer using two cell lines with variable levels of CAR on their surface. Infection of PC-3 (low CAR) cells with Ad.5-*mda*-7 was significantly less effective in reducing cell proliferation and viability than Ad.5/3-*mda*-7. This differential effect correlated with a reduced level of MDA-7/IL-24 protein being produced in PC-3 cells infected with Ad.5-*mda*-7 versus Ad.5/3-*mda*-7. At the molecular level, compared to Ad.5-*mda*-7, Ad.5/3-*mda*-7 efficiently induced apoptosis as evidenced by increased cleavage

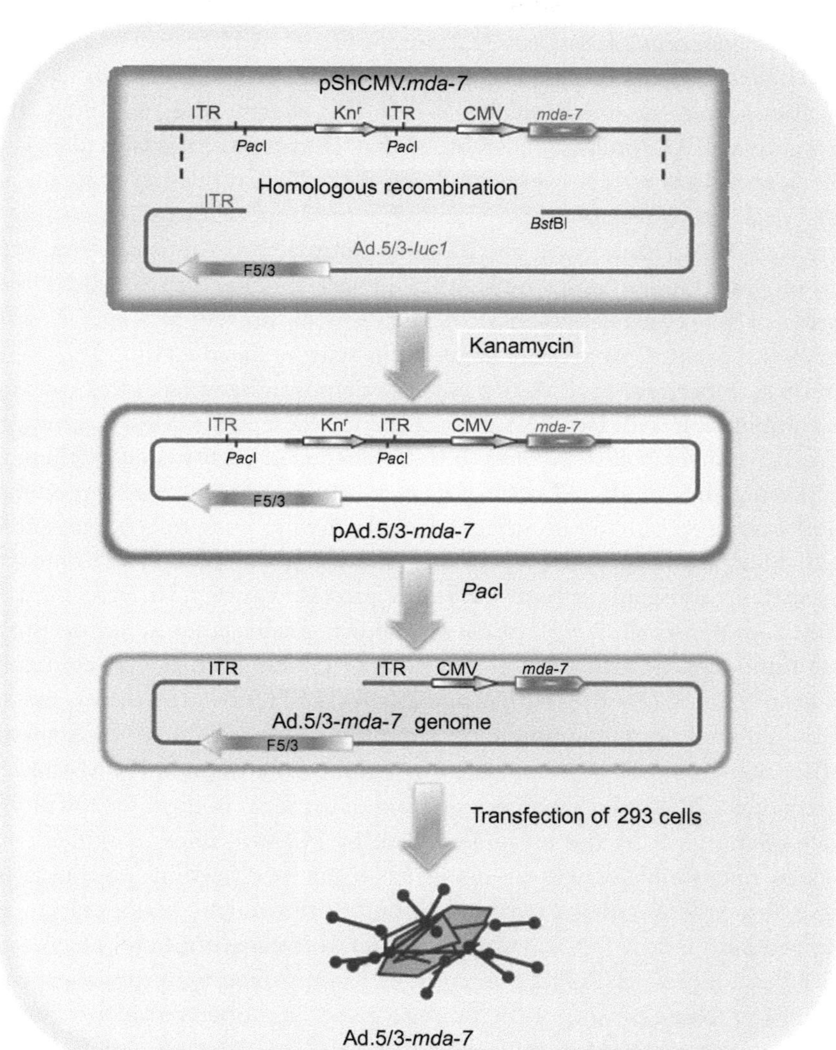


**Figure 1.3** Constructing a tropism-modified adenovirus (Ad.5/3) carrying the *mda-7*/IL-24 gene. The genome of Ad.5/3.*mda-7* was generated by homologous recombination between the linearized plasmid pShCMV.*mda-7* and *BstB*I-digested genomic DNA of Ad.5/3-Luc1, and kanamycin selection resulted in the pAd.5/3-*mda-7* genome, where the CMV promoter in place of the early viral E1 region drives *mda-7*/IL-24 expression. This plasmid was digested with *Pac*I to release viral ITRs and transfected in A549 cells to rescue the Ad.5/3-*mda-7*. *Taken from Dash, Dmitriev, et al. (2010)*. (See Page 2 in Color Section at the back of book.)

of PARP, downregulation of antiapoptotic proteins Bcl-2 and Bcl-$x_L$, as well as activation of caspase 3/7. Interestingly, Ad.5/3-*mda*-7, but not Ad.5-*mda*-7 efficiently activated the p38 MAPK pathway, known to have a crucial role in mediating *mda*-7/IL-24 effects, and inhibited the prosurvival ERK pathway suggesting that chimerism maintains the *bona fide* downstream effects exerted by *mda*-7/IL-24. Correlating with the *in vitro* data, Ad.5/3-*mda*-7 also showed a profound enhanced antitumor activity (Dash, Dmitriev, et al., 2010) as compared to Ad.5-*mda*-7 in PC-3 xenograft models using nude mice. These findings provide definitive evidence for enhanced therapeutic efficacy of the Ad.5/3-*mda*-7 virus versus Ad.5-*mda*-7 in prostate cancer cells with reduced CAR.

Dash, Dmitriev, et al. (2010) recently demonstrated that Ad.5/3-*mda*-7 in combination with BI-97C1 (Sabutoclax), which targets Mcl-1, sensitizes prostate cancer cells to *mda*-7/IL-24-mediated cytotoxicity, thereby enhancing therapeutic efficacy. Evidence was provided using two animal models, one employing immune deficient athymic nude mice xenografted with human prostate tumor cells and one using a spontaneous immunocompetent transgenic mouse model of prostate cancer (Hi-Myc mouse) (Ellwood-Yen et al., 2003). Nude mice bearing subcutaneous human prostate tumors established on both flanks were treated with the combination of *mda*-7/IL-24 (Ad.5/3-*mda*-7) and BI-97C1. Tumor growth was markedly inhibited by the combination treatment compared to treatment with either agent alone. This observation was further validated in Hi-Myc mice (Dash et al., 2011). To avoid sequestering of the virus in the liver and clearance of the virus by the immune system by Hi-Myc mice, complement-treated microbubbles (will be discussed in the next section) encapsulated Ad.5/3-*mda*-7 was delivered intravenously followed by sonoporation in the prostatic area. BI-97C1 was administered intraperitoneally. Decreased Ki-67 and increased TUNEL expression accompanied with tumor growth inhibition (based on size of the prostate gland) was observed in sections of the prostates of Hi-Myc mice treated with Ad.5/3-*mda*-7 and BI-97C1 (Sabutoclax) as compared to each single agent alone, signifying a sensitization role of BI-97C1 (Sabutoclax). Further results provide definitive evidence for enhanced therapeutic efficacy of Ad.5/3-*mda*-7 versus Ad.5-*mda*-7 in ovarian carcinoma, malignant glioma, renal carcinoma, and colorectal carcinoma with reduced CAR (Azab et al., 2011; Eulitt et al., 2011; Hamed et al., 2010). Using serotype chimerism, a novel CTV (Ad.5/3-CTV) has now been created in collaboration with Drs. I.P. Dmitriev and D.T. Curiel (WUSM, St. Louis, MO) by replacing the Ad.5 fiber knob with the Ad.3 fiber knob resulting in enhanced infection of tumor cells in a CAR-independent manner and their therapeutic efficacy against cancer from various anatomic origins is currently under investigation (unpublished data).

## 5.2. Ultrasound-targeted microbubble destruction: A novel, safe, and efficient approach for delivery of therapeutic agents

A major challenge for effective cancer therapy is the ability to specifically deliver therapeutics directly into diseased tissue. Ultrasound (US) contrast agents (microbubbles) have recently been underscored as a potential candidate for improving delivery of molecules to target tissues (Howard et al., 2006; Larina et al., 2005; Lawrie et al., 2000; Ng & Liu, 2002). It contains high-molecular weight gasses with less solubility and diffusivity, which increases microbubble endurance and allows passage through the microcirculation. Microbubbles can be injected in peripheral veins and can survive for several minutes within the bloodstream (Goldberg, Liu, & Forsberg, 1994; Howard et al., 2006). The ideal microbubble diameters are most likely between 2.5 and 4 μm. This is small enough to avert entrapment within the pulmonary capillary bed (ranging from 5 to 8 μm in diameter) but large enough to entrap and shield viral vectors such as Ads from the surrounding *milieu*. The gas-filled microspheres effectively lower the energy threshold for nonthermal cavitation. Ultrasound-targeted microbubble destruction (UTMD) enables focal release of entrapped materials as well as the creation of small shock waves that intensify cellular permeability (Pitt, Husseini, & Staples, 2004). In addition, the microbubbles safeguard the agents from rapid degradation by the immune system, thus allowing for intravenous injection rather than direct target organ delivery by catheter-based tactics or operative bed injections (Howard et al., 2006). This is especially important in cancer therapy of potentially inaccessible tumors because the microbubbles may also limit the amount of inflammatory response to the agents and should allow repeated injections. Howard et al. (2006) for the first time evaluated the feasibility of microbubbles for site-specific gene delivery in both *in vitro* and *in vivo* systems. The contrast agents were tested for their ability to enclose and to protect an Ad vector carrying the GFP marker gene (Ad.*GFP*) into the microbubbles. Systemic delivery of Ad.*GFP* enclosed in microbubbles resulted in specific targeting of the *GFP* transgene. Both fluorescence microscopy and GFP immunohistochemistry demonstrated

US-guided specific transduction in the targeted tissue only, with no uptake in heart, lungs, or liver using complement-pretreated Ad.*GFP* microbubbles.

To confirm the ability of the US contrast agent to deliver viruses efficiently and specifically to defined sites *in vivo*, prostate tumor xenografts were established in both flanks of athymic nude mice by injecting each site with DU-145, aggressive human prostate carcinoma cells. The tumor-bearing nude mice were then injected in their tail vein with contrast agent that was reconstituted with Ad.*GFP* or water as control, and a portable SonoSite Micro-Maxx US platform (SonoSite, Inc., Bothell, WA) equipped with a L25 linear array transducer set at 0.7 Mechanical Index, 1.8 MPa for 10 min, was used to sonoporate only the tumor implanted on the right side. Immunoblotting of total protein extracts from different organs such as kidney, liver, lung, heart, and tumors from both right and left flanks showed expression of *GFP* only in the left tumors, the tumors that were exposed to US (Greco et al., 2010), confirming microbubble-mediated specific delivery. This study further explored the systemic delivery of Ad.*CTV-m7* by microbubbles. Experimentally, to confirm the ability of microbubbles to deliver viruses efficiently and specifically, tumor xenografts were established in both flanks of athymic nude mice by injecting each site with DU-145 human prostate carcinoma cells overexpressing antiapoptotic proteins, Bcl-2 or Bcl-$x_L$, DU-Bcl-$x_L$. The tumor-bearing nude mice were then injected via tail vein with 100 μL of Targestar-P contrast agent that was reconstituted with Ad.*CTV-m7* treated with complement and only the tumor implanted on the left side was sonoporated (Greco et al., 2010). The US-guided delivery completely eradicated not only targeted therapy-resistant tumors, but also nontargeted distant tumors (established in the opposite flank) because of the "bystander" antitumor effect, thereby implementing a "cure" in athymic nude mice (Greco et al., 2010). Very recently, we also confirmed the utility of the UTMD approach in a spontaneous model of prostate cancer, the Hi-Myc mouse (Dash et al., 2011; Greco et al., 2010).

## 6. CONCLUSIONS AND FUTURE DIRECTIONS

Despite meaningful advances in early detection and treatment of localized cancers, no single or combinatorial therapeutic strategy has proven effective in lessening morbidity or establishing a cure of metastatic disease. Notably, it is projected that 90% of cancer-associated mortality may be due to disseminated tumor growth that impairs vital organ(s) function. There are multiple challenges and barriers to effective therapy of metastatic cancers.

One is to be able to accurately locate metastases. A second impediment to effective therapy is to deliver the therapeutic to the metastases. A third limitation is the need for a therapeutic gene product or reagent that can circulate through the body to find metastases. The fourth limitation is to provide a means of protecting the therapeutic from destruction by the immune system and nonspecific trapping in the liver or other organ sites not harboring

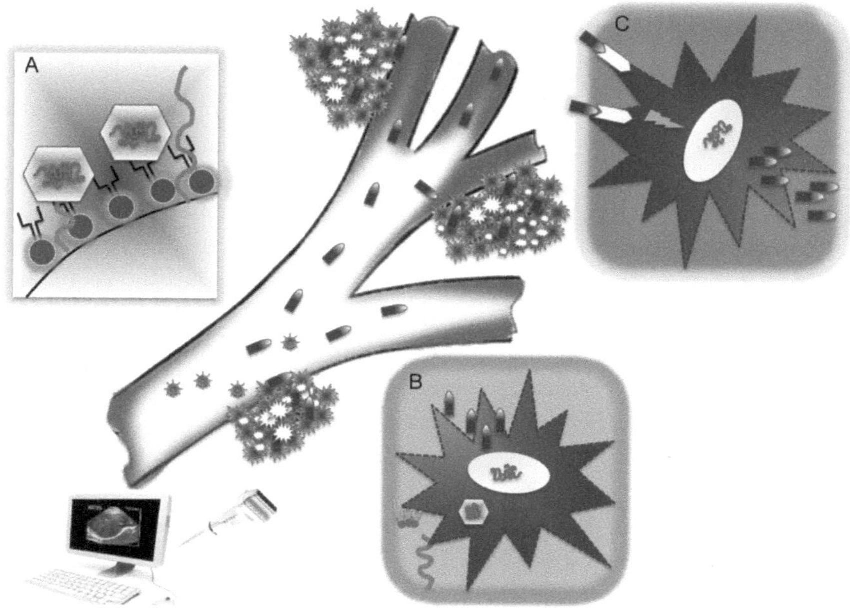


**Figure 1.4** Delivery of Cancer Terminator Viruses (CTVs) systemically following complexing with microbubbles (MB) coupled with ultrasound-targeted MB destruction, the UTMD approach. Complexes of CTVs with MBs are delivered intravenously (Box A, Ad incorporated in the lipid shell of MBs), which are released at the primary tumor site by the application of ultrasound (Box B, sonoporation of MBs in the tumor with an ultrasound probe). After intracellular entry, the CTVs replicate selectively in tumor cells, resulting in robust production of *mda-7*/IL-24 that when translated into MDA-7/IL-24 protein cause ER stress and "unfolded protein stress response" and cancer cell death. MDA-7/IL-24 is subsequently released into the circulatory system and due to virtue of its "bystander activity" (Box C, binding of MDA-7/IL-24 with IL-20R1/IL-20R2 or IL-20R1/IL-22R1 and promoting intracellular signaling leading to autocrine production of MDA-7/IL-24), which would be anticipated to induce tumor-specific apoptosis of primary and distant tumors, antiangiogenic effects in primary and distant tumor vasculatures, and immune modulatory effects targeting tumors for immune destruction. It is believed to be the "summation" of these multifaceted antitumor properties of MDA-7/IL-24 that promotes selective destruction of both primary and distant tumors. (See Page 3 in Color Section at the back of book.)

metastases. Although we are still not close to achieving the objective of producing the perfect cancer gene therapeutic, the CTV approach is appealing because it provides (i) cancer-specific replication based on expression of the PEG-Prom either in primary or distant sites and (ii) selective local and systemic killing of cancer cells by means of the cancer-specific apoptosis-inducing cytokine *mda*-7/IL-24 or IFN-$\gamma$ (Dash, Dmitriev, et al., 2010). The UTMD approach provides a means of delivering the CTV in the circulation in a protected "stealth" form with targeted delivery to sites of tumors and metastases (Dash, Dmitriev, et al., 2010). Additionally, engineering modifications in Ad tropism to enhance infectivity provides a way of increasing the efficacy of CTV delivery into tumor cells with release of *mda*-7/IL-24 or IFN-$\gamma$ into the tumor vasculature and the circulation (Fig. 1.4). In addition, detection and therapy of metastases have proven difficult and inefficient for many reasons, which emphasize the inadequacies of conventional diagnostic and therapeutic approaches. Very recently, it was confirmed that systemic delivery of the PEG-Prom linked to and regulating an imaging construct would enable tumor-specific expression of reporter genes, not only within a primary tumor but also in associated metastases in a manner broadly applicable to tumors of different tissue origin or subtype (Bhang et al., 2011). This approach holds potential for locating the enemy throughout the body. It is anticipated that, with the summation of these approaches, we will be one step closer to evoking a complete eradication of both primary and distant metastatic tumors, with minimal toxicity to normal organs, leading to a potential "*cure*" in cancer patients.

## ACKNOWLEDGMENTS

The present studies were supported in part by NIH grants R01 CA097318 (P. B. F.), R01 CA127641 (P. B. F.), P01 CA104177 (P. B. F. and P. D.), R01 CA108520 (P. D. and P. B. F.), R01 CA138540 (D. S.), R03 MH093195 (P. B. F.), DOD W81XWH-10-PCRP-SIDA (P. B. F. and X.-Y. W.), DOD W81XWH-11-1-0186 (S. S. and P. B. F.), the National Foundation for Cancer Research (NFCR) (P. B. F.), and the Samuel Waxman Cancer Research Foundation (P. B. F. and D. S.). D. S. and X.-Y. W. are Harrison Scholars in VCU Massey Cancer Center and D. S. is a Blick Scholar in VCU School of Medicine. P. B. F. holds the Thelma Newmeyer Corman Chair in Cancer Research in the VCU School of Medicine.

## REFERENCES

Alba, R., Bosch, A., & Chillon, M. (2005). Gutless adenovirus: Last-generation adenovirus for gene therapy. *Gene Therapy*, *12*(Suppl. 1), S18–S27.

Azab, B., Dash, R., Das, S. K., Bhutia, S. K., Shen, X. N., Quinn, B. A., et al. (2011). Enhanced delivery of mda-7/IL-24 using a serotype chimeric adenovirus (Ad.5/3) in

combination with the Apogossypol derivative BI-97C1 (Sabutoclax) improves therapeutic efficacy in low CAR colorectal cancer cells. *Journal of Cellular Physiology, 227*(5), 2145–2153.
Babiss, L. E., Zimmer, S. G., & Fisher, P. B. (1985). Reversibility of progression of the transformed phenotype in Ad5-transformed rat embryo cells. *Science, 228*(4703), 1099–1101.
Banerjee, N. S., Rivera, A. A., Wang, M., Chow, L. T., Broker, T. R., Curiel, D. T., et al. (2004). Analyses of melanoma-targeted oncolytic adenoviruses with tyrosinase enhancer/promoter-driven E1A, E4, or both in submerged cells and organotypic cultures. *Molecular Cancer Therapeutics, 3*(4), 437–449.
Bardeesy, N., & DePinho, R. A. (2002). Pancreatic cancer biology and genetics. *Nature Reviews. Cancer, 2*(12), 897–909.
Bhang, H. E. C., Gabrielson, K. L., Laterra, J., Fisher, P. B., & Pomper, M. G. (2011). Tumor-specific imaging through progression elevated gene-3 promoter-driven gene expression. *Nature Medicine, 17*(1), 123–U302.
Bhutia, S. K., Das, S. K., Kegelman, T. P., Azab, B., Dash, R., Su, Z. Z., et al. (2011). mda-7/IL-24 differentially regulates soluble and nuclear clusterin in prostate cancer. *Journal of Cellular Physiology, 227*, 1805–1813.
Bhutia, S. K., Dash, R., Das, S. K., Azab, B., Su, Z. Z., Lee, S. G., et al. (2010). Mechanism of autophagy to apoptosis switch triggered in prostate cancer cells by antitumor cytokine melanoma differentiation-associated gene 7/interleukin-24. *Cancer Research, 70*(9), 3667–3676.
Bischoff, J. R., Kirn, D. H., Williams, A., Heise, C., Horn, S., Muna, M., et al. (1996). An adenovirus mutant that replicates selectively in p53-deficient human tumor cells. *Science, 274*(5286), 373–376.
Borden, E. C., Lotan, R., Levens, D., Young, C. W., & Waxman, S. (1993). Differentiation therapy of cancer: Laboratory and clinical investigations. *Cancer Research, 53*(17), 4109–4115.
Cancer Figurs and Facts (2012). January 4, 2012. http://www.cancer.org.
Cao, X., Wei, R., Liu, X., Zeng, Y., Huang, H., Ding, M., et al. (2011). Cancer targeting gene-viro-therapy specific for liver cancer by alpha-fetoprotein-controlled oncolytic adenovirus expression of SOCS3 and IL-24. *Acta Biochimica et Biophysica Sinica (Shanghai), 43*(10), 813–821.
Cascallo, M., Capella, G., Mazo, A., & Alemany, R. (2003). Ras-dependent oncolysis with an adenovirus VAI mutant. *Cancer Research, 63*(17), 5544–5550.
Caudell, E. G., Mumm, J. B., Poindexter, N., Ekmekcioglu, S., Mhashilkar, A. M., Yang, X. H., et al. (2002). The protein product of the tumor suppressor gene, melanoma differentiation-associated gene 7, exhibits immunostimulatory activity and is designated IL-24. *The Journal of Immunology, 168*(12), 6041–6046.
Chada, S., Mhashilkar, A. M., Liu, Y., Nishikawa, T., Bocangel, D., Zheng, M., et al. (2006). mda-7 gene transfer sensitizes breast carcinoma cells to chemotherapy, biologic therapies and radiotherapy: Correlation with expression of bcl-2 family members. *Cancer Gene Therapy, 13*(5), 490–502.
Chada, S., Sutton, R. B., Ekmekcioglu, S., Ellerhorst, J., Mumm, J. B., Leitner, W. W., et al. (2004). MDA-7/IL-24 is a unique cytokine—Tumor suppressor in the IL-10 family. *International Immunopharmacology, 4*(5), 649–667.
Chen, L., Chen, D., Manome, Y., Dong, Y., Fine, H. A., & Kufe, D. W. (1995). Breast cancer selective gene expression and therapy mediated by recombinant adenoviruses containing the DF3/MUC1 promoter. *The Journal of Clinical Investigation, 96*(6), 2775–2782.
Chen, M. J., Green, N. K., Reynolds, G. M., Flavell, J. R., Mautner, V., Kerr, D. J., et al. (2004). Enhanced efficacy of Escherichia coli nitroreductase/CB1954 prodrug activation

gene therapy using an E1B-55K-deleted oncolytic adenovirus vector. *Gene Therapy*, *11* (14), 1126–1136.

Choi, K. J., Kim, J. H., Lee, Y. S., Kim, J., Suh, B. S., Kim, H., et al. (2006). Concurrent delivery of GM-CSF and B7-1 using an oncolytic adenovirus elicits potent antitumor effect. *Gene Therapy*, *13*(13), 1010–1020.

Colombel, M., Symmans, F., Gil, S., O'Toole, K. M., Chopin, D., Benson, M., et al. (1993). Detection of the apoptosis-suppressing oncoprotein bc1-2 in hormone-refractory human prostate cancers. *The American Journal of Pathology*, *143*(2), 390–400.

Cunningham, C. C., Chada, S., Merritt, J. A., Tong, A., Senzer, N., Zhang, Y., et al. (2005). Clinical and local biological effects of an intratumoral injection of mda-7 (IL24; INGN 241) in patients with advanced carcinoma: A phase I study. *Molecular Therapy*, *11*(1), 149–159.

Dash, R., Azab, B., Quinn, B. A., Shen, X., Wang, X. Y., Das, S. K., et al. (2011). Apogossypol derivative BI-97C1 (Sabutoclax) targeting Mcl-1 sensitizes prostate cancer cells to mda-7/IL-24-mediated toxicity. *Proceedings of the National Academy of Sciences of the United States of America*, *108*(21), 8785–8790.

Dash, R., Bhutia, S. K., Azab, B., Su, Z. Z., Quinn, B. A., Kegelmen, T. P., et al. (2010). mda-7/IL-24: A unique member of the IL-10 gene family promoting cancer-targeted toxicity. *Cytokine & Growth Factor Reviews*, *21*(5), 381–391.

Dash, R., Dmitriev, I., Su, Z. Z., Bhutia, S. K., Azab, B., Vozhilla, N., et al. (2010). Enhanced delivery of mda-7/IL-24 using a serotype chimeric adenovirus (Ad.5/3) improves therapeutic efficacy in low CAR prostate cancer cells. *Cancer Gene Therapy*, *17*(7) 447–456.

Dedieu, J. F., Vigne, E., Torrent, C., Jullien, C., Mahfouz, I., Caillaud, J. M., et al. (1997). Long-term gene delivery into the livers of immunocompetent mice with E1/E4-defective adenoviruses. *Journal of Virology*, *71*(6), 4626–4637.

DeWeese, T. L., van der Poel, H., Li, S., Mikhak, B., Drew, R., Goemann, M., et al. (2001). A phase I trial of CV706, a replication-competent, PSA selective oncolytic adenovirus, for the treatment of locally recurrent prostate cancer following radiation therapy. *Cancer Research*, *61*(20), 7464–7472.

Di Lorenzo, G., & De Placido, S. (2006). Hormone refractory prostate cancer (HRPC): Present and future approaches of therapy. *International Journal of Immunopathology and Pharmacology*, *19*(1), 11–34.

Donahue, T. R., & Reber, H. A. (2010). Pancreatic surgery. *Current Opinion in Gastroenterology*, *26*(5), 499–505.

Eager, R., Harle, L., & Nemunaitis, J. (2008). Ad-MDA-7; INGN 241: A review of preclinical and clinical experience. *Expert Opinion on Biological Therapy*, *8*(10), 1633–1643.

Ellwood-Yen, K., Graeber, T. G., Wongvipat, J., Iruela-Arispe, M. L., Zhang, J., Matusik, R., et al. (2003). Myc-driven murine prostate cancer shares molecular features with human prostate tumors. *Cancer Cell*, *4*(3), 223–238.

Emdad, L., Lebedeva, I. V., Su, Z. Z., Gupta, P., Sauane, M., Dash, R., et al. (2009). Historical perspective and recent insights into our understanding of the molecular and biochemical basis of the antitumor properties of mda-7/IL-24. *Cancer Biology & Therapy*, *8*(5), 391–400.

Emdad, L., Sarkar, D., Su, Z. Z., Boukerche, H., Bar-Eli, M., & Fisher, P. B. (2005). Progression elevated gene-3 (PEG-3) induces pleiotropic effects on tumor progression: Modulation of genomic stability and invasion. *Journal of Cellular Physiology*, *202*(1), 135–146.

Enders, J. F., Bell, J. A., Dingle, J. H., Francis, T., Jr., Hilleman, M. R., Huebner, R. J., et al. (1956). Adenoviruses: Group name proposed for new respiratory-tract viruses. *Science*, *124*(3212), 119–120.

Eulitt, P. J., Park, M. A., Hossein, H., Cruikshanks, N., Yang, C., Dmitriev, I. P., et al. (2011). Enhancing mda-7/IL-24 therapy in renal carcinoma cells by inhibiting multiple protective signaling pathways using sorafenib and by Ad5/3 gene delivery. *Cancer Biology & Therapy*, *10*(12), 1290–1305.

Fisher, P. B. (2005). Is mda-7/IL-24 a "magic bullet" for cancer? *Cancer Research*, *65*(22), 10128–10138.

Fisher, P. B., Bozzone, J. H., & Weinstein, I. B. (1979). Tumor promoters and epidermal growth factor stimulate anchorage-independent growth of adenovirus-transformed rat embryo cells. *Cell*, *18*(3), 695–705.

Fisher, P. B., Gopalkrishnan, R. V., Chada, S., Ramesh, R., Grimm, E. A., Rosenfeld, M. R., et al. (2003). mda-7/IL-24, a novel cancer selective apoptosis inducing cytokine gene: From the laboratory into the clinic. *Cancer Biology & Therapy*, *2*(4 Suppl. 1), S23–S37.

Fisher, P. B., Hermo, H., Jr., Solowey, W. E., Dietrich, M. C., Edwalds, G. M., Weinstein, I. B., Langer, J. A., Pestka, S., Giacomini, P., Kusama, M., et al. (1986). Effect of recombinant human fibroblast interferon and mezerein on growth, differentiation, immune interferon binding and tumor associated antigen expression in human melanoma cells. *Anticancer Res*, *6*(4), 765–774.

Fisher, P. B., Prignoli, D. R., Hermo, H., Jr., Weinstein, I. B., & Pestka, S. (1985). Effects of combined treatment with interferon and mezerein on melanogenesis and growth in human melanoma cells. *Journal of Interferon Research*, *5*(1), 11–22.

Freytag, S. O., Khil, M., Stricker, H., Peabody, J., Menon, M., DePeralta-Venturina, M., et al. (2002). Phase I study of replication-competent adenovirus-mediated double suicide gene therapy for the treatment of locally recurrent prostate cancer. *Cancer Research*, *62*(17), 4968–4976.

Freytag, S. O., Stricker, H., Pegg, J., Paielli, D., Pradhan, D. G., Peabody, J., et al. (2003). Phase I study of replication-competent adenovirus-mediated double-suicide gene therapy in combination with conventional-dose three-dimensional conformal radiation therapy for the treatment of newly diagnosed, intermediate- to high-risk prostate cancer. *Cancer Research*, *63*(21), 7497–7506.

Fuerer, C., & Iggo, R. (2002). Adenoviruses with Tcf binding sites in multiple early promoters show enhanced selectivity for tumour cells with constitutive activation of the wnt signalling pathway. *Gene Therapy*, *9*(4), 270–281.

Fueyo, J., Gomez-Manzano, C., Alemany, R., Lee, P. S., McDonnell, T. J., Mitlianga, P., et al. (2000). A mutant oncolytic adenovirus targeting the Rb pathway produces anti-glioma effect in vivo. *Oncogene*, *19*(1), 2–12.

Fujiwara, T., Shirakawa, Y., & Kagawa, S. (2011). Telomerase-specific oncolytic virotherapy for human gastrointestinal cancer. *Expert Review of Anticancer Therapy*, *11*(4), 525–532.

Fujiwara, T., Urata, Y., & Tanaka, N. (2007). Telomerase-specific oncolytic virotherapy for human cancer with the hTERT promoter. *Current Cancer Drug Targets*, 7(2), 191–201.

Fukazawa, T., Matsuoka, J., Yamatsuji, T., Maeda, Y., Durbin, M. L., & Naomoto, Y. (2010). Adenovirus-mediated cancer gene therapy and virotherapy (Review). *International Journal of Molecular Medicine*, *25*(1), 3–10.

Goldberg, B. B., Liu, J. B., & Forsberg, F. (1994). Ultrasound contrast agents: A review. *Ultrasound in Medicine & Biology*, *20*(4), 319–333.

Gomez-Manzano, C., Fueyo, J., Kyritsis, A. P., Steck, P. A., Roth, J. A., McDonnell, T. J., et al. (1996). Adenovirus-mediated transfer of the p53 gene produces rapid and generalized death of human glioma cells via apoptosis. *Cancer Research*, *56*(4), 694–699.

Gopalkrishnan, R. V., Sauane, M., & Fisher, P. B. (2004). Cytokine and tumor cell apoptosis inducing activity of mda-7/IL-24. *International Immunopharmacology*, *4*(5), 635–647.

Greco, A., Di Benedetto, A., Howard, C. M., Kelly, S., Nande, R., Dementieva, Y., et al. (2010). Eradication of therapy-resistant human prostate tumors using an ultrasound-guided site-specific cancer terminator virus delivery approach. *Molecular Therapy, 18* (2), 295–306.
Gupta, P., Emdad, L., Lebedeva, I. V., Sarkar, D., Dent, P., Curiel, D. T., et al. (2006). mda-7/IL-24: Multifunctional cancer-specific apoptosis-inducing cytokine. *Pharmacology and Therapeutics, 111*(3), 596–628.
Hall, A. R., Dix, B. R., O'Carroll, S. J., & Braithwaite, A. W. (1998). p53-dependent cell death/apoptosis is required for a productive adenovirus infection. *Nature Medicine, 4*(9), 1068–1072.
Hallenbeck, P. L., Chang, Y. N., Hay, C., Golightly, D., Stewart, D., Lin, J., et al. (1999). A novel tumor-specific replication-restricted adenoviral vector for gene therapy of hepatocellular carcinoma. *Human Gene Therapy, 10*(10), 1721–1733.
Hamed, H. A., Yacoub, A., Park, M. A., Eulitt, P. J., Dash, R., Sarkar, D., et al. (2010). Inhibition of multiple protective signaling pathways and Ad.5/3 delivery enhances mda-7/IL-24 therapy of malignant glioma. *Molecular Therapy, 18*(6), 1130–1142.
Haviv, Y. S., Blackwell, J. L., Kanerva, A., Nagi, P., Krasnykh, V., Dmitriev, I., et al. (2002). Adenoviral gene therapy for renal cancer requires retargeting to alternative cellular receptors. *Cancer Research, 62*(15), 4273–4281.
Heise, C., Hermiston, T., Johnson, L., Brooks, G., Sampson-Johannes, A., Williams, A., et al. (2000). An adenovirus E1A mutant that demonstrates potent and selective systemic anti-tumoral efficacy. *Nature Medicine, 6*(10), 1134–1139.
Heise, C., & Kirn, D. H. (2000). Replication-selective adenoviruses as oncolytic agents. *The Journal of Clinical Investigation, 105*(7), 847–851.
Heise, C., Sampson-Johannes, A., Williams, A., McCormick, F., Von Hoff, D. D., & Kirn, D. H. (1997). ONYX-015, an E1B gene-attenuated adenovirus, causes tumor-specific cytolysis and antitumoral efficacy that can be augmented by standard chemotherapeutic agents. *Nature Medicine, 3*(6), 639–645.
Heise, C. C., Williams, A. M., Xue, S., Propst, M., & Kirn, D. H. (1999). Intravenous administration of ONYX-015, a selectively replicating adenovirus, induces antitumoral efficacy. *Cancer Research, 59*(11), 2623–2628.
Hoffmann, D., & Wildner, O. (2006). Restriction of adenoviral replication to the transcriptional intersection of two different promoters for colorectal and pancreatic cancer treatment. *Molecular Cancer Therapeutics, 5*(2), 374–381.
Hollander, M. C., Poola-Kella, S., & Fornace, A. J., Jr. (2003). Gadd34 functional domains involved in growth suppression and apoptosis. *Oncogene, 22*(25), 3827–3832.
Howard, C. M., Forsberg, F., Minimo, C., Liu, J. B., Merton, D. A., & Claudio, P. P. (2006). Ultrasound guided site specific gene delivery system using adenoviral vectors and commercial ultrasound contrast agents. *Journal of Cellular Physiology, 209*(2), 413–421.
Huang, E. Y., Madireddi, M. T., Gopalkrishnan, R. V., Leszczyniecka, M., Su, Z., Lebedeva, I. V., et al. (2001). Genomic structure, chromosomal localization and expression profile of a novel melanoma differentiation associated (mda-7) gene with cancer specific growth suppressing and apoptosis inducing properties. *Oncogene, 20*(48), 7051–7063.
Huang, T. G., Savontaus, M. J., Shinozaki, K., Sauter, B. V., & Woo, S. L. (2003). Telomerase-dependent oncolytic adenovirus for cancer treatment. *Gene Therapy, 10* (15), 1241–1247.
Inoue, S., Shanker, M., Miyahara, R., Gopalan, B., Patel, S., Oida, Y., et al. (2006). MDA-7/IL-24-based cancer gene therapy: Translation from the laboratory to the clinic. *Current Gene Therapy, 6*(1), 73–91.
Irving, J., Wang, Z., Powell, S., O'Sullivan, C., Mok, M., Murphy, B., et al. (2004). Conditionally replicative adenovirus driven by the human telomerase promoter provides

broad-spectrum antitumor activity without liver toxicity. *Cancer Gene Therapy, 11*(3), 174–185.
Jaffee, E. M., Hruban, R. H., Canto, M., & Kern, S. E. (2002). Focus on pancreas cancer. *Cancer Cell, 2*(1), 25–28.
Jemal, A., Siegel, R., Xu, J., & Ward, E. (2010). Cancer statistics, 2010. *CA: A Cancer Journal for Clinicians, 60*(5), 277–300.
Jiang, H., & Fisher, P. B. (1993). Use of a sensitive and efficient subtraction hybridization protocol for the identification of genes differentially regulated during the induction of differentiation in human melanoma cells. *Molecular and Cellular Differentiation, 1*(3), 285–299.
Jiang, H., Lin, J. J., Su, Z. Z., Goldstein, N. I., & Fisher, P. B. (1995). Subtraction hybridization identifies a novel melanoma differentiation associated gene, mda-7, modulated during human melanoma differentiation, growth and progression. *Oncogene, 11*(12), 2477–2486.
Jiang, H., Shah, S., & Hilt, D. C. (1993). Organization, sequence, and expression of the murine S100 beta gene. Transcriptional regulation by cell type-specific cis-acting regulatory elements. *The Journal of Biological Chemistry, 268*(27), 20502–20511.
Jiang, H., Su, Z. Z., Lin, J. J., Goldstein, N. I., Young, C. S., & Fisher, P. B. (1996). The melanoma differentiation associated gene mda-7 suppresses cancer cell growth. *Proceedings of the National Academy of Sciences of the United States of America, 93*(17), 9160–9165.
Johnson, L., Shen, A., Boyle, L., Kunich, J., Pandey, K., Lemmon, M., et al. (2002). Selectively replicating adenoviruses targeting deregulated E2F activity are potent, systemic antitumor agents. *Cancer Cell, 1*(4), 325–337.
Kanai, F., Shiratori, Y., Yoshida, Y., Wakimoto, H., Hamada, H., Kanegae, Y., et al. (1996). Gene therapy for alpha-fetoprotein-producing human hepatoma cells by adenovirus-mediated transfer of the herpes simplex virus thymidine kinase gene. *Hepatology, 23* (6), 1359–1368.
Kanerva, A., Mikheeva, G. V., Krasnykh, V., Coolidge, C. J., Lam, J. T., Mahasreshti, P. J., et al. (2002). Targeting adenovirus to the serotype 3 receptor increases gene transfer efficiency to ovarian cancer cells. *Clinical Cancer Research, 8*(1), 275–280.
Kawakami, Y., Li, H., Lam, J. T., Krasnykh, V., Curiel, D. T., & Blackwell, J. L. (2003). Substitution of the adenovirus serotype 5 knob with a serotype 3 knob enhances multiple steps in virus replication. *Cancer Research, 63*(6), 1262–1269.
Kim, J., Cho, J. Y., Kim, J. H., Jung, K. C., & Yun, C. O. (2002). Evaluation of E1B gene-attenuated replicating adenoviruses for cancer gene therapy. *Cancer Gene Therapy, 9*(9), 725–736.
Kim, E., Kim, J. H., Shin, H. Y., Lee, H., Yang, J. M., Kim, J., et al. (2003). Ad-mTERT-delta19, a conditional replication-competent adenovirus driven by the human telomerase promoter, selectively replicates in and elicits cytopathic effect in a cancer cell-specific manner. *Human Gene Therapy, 14*(15), 1415–1428.
Kim, M., Zinn, K. R., Barnett, B. G., Sumerel, L. A., Krasnykh, V., Curiel, D. T., et al. (2002). The therapeutic efficacy of adenoviral vectors for cancer gene therapy is limited by a low level of primary adenovirus receptors on tumour cells. *European Journal of Cancer, 38*(14), 1917–1926.
Kirn, D. H., & McCormick, F. (1996). Replicating viruses as selective cancer therapeutics. *Molecular Medicine Today, 2*(12), 519–527.
Krajewska, M., Krajewski, S., Epstein, J. I., Shabaik, A., Sauvageot, J., Song, K., et al. (1996). Immunohistochemical analysis of bcl-2, bax, bcl-X, and mcl-1 expression in prostate cancers. *The American Journal of Pathology, 148*(5), 1567–1576.
Krajewski, S., Krajewska, M., Shabaik, A., Wang, H. G., Irie, S., Fong, L., et al. (1994). Immunohistochemical analysis of in vivo patterns of Bcl-X expression. *Cancer Research, 54*(21), 5501–5507.

Krasnykh, V. N., Mikheeva, G. V., Douglas, J. T., & Curiel, D. T. (1996). Generation of recombinant adenovirus vectors with modified fibers for altering viral tropism. *Journal of Virology, 70*(10), 6839–6846.

Lanson, N. A., Jr., Friedlander, P. L., Schwarzenberger, P., Kolls, J. K., & Wang, G. (2003). Replication of an adenoviral vector controlled by the human telomerase reverse transcriptase promoter causes tumor-selective tumor lysis. *Cancer Research, 63*(22), 7936–7941.

Larina, I. V., Evers, B. M., Ashitkov, T. V., Bartels, C., Larin, K. V., & Esenaliev, R. O. (2005). Enhancement of drug delivery in tumors by using interaction of nanoparticles with ultrasound radiation. *Technology in Cancer Research & Treatment, 4*(2), 217–226.

Lawrie, A., Brisken, A. F., Francis, S. E., Cumberland, D. C., Crossman, D. C., & Newman, C. M. (2000). Microbubble-enhanced ultrasound for vascular gene delivery. *Gene Therapy, 7*(23), 2023–2027.

Lebedeva, I. V., Emdad, L., Su, Z. Z., Gupta, P., Sauane, M., Sarkar, D., et al. (2007). mda-7/IL-24, novel anticancer cytokine: Focus on bystander antitumor, radiosensitization and antiangiogenic properties and overview of the phase I clinical experience (Review). *International Journal of Oncology, 31*(5), 985–1007.

Lebedeva, I. V., Sarkar, D., Su, Z. Z., Kitada, S., Dent, P., Stein, C. A., et al. (2003). Bcl-2 and Bcl-x(L) differentially protect human prostate cancer cells from induction of apoptosis by melanoma differentiation associated gene-7, mda-7/IL-24. *Oncogene, 22*(54), 8758–8773.

Lebedeva, I. V., Su, Z. Z., Sarkar, D., Kitada, S., Dent, P., Waxman, S., et al. (2003). Melanoma differentiation associated gene-7, mda-7/interleukin-24, induces apoptosis in prostate cancer cells by promoting mitochondrial dysfunction and inducing reactive oxygen species. *Cancer Research, 63*(23), 8138–8144.

Lee, B., Choi, J., Kim, J., Kim, J. H., Joo, C. H., Cho, Y. K., et al. (2002). Oncolysis of human gastric cancers by an E1B 55 kDa-deleted YKL-1 adenovirus. *Cancer Letters, 185*(2), 225–233.

Li, Y., Chen, Y., Dilley, J., Arroyo, T., Ko, D., Working, P., et al. (2003). Carcinoembryonic antigen-producing cell-specific oncolytic adenovirus, OV798, for colorectal cancer therapy. *Molecular Cancer Therapeutics, 2*(10), 1003–1009.

Liang, J., Huang, R. L., Huang, Q., Peng, Z., Zhang, P. H., & Wu, Z. X. (2011). Adenovirus-mediated human interleukin 24 (MDA-7/IL-24) selectively suppresses proliferation and induces apoptosis in keloid fibroblasts. *Annals of Plastic Surgery, 66*(6), 660–666.

Mathis, J. M., Stewart, P. L., Zhu, Z. B., & Curiel, D. T. (2006). Advanced generation adenoviral virotherapy agents embody enhanced potency based upon CAR-independent tropism. *Clinical Cancer Research, 12*(9), 2651–2656.

Miao, J., Chen, G. G., Chun, S. Y., Yun, J. P., Chak, E. C., Ho, R. L., et al. (2006). Adenovirus-mediated tBid overexpression results in therapeutic effects on p53-resistant hepatocellular carcinoma. *International Journal of Cancer, 119*(8), 1985–1993.

Nettelbeck, D. M. (2003). Virotherapeutics: Conditionally replicative adenoviruses for viral oncolysis. *Anti-Cancer Drugs, 14*(8), 577–584.

Nettelbeck, D. M., Rivera, A. A., Balague, C., Alemany, R., & Curiel, D. T. (2002). Novel oncolytic adenoviruses targeted to melanoma: Specific viral replication and cytolysis by expression of E1A mutants from the tyrosinase enhancer/promoter. *Cancer Research, 62*(16), 4663–4670.

Ng, K. Y., & Liu, Y. (2002). Therapeutic ultrasound: Its application in drug delivery. *Medicinal Research Reviews, 22*(2), 204–223.

O'Shea, C. C., Johnson, L., Bagus, B., Choi, S., Nicholas, C., Shen, A., et al. (2004). Late viral RNA export, rather than p53 inactivation, determines ONYX-015 tumor selectivity. *Cancer Cell, 6*(6), 611–623.

Park, J. S., Qiao, L., Su, Z. Z., Hinman, D., Willoughby, K., McKinstry, R., et al. (2008). PERK-dependent regulation of MDA-7/IL-24-induced autophagy in primary human glioma cells. *Autophagy*, *4*(4), 513–515.

Pitt, W. G., Husseini, G. A., & Staples, B. J. (2004). Ultrasonic drug delivery—A general review. *Expert Opinion on Drug Delivery*, *1*(1), 37–56.

Ramesh, R., Mhashilkar, A. M., Tanaka, F., Saito, Y., Branch, C. D., Sieger, K., et al. (2003). Melanoma differentiation-associated gene 7/interleukin (IL)-24 is a novel ligand that regulates angiogenesis via the IL-22 receptor. *Cancer Research*, *63*(16), 5105–5113.

Rein, D. T., Breidenbach, M., Kirby, T. O., Han, T., Siegal, G. P., Bauerschmitz, G. J., et al. (2005). A fiber-modified, secretory leukoprotease inhibitor promoter-based conditionally replicating adenovirus for treatment of ovarian cancer. *Clinical Cancer Research*, *11*(3), 1327–1335.

Rivera, A. A., Davydova, J., Schierer, S., Wang, M., Krasnykh, V., Yamamoto, M., et al. (2004). Combining high selectivity of replication with fiber chimerism for effective adenoviral oncolysis of CAR-negative melanoma cells. *Gene Therapy*, *11*(23), 1694–1702.

Rodriguez, R., Schuur, E. R., Lim, H. Y., Henderson, G. A., Simons, J. W., & Henderson, D. R. (1997). Prostate attenuated replication competent adenovirus (ARCA) CN706: A selective cytotoxic for prostate-specific antigen-positive prostate cancer cells. *Cancer Research*, *57*(13), 2559–2563.

Roth, J. A., & Cristiano, R. J. (1997). Gene therapy for cancer: What have we done and where are we going? *Journal of the National Cancer Institute*, *89*(1), 21–39.

Russo, S., Butler, J., Ove, R., & Blackstock, A. W. (2007). Locally advanced pancreatic cancer: A review. *Seminars in Oncology*, *34*(4), 327–334.

Sachs, L. (1978). Control of normal cell differentiation and the phenotypic reversion of malignancy in myeloid leukaemia. *Nature*, *274*(5671), 535–539.

Sachs, L. (1987). Cell differentiation and bypassing of genetic defects in the suppression of malignancy. *Cancer Research*, *47*(8), 1981–1986.

Sachs, L. (1990). The control of growth and differentiation in normal and leukemic blood cells. *Cancer*, *65*(10), 2196–2206.

Sarkar, D., Lebedeva, I. V., Gupta, P., Emdad, L., Sauane, M., Dent, P., et al. (2007). Melanoma differentiation associated gene-7 (mda-7)/IL-24: A 'magic bullet' for cancer therapy? *Expert Opinion on Biological Therapy*, 7(5), 577–586.

Sarkar, D., Su, Z. Z., & Fisher, P. B. (2006). Unique conditionally replication competent bipartite adenoviruses-cancer terminator viruses (CTV): Efficacious reagents for cancer gene therapy. *Cell Cycle*, *5*(14), 1531–1536.

Sarkar, D., Su, Z. Z., Lebedeva, I. V., Sauane, M., Gopalkrishnan, R. V., Valerie, K., et al. (2002). mda-7 (IL-24) Mediates selective apoptosis in human melanoma cells by inducing the coordinated overexpression of the GADD family of genes by means of p38 MAPK. *Proceedings of the National Academy of Sciences of the United States of America*, *99* (15), 10054–10059.

Sarkar, D., Su, Z. Z., Park, E. S., Vozhilla, N., Dent, P., Curiel, D. T., et al. (2008). A cancer terminator virus eradicates both primary and distant human melanomas. *Cancer Gene Therapy*, *15*(5), 293–302.

Sarkar, D., Su, Z. Z., Vozhilla, N., Park, E. S., Gupta, P., & Fisher, P. B. (2005). Dual cancer-specific targeting strategy cures primary and distant breast carcinomas in nude mice. *Proceedings of the National Academy of Sciences of the United States of America*, *102*(39), 14034–14039.

Sarkar, D., Su, Z. Z., Vozhilla, N., Park, E. S., Randolph, A., Valerie, K., et al. (2005). Targeted virus replication plus immunotherapy eradicates primary and distant pancreatic tumors in nude mice. *Cancer Research*, *65*(19), 9056–9063.

Sauane, M., Gopalkrishnan, R. V., Lebedeva, I., Mei, M. X., Sarkar, D., Su, Z. Z., et al. (2003). MDA-7/IL-24: Novel cancer growth suppressing and apoptosis inducing cytokine. *Cytokine & Growth Factor Reviews, 14*(1), 35–51.

Sauane, M., Lebedeva, I. V., Su, Z. Z., Choo, H. T., Randolph, A., Valerie, K., et al. (2004). Melanoma differentiation associated gene-7/interleukin-24 promotes tumor cell-specific apoptosis through both secretory and nonsecretory pathways. *Cancer Research, 64*(9), 2988–2993.

Sauane, M., Su, Z. Z., Gupta, P., Lebedeva, I. V., Dent, P., Sarkar, D., et al. (2008). Autocrine regulation of mda-7/IL-24 mediates cancer-specific apoptosis. *Proceedings of the National Academy of Sciences of the United States of America, 105*(28), 9763–9768.

Smith, A. E. (1995). Viral vectors in gene therapy. *Annual Review of Microbiology, 49*, 807–838.

Smith, R. R., Huebner, R. J., Rowe, W. P., Schatten, W. E., & Thomas, L. B. (1956). Studies on the use of viruses in the treatment of carcinoma of the cervix. *Cancer, 9*(6), 1211–1218.

Staudt, M. R., Depass, A. L., Sarkar, D., & Fisher, P. B. (2009). Model cell culture system for defining the molecular and biochemical events mediating terminal differentiation of human melanoma cells. *Journal of Cellular Physiology, 218*(2), 304–314.

Staudt, M. R., & Dittmer, D. P. (2007). The Rta/Orf50 transactivator proteins of the gamma-herpesviridae. *Current Topics in Microbiology and Immunology, 312*, 71–100.

Sternberg, C. N. (2002). Highlights of contemporary issues in the medical management of prostate cancer. *Critical Reviews in Oncology/Hematology, 43*(2), 105–121.

Su, Z. Z., Emdad, L., Sarkar, D., Randolph, A., Valerie, K., Yacoub, A., et al. (2005). Potential molecular mechanism for rodent tumorigenesis: Mutational generation of Progression Elevated Gene-3 (PEG-3). *Oncogene, 24*(13), 2247–2255.

Su, Z., Emdad, L., Sauane, M., Lebedeva, I. V., Sarkar, D., Gupta, P., et al. (2005). Unique aspects of mda-7/IL-24 antitumor bystander activity: Establishing a role for secretion of MDA-7/IL-24 protein by normal cells. *Oncogene, 24*(51), 7552–7566.

Su, Z. Z., Gopalkrishnan, R. V., Narayan, G., Dent, P., & Fisher, P. B. (2002). Progression elevated gene-3, PEG-3, induces genomic instability in rodent and human tumor cells. *Journal of Cellular Physiology, 192*(1), 34–44.

Su, Z. Z., Lebedeva, I. V., Sarkar, D., Gopalkrishnan, R. V., Sauane, M., Sigmon, C., et al. (2003). Melanoma differentiation associated gene-7, mda-7/IL-24, selectively induces growth suppression, apoptosis and radiosensitization in malignant gliomas in a p53-independent manner. *Oncogene, 22*(8), 1164–1180.

Su, C., Peng, L., Sham, J., Wang, X., Zhang, Q., Chua, D., et al. (2006). Immune gene-viral therapy with triplex efficacy mediated by oncolytic adenovirus carrying an interferon-gamma gene yields efficient antitumor activity in immunodeficient and immunocompetent mice. *Molecular Therapy, 13*(5), 918–927.

Su, Z. Z., Sarkar, D., Emdad, L., Duigou, G. J., Young, C. S., Ware, J., et al. (2005). Targeting gene expression selectively in cancer cells by using the progression-elevated gene-3 promoter. *Proceedings of the National Academy of Sciences of the United States of America, 102*(4), 1059–1064.

Su, Z. Z., Shi, Y., & Fisher, P. B. (1997). Subtraction hybridization identifies a transformation progression-associated gene PEG-3 with sequence homology to a growth arrest and DNA damage-inducible gene. *Proceedings of the National Academy of Sciences of the United States of America, 94*(17), 9125–9130.

Su, Z., Shi, Y., & Fisher, P. B. (2000). Cooperation between AP1 and PEA3 sites within the progression elevated gene-3 (PEG-3) promoter regulate basal and differential expression of PEG-3 during progression of the oncogenic phenotype in transformed rat embryo cells. *Oncogene, 19*(30), 3411–3421.

Su, Z., Shi, Y., Friedman, R., Qiao, L., McKinstry, R., Hinman, D., et al. (2001). PEA3 sites within the progression elevated gene-3 (PEG-3) promoter and mitogen-activated protein kinase contribute to differential PEG-3 expression in Ha-ras and v-raf oncogene transformed rat embryo cells. *Nucleic Acids Research*, *29*(8), 1661–1671.

Thomas, C. E., Ehrhardt, A., & Kay, M. A. (2003). Progress and problems with the use of viral vectors for gene therapy. *Nature Reviews Genetics*, *4*(5), 346–358.

Tong, A. W., Nemunaitis, J., Su, D., Zhang, Y., Cunningham, C., Senzer, N., et al. (2005). Intratumoral injection of INGN 241, a nonreplicating adenovector expressing the melanoma-differentiation associated gene-7 (mda-7/IL24): Biologic outcome in advanced cancer patients. *Molecular Therapy*, *11*(1), 160–172.

Tooze, J. (1980). *Molecular biology of tumor viruses (Part 2)*. New York: Cold Spring Harbor Laboratory.

Turnell, A. S., Grand, R. J., & Gallimore, P. H. (1999). The replicative capacities of large E1B-null group A and group C adenoviruses are independent of host cell p53 status. *Journal of Virology*, *73*(3), 2074–2083.

Van Houdt, W. J., Haviv, Y. S., Lu, B., Wang, M., Rivera, A. A., Ulasov, I. V., et al. (2006). The human survivin promoter: A novel transcriptional targeting strategy for treatment of glioma. *Journal of Neurosurgery*, *104*(4), 583–592.

Volk, A. L., Rivera, A. A., Kanerva, A., Bauerschmitz, G., Dmitriev, I., Nettelbeck, D. M., et al. (2003). Enhanced adenovirus infection of melanoma cells by fiber-modification: Incorporation of RGD peptide or Ad5/3 chimerism. *Cancer Biology & Therapy*, *2*(5), 511–515.

Whyte, P., Ruley, H. E., & Harlow, E. (1988). Two regions of the adenovirus early region 1A proteins are required for transformation. *Journal of Virology*, *62*(1), 257–265.

Wildner, O., Blaese, R. M., & Morris, J. C. (1999). Therapy of colon cancer with oncolytic adenovirus is enhanced by the addition of herpes simplex virus-thymidine kinase. *Cancer Research*, *59*(2), 410–413.

Wirth, T., Zender, L., Schulte, B., Mundt, B., Plentz, R., Rudolph, K. L., et al. (2003). A telomerase-dependent conditionally replicating adenovirus for selective treatment of cancer. *Cancer Research*, *63*(12), 3181–3188.

Wu, Y. M., Zhang, K. J., Yue, X. T., Wang, Y. Q., Yang, Y., Li, G. C., et al. (2009). Enhancement of tumor cell death by combining cisplatin with an oncolytic adenovirus carrying MDA-7/IL-24. *Acta Pharmacologica Sinica*, *30*(4), 467–477.

Wu, T. L., & Zhou, D. (2011). Viral delivery for gene therapy against cell movement in cancer. *Advanced Drug Delivery Reviews*, *63*(8), 671–677.

Xiao, L., Li, X., Niu, N., Qian, J., Xie, G., & Wang, Y. (2010). Dichloroacetate (DCA) enhances tumor cell death in combination with oncolytic adenovirus armed with MDA-7/IL-24. *Molecular and Cellular Biochemistry*, *340*(1–2), 31–40.

Xue, X. B., Xiao, C. W., Zhang, H., Lu, A. G., Gao, W., Zhou, Z. Q., et al. (2010). Oncolytic adenovirus SG600-IL24 selectively kills hepatocellular carcinoma cell lines. *World Journal of Gastroenterology*, *16*(37), 4677–4684.

Yamamoto, M., & Curiel, D. T. (2010). Current issues and future directions of oncolytic adenoviruses. *Molecular Therapy*, *18*(2), 243–250.

Yang, B. X., Duan, Y. J., Dong, C. Y., Zhang, F., Gao, W. F., Cui, X. Y., et al. (2011). Novel functions for mda-7/IL-24 and IL-24 delE5: Regulation of differentiation of acute myeloid leukemic cells. *Molecular Cancer Therapeutics*, *10*(4), 615–625.

Yang, Y., Nunes, F. A., Berencsi, K., Gonczol, E., Engelhardt, J. F., & Wilson, J. M. (1994). Inactivation of E2a in recombinant adenoviruses improves the prospect for gene therapy in cystic fibrosis. *Nature Genetics*, *7*(3), 362–369.

Yang, Y., Su, Q., & Wilson, J. M. (1996). Role of viral antigens in destructive cellular immune responses to adenovirus vector-transduced cells in mouse lungs. *Journal of Virology*, *70*(10), 7209–7212.

Yang, Y., & Wilson, J. M. (1995). Clearance of adenovirus-infected hepatocytes by MHC class I-restricted CD4+ CTLs in vivo. *The Journal of Immunology, 155*(5), 2564–2570.

Ye, X., Lu, Q., Zhao, Y., Ren, Z., Ren, X. W., Qiu, Q. H., et al. (2005). Conditionally replicative adenovirus vector carrying TRAIL gene for enhanced oncolysis of human hepatocellular carcinoma. *International Journal of Molecular Medicine, 16*(6), 1179–1184.

Zhaohui, P. (2005). Current status of gendicine in China: Recombinant human Ad-p53 agent for treatment of cancers. *Human Gene Therapy, 16*, 1016–1027.

Zou, W., Luo, C., Zhang, Z., Liu, J., Gu, J., Pei, Z., et al. (2004). A novel oncolytic adenovirus targeting to telomerase activity in tumor cells with potent. *Oncogene, 23* (2), 457–464.

CHAPTER TWO

# Adenovirus Strategies for Tissue-Specific Targeting

**Matthew S. Beatty**[*,†], **David T. Curiel**[*,1]
[*]Division of Cancer Biology, Department of Radiation Oncology, Washington University School of Medicine in St. Louis, St. Louis, Missouri, USA
[†]Division of Molecular and Cellular Pathology, Department of Pathology, University of Alabama at Birmingham, Birmingham, Alabama, USA
[1]Corresponding author: e-mail address: dcuriel@wustl.edu

## Contents



## Abstract

Cancer gene therapy approaches have benefited greatly from the utilization of molecular-based therapeutics. Of these, adenovirus-based interventions hold much promise as a platform for targeted therapeutic delivery to tumors. However, a barrier to this progression is the lack of native adenovirus receptor expression on a variety of cancer types. As such, any adenovirus-based cancer therapy must take into consideration retargeting the vector to nonnative cellular surface receptors. Predicated upon the

*Advances in Cancer Research*, Volume 115
ISSN 0065-230X
http://dx.doi.org/10.1016/B978-0-12-398342-8.00002-1

knowledge gained in native adenovirus biology, several strategies to transductionally retarget adenovirus have emerged. Herein, we describe the biological hurdles as well as strategies utilized in adenovirus transductional targeting, covering the progress of both adapter-based and genetic manipulation-based targeting. Additionally, we discuss recent translation of these targeting strategies into a clinical setting.

## 1. INTRODUCTION

The development of rationally designed cancer interventions has followed the progress of the molecular understanding of cancer development and progression. To this end, gene therapy has endeavored multiple strategies for molecular targeted therapeutics. Of these strategies, adenovirus (Ad)-based vectors have been used prevalently, especially in the field of cancer. Ad vectors entail many characteristics that make it an ideal choice compared to other vectors. Biologically, Ad is able to efficiently transduce a variety of both dividing and quiescent cell types *in vitro* and *in vivo*. Additionally, the amenability to genetic modification, large genetic payload capacity, and the ability to produce high titers of good manufacturing practice quality are all factors that favor the use of Ad-based vectors as cancer therapeutics. Of importance, Ad-based vectors have shown an impressive safety record in the preclinical and clinical setting. However, despite the safety profile and preclinical efficacy, these vectors have failed to achieve therapeutic efficacy in the clinical setting. Thus, efforts have been refocused on basic vector design, especially on maximizing gene delivery by specifically transducing the target cell population. Achievements in Ad transductional targeting technology have steadily progressed and show great promise as a therapeutic for cancer treatment.

## 2. AD BIOLOGY

The most common Ad used in gene therapy, human Ad serotype 5, is a member of the Adenoviridae family. This family comprises 51 Ad serotypes originally classified by their ability to be neutralized by animal antisera. These serotypes are then divided into six species based upon hemagglutination properties, oncogenicity, and genomic structure (Davison, Benko, & Harrach, 2003). In addition, there are some correlations between species and tissue target and clinical presentation. Of importance to current Ad-based gene therapy vectors are species B, C, and D. These Ad species comprise the serotypes most commonly used in gene therapy.

## 3. AD CAPSID STRUCTURE

The human Ad is a 36-kb double-stranded DNA virus encapsulated by an icosahedral protein particle (Fig. 2.1). Progress in X-ray crystallography and cryo-electron microscopy has furthered our understanding of the structural components of the Ad capsid (Fabry et al., 2005; Reddy, Natchiar, Stewart, & Nemerow, 2010; Saban, Silvestry, Nemerow, & Stewart, 2006). Three major proteins comprise most of the Ad capsid. Hexon is a homotrimeric protein found in 720 copies and comprises most of the Ad capsid, playing a structural role as the main coating protein. The Ad hexon is highly conserved among human serotypes; however, hexon contains nine surface exposed hypervariable regions (HVR1–HVR9) which contain serotype variation (Burnett, 1985). Penton forms a homopentamer which makes up the penton base. This pentameric structure is located at each of the 12 vertices and plays a structural role, interacting with each of the Ad capsid 12 fibers. In addition, penton contains an Arg-Gly-Asp (RGD) motif which is responsible for virion internalization. Fiber is found as a homotrimer at each vertex and noncovalently binds to its corresponding penton base via its N-terminus. Fiber is composed of a shaft domain and a globular knob

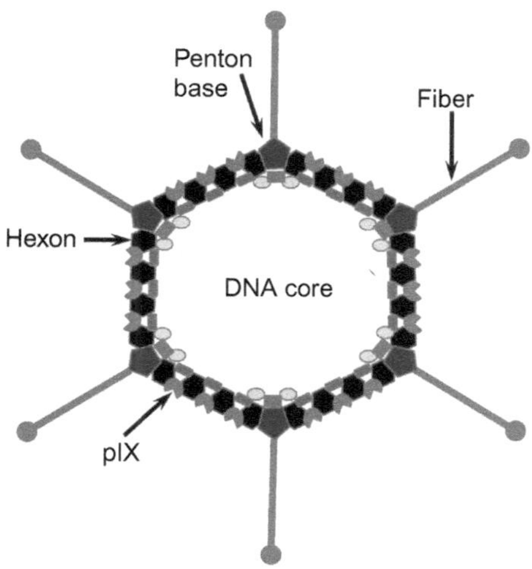


**Figure 2.1** Adenovirus capsid structure: The adenovirus capsid structure is composed of both major capsid proteins and minor proteins. Labeled here are capsid proteins hexon, penton base, fiber, and polypeptide IX (pIX). These capsid structures have all been utilized as locales for retargeting Ad strategies.

domain found at the distal tip, which plays a role as the primary cellular attachment site (Zubieta, Schoehn, Chroboczek, & Cusack, 2005). In addition to the major capsid proteins, the Ad capsid comprises of an assortment of minor proteins: IIIa, VI, VIII, and IX. Of note, polypeptide IX (pIX), which has gained favor as an alternative locus for targeting incorporation (discussed later), is a structural protein that is believed to help stabilize the Ad capsid (Vellinga, Van der Heijdt, & Hoeben, 2005). Four monomers of pIX interact to form a four-helix bundle with a surface exposed C-terminus (Marsh et al., 2006; Saban et al., 2006).

## 4. AD ENTRY BIOLOGY

Ad entry biology comprises two distinct steps: attachment of the virus to its primary receptor subsequently followed by molecular interactions that lead to internalization of the virus. Initially, high-affinity interactions occur when the fiber knob domain binds to its cognate primary receptor. In general, for serotypes from species A, C, E, and F, this receptor is the coxsackie and Ad receptor (CAR), while serotypes from species B and D tend to utilize alternative receptors. CAR is a 46-kDa protein that is a member of the immunoglobulin superfamily and is involved in the formation of tight junctions (Coyne & Bergelson, 2006; Philipson & Pettersson, 2004). Once the Ad virion has attached, cellular integrins including $\alpha v\beta 3$, $\alpha v\beta 5$ (Wickham, Mathias, Cheresh, & Nemerow, 1993), $\alpha v\beta 1$ (Li et al., 2001), $\alpha 3\beta 1$, and $\alpha 5\beta 1$ (Davison, Diaz, Hart, Santis, & Marshall, 1997) interact with RGD motifs in the penton base. This interaction induces cellular responses that lead to cytoskeleton alterations which aid in internalization (Li, Stupack, Bokoch, & Nemerow, 1998; Li, Stupack, Klemke, Cheresh, & Nemerow, 1998). Ultimately, virus internalization occurs via clathrin-coated vesicles, and the Ad virion is transported to the endosome (Meier et al., 2002). Upon endosomal acidification, the Ad virion disassembles and is released into the cytoplasm where it ultimately travels to the infected cells nucleus for viral replication.

## 5. TRANSDUCTIONAL TARGETING OF AD VECTORS

Knowledge gained from studies concerning native Ad entry biology has predicated understanding of findings that non-CAR-expressing cancer cells are refractory to Ad infection and gene delivery. Thus, if target cancer cells exhibit low levels of CAR, we are left with a scenario whereby high-CAR-expressing nontarget cells are effectively transduced, while low-CAR cancer cells show poor transduction. Higher expression of CAR also appears to be growth inhibitory in some cancers (Okegawa et al., 2001). In summary, targeting CAR appears to be strategically incompatible with Ad-based cancer therapeutics.

Biodistribution of Ad, although effected by CAR distribution, is not solely determined by expression profiles *in vivo* (Fechner et al., 1999). Systemic administration of Ad, intravenously, results in the majority of transduction occurring in the liver followed by the spleen, heart, lung, and kidneys of mice. This profile, however, does not correlate with the highest levels of CAR expression (Wood et al., 1999). This is especially true in regard to liver transduction which absorbs the vast majority of systemic Ad vector via hepatic kupffer cell uptake (Tao et al., 2001) and hepatocyte transduction, potentially resulting in liver toxicity. Due to the toxicity issues regarding liver transduction, this biological interaction *in vivo* has been given great scrutiny.

Early strategies to retarget Ad to non-CAR pathways were initially thought to also detarget the liver, as the initial hypothesis was that liver transduction was CAR and integrin dependent. However, studies that ablated CAR and integrin binding in the Ad capsid had little effect on biodistribution profiles (Alemany & Curiel, 2001; Smith et al., 2002). Thus, Ad liver tropism was shown to be linked to a novel pathway. Following initial studies implicating motifs in the fiber shaft (Breidenbach et al., 2004; Smith et al., 2003; Vigne et al., 2003), Shayakhmetov, Gaggar, Ni, Li, and Lieber (2005) reported a major role for fiber interactions with blood coagulation factors and complement component C4 binding protein in hepatocyte and kupffer cell transduction. Modification of the Ad5 fiber to ablate this interaction resulted in a 50-fold decrease in liver transduction along with reduced levels of liver toxicity. Analysis determined that this *in vivo* tropism was due to Ad associating with hepatocellular heparin sulfate proteoglycan and low-density lipoprotein receptor-related protein (Shayakhmetov et al., 2005).

More recent studies, however, have shown that fiber structure and motifs do not play a role in liver sequestration and have elucidated hexon interactions with blood coagulation factors as the major pathway directing hepatocyte transduction by systemic delivery of Ad. Kalyuzhniy et al. (2008) and Waddington et al. (2008) defined the specific interaction between blood coagulation factor X (FX) and hexon. In addition, utilizing structural studies,

FX was shown to interact with hypervariable regions 3, 5, and 7 of hexon. Both groups showed that this interaction could be inhibited by mutated forms of hexon or by pharmacological methods involving warfarin or snake venom protein X-bp (Kalyuzhniy et al., 2008; Waddington et al., 2008). Additionally, Waddington et al. elucidated that different serotypes of Ad interact with FX with different affinities, some to the point of not binding at all. Of note, those Ad serotypes that did not bind FX were all from species D. Following this, candidate viruses from low (Ad35) and nonbinding (Ad26 and Ad48) groups were examined *in vivo* by intravenous injection with or without X-bp protein. These viruses showed a lack of liver transduction (Waddington et al., 2008).

From this work, several groups have reported success with genetic manipulations of hexon to ablate liver sequestration. These strategies fall into two different categories. From the original studies, hexon mutations blocking the FX/hexon interaction have been utilized and shown to drastically reduce liver sequestration and transduction. This strategy was further developed by Alba et al., identifying the exact amino acids in hypervariable regions 5 and 7 responsible for FX binding. Altering either of these points resulted in a drastic decrease in FX binding and FX-mediated gene transduction (Alba et al., 2009). These FX-ablated Ad vectors also possessed altered biodistribution, with decreased liver transduction and greater vector accumulation in the spleen, especially following macrophage depletion (Alba et al., 2010). Additionally, since different serotypes have different affinities for FX, some labs have reported that either whole hexon swaps or hypervariable region swaps with lesser binding serotypes have also been successful in preventing liver uptake. Short et al. (2010) utilized the swapping of Ad5 hexon with that of Ad3 hexon. This modification was shown to block FX binding to Ad virions by surface plasmon resonance (SPR) analysis and prevent FX-mediated gene transduction *in vitro*. Interestingly, ablation of FX binding in an oncolytic Ad vector provided increased tumor killing and prolonged viral replication in a skov3.ip1 subcutaneous flank tumor model (Short et al., 2010).

In summary, primary biology determined by *in vitro* experimentation as well as host interactions outside of this primary pathway plays a role in the overall *in vivo* biodistribution of Ad. It is clear that alternative targets for cell transduction must be explored as a wide variety of target cells including cancer are not amenable to CAR-targeted Ad vectors. Two distinct approaches have been utilized to transductionally target Ad-based vectors: (1) adapter based and (2) genetically capsid modification. While these strategies show great efficacy in retargeting Ad-based therapeutics, it is also clear that liver detargeting must also be considered in any Ad vector as these external biological forces will effect overall target cell transduction *in vivo*. Thus, any Ad-based therapeutic must entail both detargeting strategies and retargeting strategies in order to reach its full therapeutic efficacy.

## 6. AD TRANSDUCTIONAL TARGETING: ADAPTER-BASED STRATEGIES

The adapter-based strategy (Fig. 2.2A) involves the use of a molecular bridge to retarget the Ad from its native primary receptor to a different cell surface receptor. This function is performed due to the bispecific nature of adapter molecules. One end of the adapter binds specifically to the Ad capsid, while the alternative end redirects Ad away from its native CAR binding and interacts with the alternative cellular receptor. Of note, this process does not impede with Ad infection as entry biology is determined by a two-step

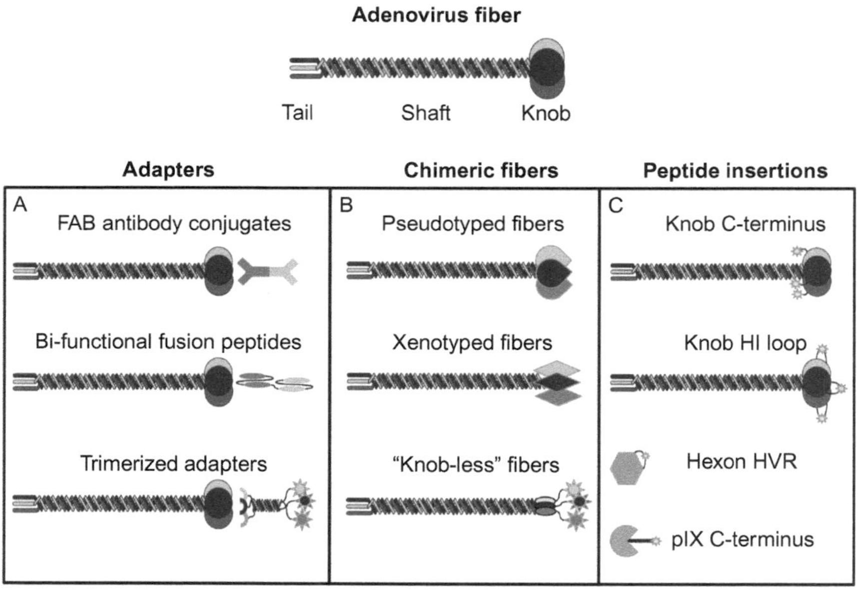


**Figure 2.2** Transductional retargeting modifications: Adenoviral vectors can be retargeted to specific cell surface receptors using a variety of strategies. (A) Heterologous retargeting ligands (adapters) composed of a variety of ligands including FAB antibodies, single-chain antibodies, and biological ligands. (B) Chimeric Ads composed of fiber/knob domains from alternative serotypes and "knob-less" fiber modifications. (C) Genetic incorporation of peptides into fiber, hexon, and pIX.

process, with binding being a separate mechanism from internalization. Conceptually, the most elegant location for conjugating an adapter molecule to the Ad virion is the fiber knob domain. This interaction allows for the retargeting of Ad to alternative receptors while also detargeting Ad from its primary receptor CAR. As such, within this retargeting strategy, the majority of adapters interact with the Ad knob domain. However, alternative capsid sites have been used as adapter interaction locales and have shown to be effective in retargeting Ad.

## 7. FAB ANTIBODY ADAPTERS

The first *in vitro* manifestations of the adapter strategy were accomplished by conjugating an anti-knob FAB antibody to a retargeting ligand. This strategy was first shown by chemically conjugating an anti-knob FAB antibody to folate. Conjugated to Ad virions, this adapter showed CAR-independent targeting to cancer cells overexpressing the folate receptor (Douglas et al., 1996). FAB antibody adapters against fibroblast growth factor 2 (FGF2) receptor have also been widely used and shown to retarget Ad-based vectors to a variety of cancers including pancreatic cancer (Huch et al., 2006), melanoma (Gu et al., 1999), Kaposi's sarcoma (Goldman et al., 1997), ovarian cancer (Printz et al., 2000), and head and neck cancer (Araki et al., 2010; Figures et al., 2009). In all, a wide variety of alternative cancer-relevant targeting ligands have been conjugated to anti-knob FAB antibodies and used to retarget Ad vectors to cellular receptors such as CD40 (Tillman et al., 1999), epidermal growth factor (EGF) receptor (Miller et al., 1998), epithelial cell adhesion molecule (EPCAM) (Haisma et al., 1999; Kraaij, van Rijswijk, Oomen, Haisma, & Bangma, 2005), prostate-specific membrane antigen (PSMA) (Kraaij et al., 2005), and Tag-72 (Kelly et al., 2000), resulting in similar transductional gains over untargeted Ad. Full utility of this strategy was progressed when Reynolds et al. (2000) employed an anti-knob FAB antibody chemically conjugated to monoclonal antibody (9B9) against angiotensin-converting enzyme (ACE). This cellular marker is found in the pulmonary epithelium and upregulated in various pulmonary diseases. Of note, this study showed a 20-fold increase in lung gene expression while also reducing liver expression by 83% when compared to an untargeted Ad vector. More recently, an anti-knob FAB antibody fused to FGF2 showed increased transduction efficiency in a mouse model of head and neck carcinoma. The targeted Ad, expressing a mutant Rad50 protein, also demonstrated an increase in tumor suppression compared to untargeted Ad when combined with cisplatinum, resulting in greater DNA double-strand breaks and reduced angiogenesis (Araki et al., 2010).

## 8. RECOMBINANT FUSION ADAPTERS

Studies have shown FAB antibody-based adapters to be effective in providing a proof-of-principle strategy for retargeting Ad vectors. However, the production of these molecules has been hampered by the randomness of chemical conjugation, producing heterogeneous populations of molecules. As such, strategies that utilize single-component molecules could be advantageous. Thus, more recent efforts have focused on the development of single-component fusion proteins that can be easily expressed and whose population is genetically homogenous. Building on the use of anti-knob antibodies, Haisma et al. (2000) developed a recombinant fusion protein composed of a single-chain antibody (scFv) against Ad fiber knob domain fused to an scFv against EGF receptor. This adapter could be easily purified without loss of function and increased Ad gene transfer compared to untargeted Ad in EGF receptor expressing cell lines (Haisma et al., 2000). Later, Haisma et al. showed selective targeting to a variety of angiogenesis-related markers including $\alpha v\beta 3$ integrins, vascular endothelial growth factor receptor 2, and the angiopoietin receptor TIE-2. These adapters retargeted Ad to both mouse and human endothelial cell lines. In addition, they showed *in vivo* tumor-specific targeting to a subcutaneous C26 colon carcinoma (Haisma et al., 2010). These scFv diabodies have also shown efficacy in a variety of studies retargeting Ad to cellular targets such as EGF receptor (Carette et al., 2007; Grill et al., 2001), EPCAM (Heideman et al., 2002), human epidermal growth factor receptor 2 (HER2/neu), carcinoembryonic antigen (CEA), endoglin (CD105) (Nettelbeck et al., 2001), and high molecular weight melanoma antigen (Nettelbeck et al., 2004).

In addition to scFv diabody-based adapters, a novel adapter utilizing a truncated, soluble portion of CAR (sCAR) fused to a targeting molecule was developed. Utilizing the first sCAR-based adapters, Dmitriev, Kashentseva, Rogers, Krasnykh, and Curiel (2000) created sCAR adapters fused to either an anti-CD40 antibody or EGF. With this, they demonstrated in several EGF-overexpressing cancer cell lines a ninefold increase in gene expression when compared to nontargeted Ad vectors. In addition to Ad-targeting specificity, Dmitriev et al. questioned the stability of these adapters when complexed with Ad virions. They showed through comparing

Ad/sCAR-EGF and Ad/sCAR-EGF purified through gel filtration that there was no difference in targeting profile, confirming the stability of these complexes. Similar to scFv diabody-based adapters, studies have shown this strategy to be efficacious in retargeting Ad vectors utilizing a wide variety of targeting ligands against EGF receptor (Harvey et al., 2010), urokinase-type plasminogen activator receptor (Harvey et al., 2010), CEA (Everts et al., 2005; Li, Everts, Yamamoto, Curiel, & Herschman, 2009), HER2/neu (Barker et al., 2003), CD40 (Hakkarainen et al., 2003), and high-affinity Fcgamma receptor I (CD64) (Ebbinghaus et al., 2001).

Kashentseva, Seki, Curiel, and Dmitriev (2002) furthered this strategy by incorporating a novel trimerization domain. This trimeric adapter contained sCAR fused to a HER2/neu-specific scFv and displayed increased affinity for the Ad fiber knob. In addition, the adapter augmented gene transduction 17-fold in HER2/neu positive breast and ovarian cancer cell lines. Kim et al. (2002) additionally reported that adapter trimerization provided drastic increase (100-fold) in gene transduction over its identical monomeric adapter. These studies have proven the targeting efficacy and stability of adapter virus complexes and have paved the way for these complexes to be utilized in a clinical setting, where complex stability could affect targeting efficacy and safety.

## 9. ALTERNATIVE ADAPTER BINDING LOCALES

In addition to bispecific adapters that utilize the Ad fiber knob domain for presentation, penton base and hexon have also been explored as sites for adapter-based retargeting. Li, Brown, Von Seggern, Brown, and Nemerow (2000) constructed a bispecific adapter molecule composed of an anti-penton monoclonal antibody fused to tumor necrosis factor α, insulin-like growth factor 1, and EGF. These adapters when complexed with Ad provided increased gene transduction in M21-L12 melanoma cells. The use of hexon as a retargeting local is predicated on newly identified *in vivo* biology regarding Ad interaction with blood coagulation factors. As previously described, upon entering circulation, Ad hexon interacts with FX. Chen, May, and Barry (2010) utilized this interaction to create a novel retargeting ligand by fusing the GLA domain of FX to an scFv against HER2/neu. When conjugated to Ad, this novel adapter promoted increased transduction of HER2/neu positive cells versus cells lacking HER2/neu expression. They additionally showed that this strategy could be used to target other cell surface receptors such as EGF receptor and ATP-binding cassette protein

G2. Since this technology utilizes the same binding interaction that leads to liver transduction, gains in retargeting Ad to an alternative receptor also yield gains in detargeting the liver.

## 10. ADAPTERS UTILIZING GENETICALLY MODIFIED AD CAPSID

Alternative to bispecific adapters, approaches that genetically modify the Ad virion to bind retargeting ligands have been developed. The benefit of this strategy is that a variety of off the shelf, commercially available ligands can be used without any additional modification. This strategy is best portrayed by incorporation of the immunoglobulin (Ig)-binding domain of *Staphylococcus aureus* protein A into the fiber HI loop or C-terminus. This fiber-modified Ad vector is thus able to bind a wide variety of targeting moieties that contain the Fc region of Ig. Most recently, Takahashi et al. (2011) utilized this technology to screen antibody libraries against the prostate cancer cell line LNCaP. Identifying an antibody against neural adhesion molecule 2, they showed that this antibody when conjugated to Ad increases gene transduction in prostate and breast cancer cell lines. A variety of other targets including CD40 (Korokhov et al., 2005), mesothelin (Breidenbach et al., 2005), EGF2 receptor, HER2/neu, CA242 antigen, and PSMA (Henning et al., 2005) have been explored. Utilizing the same concept, a biotin acceptor peptide (BAP) has been inserted into fiber (Parrott et al., 2003; Pereboeva, Komarova, Roth, Ponnazhagan, & Curiel, 2007). During virus production, BAP is biotinylated by the endogenous biotin ligase in 239 cells. Once purified, this virus can be conjugated to biotin-labeled ligand. pIX has also been explored as a site for BAP insertion. Campos et al. fused BAP to the terminus of pIX and compared targeting to BAP inserted into fiber. Of interest, using an anti-CD71 (transferin receptor) antibody as ligand, BAP fused to pIX failed to retarget Ad vectors unlike BAP inserted into fiber. However, when pIX-BAP Ad was retargeting with biotinylated transferrin, transduction was successful (Campos, Parrott, & Barry, 2004).

## 11. AD SECRETION OF ADAPTER

Although the adapter strategy for retargeting Ad vectors allows for efficient and specific retargeting, it is reliant upon a two-part system. The Ad vector and the adapter molecule are produced separately and are conjugated before being introduced to their cellular target. However, genetic incorporation of targeting ligands is biologically incompatible with a variety of ligands utilized in adapter-based strategies, such as scFvs. This is due to the fact that most scFv molecules must be processed through the endoplasmic reticulum (ER) in order to fold and function correctly, while Ad assembly takes place in the cytosol. Bridging this gap between cytosolically unstable targeting ligands and single-component Ad vector retargeting has been reports of Ad vectors that secrete their own adapter. With this, single-component Ad vectors can utilize targeting ligands which are not biologically compatible with viral assembly. In addition, while adapter-conjugated Ad vectors provide efficient targeting, this strategy only allows for a single round of targeting. Any viral progeny will regress to their native cellular receptor CAR. As a result, Ads that secrete their own adapter have targeting advantages over traditional adapter strategies in regard to replicating Ad vectors.

A novel method for achieving this was explored by Glasgow et al. This strategy utilizes a unique leucine zipper-based binding motif derived from vitellogenin gene-binding protein to allow for adapter conjugation. One zipper domain was incorporated into a knob-less fiber, while its corresponding zipper domain was fused to the retargeting ligand and secreted. Utilizing an scFv against CD40, Glasgow et al. showed that these novel structures could be incorporated into the Ad capsid and corresponding secreted adapter. Most importantly, upon completion of the Ad-native life cycle and Ad release, crude viral lysate was shown to have CD40-specific targeting, thus showing that the virus could interact and bind with its cognate adapter in the intercellular space (Glasgow, Mikheeva, Krasnykh, & Curiel, 2009).

Overall, adapter-based Ad-targeting studies have provided vast evidence that native Ad tropism can be retargeted to alternative cellular receptors and increase gene transfer in non-CAR-expressing cells *in vitro*. Adapter-targeted vectors have also shown to be efficacious in *in vivo* studies, showing great stability and transductional efficacy. With further development of expression systems and rigorous analysis of the stability and kinetics of vector adapter complexes, adapter-based strategies could progress to clinical translation.

## 12. AD TRANSDUCTIONAL TARGETING: GENETICALLY INCORPORATED STRATEGIES

With greater understanding of Ad virion structure, genetic manipulation of capsid proteins has yielded great strides in Ad targeting. Conceptually, genetic manipulation of the Ad capsid to incorporate novel targeting ligands could yield a multitude of targeting strategies. However, any genetic manipulation must function within the structural and biological constraints of the native Ad capsid. Based on this, most genetic manipulation of the Ad capsid has focused on the fiber, as this domain is the primary determinant of native Ad tropism and can be modified without interfering with capsid assembly.

## 13. CHIMERIC AD

Although Ad5's primary receptor is CAR and thus nonamenable to a variety of cancer therapeutic strategies, other serotypes of Ad do not use CAR as their primary binding receptor. As such, one genetic strategy for retargeting Ad, termed pseudotyping, developed as a mechanism to utilize non-CAR-targeting serotypes while not abandoning the vast knowledge of Ad5 biology (Fig. 2.2B). These pseudotyped virions have shown great transductional efficacy in a variety of Ad5 refractory cell types such as ovarian carcinoma (Rein et al., 2011; Rocconi et al., 2007), prostate cancer (Murakami, Ugai, Belousova, et al., 2010), breast cancer (Stoff-Khalili et al., 2007), colon carcinoma (Silver & Mei, 2011), glioblastoma (Hoffmann, Meyer, & Wildner, 2007), and others. With pseudotyped Ad vectors, the fiber knob domain or the entire fiber is genetically replaced with its structural counterpart from a different human serotype that recognizes an alternative cellular surface receptor. These alternative serotypes are primarily developed from species B and species D Ads. Species B viruses have been shown to interact with a variety of non-CAR receptors including CD46 (Gaggar, Shayakhmetov, & Lieber, 2003), CD80, and CD86 (Short et al., 2004). Additionally, subgroup D serotypes have been shown to interact with CD46 and the glycoprotein component $\alpha$(2–3)-linked sialic acid (Arnberg, Edlund, Kidd, & Wadell, 2000). With the great progress developed in pseudotyped Ad vectors, the strategy was also expanded to include the insertion of fiber elements from nonhuman Ad serotypes. This strategy, termed xenotyping, has yielded a variety of non-CAR-targeted Ad vectors including vectors with fiber elements from avian, bovine, canine, murine, and porcine Ad vectors (Bangari & Mittal, 2005; Bangari, Shukla, & Mittal, 2005; Glasgow et al., 2004; Nakayama et al., 2006; Renaut, Colin, Leite, Benko, & D'Halluin, 2004). Of note, most of the receptor targets for these vectors are undetermined as of today. However, the fiber element from porcine Ad serotype 4 has recently been shown to interact with glycan chains containing repeats of *N*-acetyllactosamine through evidence that this interaction leads to cellular uptake is unknown (Guardado-Calvo et al., 2010). Bovine Ad serotype 4 fiber has also been elucidated recently. This nonhuman Ad was shown to interact with two different immunoevasion molecules of the B7 family of proteins, B7-1 and B7-H1, in murine leukemia cells. Of interest, this fiber requires both interactions in order to result in cellular uptake (Grellier et al., 2011). In addition to nonhuman Ads, structurally similar binding domains from other virus species have also been incorporated into the Ad fiber. This was first shown by the incorporation of the fiber-like σ1 reovirus attachment protein into the Ad fiber. This allowed for effective transduction of target cells expressing junctional adhesion molecule (Tsuruta et al., 2005).

## 14. PEPTIDE-TARGETED AD

Although Ad pseudotyping has shown great success, it is predicated by the discovery of novel non-CAR targeting Ads. As such, development of retargeting Ad vectors has progressed into rationally designed targeted Ads. Meticulous structural studies of the knob domain of fiber have yielded two separate locations within the knob that can be exploited for genetic peptide presentation without disrupting fiber function, the C-terminus, and a region termed the HI loop (Fig. 2.2C). Conceptually, the C-terminus is an ideal location for peptide insertion. Successful genetic insertions of an integrin-binding RGD motif or polylysine peptides have yielded positive *in vitro* and *in vivo* results (Wickham et al., 1997). However, other peptide insertions have shown no effect possibly due to steric hindrances. Structural studies using a genetically inserted FLAG tag into the HI loop, an exposed loop structure connecting β-sheets H and I in the Ad knob domain, showed proof-of-principle evidence that this location is structurally amenable to peptide insertion (Krasnykh et al., 1998). In fact, further studies have shown that this location can handle peptide insertions of up to 100 amino acids without detriment to fiber function (Belousova, Krendelchtchikova, Curiel, & Krasnykh, 2002). As such, Dmitriev et al. inserted an integrin-binding RGD motif into this location and showed that this virus, AdlucRGD, has enhanced transductional efficacy and gene delivery in ovarian cancer cell lines and primary tumors versus nontargeted Ad (Dmitriev et al., 1998; Hemminki et al., 2001). This tropism-expanded Ad has been utilized widely in the field and shown to be efficacious in gene delivery to a wide variety of cancers including ovarian (Murugesan, Akiyama, Einfeld,

Wickham, & King, 2007), cervical (Rein et al., 2004), colon (Lavilla-Alonso et al., 2010), melanoma (Okada et al., 2005), and others.

Progressing beyond the tropism expansion seen in RGD and polylysine motif insertions, several groups have inserted cellular-specific targeting peptides into the HI loop. These peptides developed by traditional phage display biopanning or more novel strategies utilizing peptide incorporated Ad libraries (Bockmann, Drosten, & Putzer, 2005; Miura et al., 2007; Nishimoto et al., 2009) have been proven to be highly specific and generally amenable to Ad insertion. Nicklin et al. (2003) showed that the vascular endothelial cell-targeting peptide, SY-GYLPLP, provided increased transduction in a variety of cancer cell lines. In addition, peptides have been inserted to target a variety of cancers including head and neck (Li et al., 2008), medullary thyroid carcinoma (Schmidt et al., 2011), glioma (Piao et al., 2009), and renal cell carcinoma (Diaconu et al., 2009). In addition to classical target-specific short peptides, Myhre et al. (2009) inserted an Affibody, a small antibody mimetic, into the HI loop and showed HER2/neu and Taq polymerase-specific targeting. They also showed that HER2/neu-specific, Affibody-targeted oncolytic Ad provided increased transduction and killing in prostate cancer cells *in vitro* and increased survival time while decreasing serum prostate-specific antigen in an orthotopic mouse model of prostate cancer (Magnusson et al., 2011).

## 15. "KNOB-LESS"-TARGETED AD

While peptide insertion has shown to be a successful strategy in retargeting Ad, structural conflicts have emerged from fiber knob modifications. As such, a platform by which a wider variety of targeting ligands could be utilized would be a rational goal in further progressing Ad retargeting. The observation that Ads lacking various portions of their knob domain could be rescued leads to the concept of utilizing a knob-less fiber as a platform for ligand presentation (Fig. 2.2B). Limiting this concept, however, was the fact that the knob domain contained the trimerization domain for the fiber that is required for fiber function and insertion into the Ad capsid. Overcoming this structural problem, a foreign trimerization domain, the foldon domain of T4 fibritin, was fused to the native Ad fiber shaft to replace that which was lost by deletion of the knob domain (Papanikolopoulou, Forge, Goeltz, & Mitraki, 2004). Krasnykh, Belousova, Korokhov, Mikheeva, and Curiel (2001) replaced the fiber and knob with the bacteriophage T4 fibritin and showed that this platform could present a 6-histidine

(6-His) motif inserted into the C-terminus. This vector showed a 100-fold increase in gene expression in cells expressing an artificial 6-His binding receptor. Variant "deknobing" strategies have also been explored by Magnusson, Hong, Boulanger, and Lindholm (2001), demonstrating that an RGD motif could target integrin-expressing cells. Further, labs have progressed toward larger peptide displays such as small peptides and Affibodies. Belousova, Mikheeva, Gelovani, and Krasnykh (2008) incorporated a HER2/neu-specific Affibody into a knob-less fiber, showing that the novel Affibody technology was compatible with knob-less fiber platforms. This HER2/neu-targeted vector also showed increased gene delivery in HER2/neu-expressing cancer cells (Belousova et al., 2008). The knob-less Ad platform provides the ability to move beyond small ligands and into the use of proteins as targeting ligands. Previously, this would be a very problematic strategy as large protein insertions are much more likely to interfere with native Ad assembly and function. Notably, trimeric CD40 has been fused to this fiber providing evidence that this platform can be amenable to large protein ligands. The CD40 incorporated Ad provided CD40-specific gene delivery *in vivo* following systemic delivery (Izumi et al., 2005).

## 16. ALTERNATIVE CAPSID LOCATIONS

Although fiber is the most developed capsid protein for retargeting Ad, the difficulty of incorporating ligands into the Ad capsid has furthered the development of alternative locales. Potentially, alternative sites could provide increased presentation of the targeting ligand through increased copy number per virion and could also allow for multiple targeting ligands to be utilized on the same capsid. To date, a variety of alternative sites have been proposed and explored including hexon, pIX, and pIIIa (Fig. 2.2C). The first two have been shown to be compatible with ligand presentation, while the latter was shown incompatible due to its current structural location within the capsid (San Martin et al., 2008).

Hexon is the most abundant protein in the Ad capsid and as such is an ideal candidate for ligand incorporation. The potential 720 copies of hexon could allow for a "coating" of the Ad capsid in any incorporated ligand. Although most of the hexon sequence is highly conserved among serotypes, nine hypervariable regions are found within the hexon and have solvent-exposed loops. As such, these loops lay in an ideal location for modification. Vigne et al. (1999) genetically modified hypervariable region 5 (HVR5) and inserted an integrin-binding RGD domain. This RGD motif had no effect on hexon structure or capsid stability but increased CAR-independent transduction of vascular smooth muscle cells. Further HVRs 2, 3, and 5–7 were found to be amenable to insertion of a 6-His motif (Wu et al., 2005). In addition, anti-6-His antibodies recognized Ad vectors with 6-His inserted into HVRs 2 and 5. The rescue of Ad vectors with peptides inserted in various hypervariable regions provides us with a potential platform for various downstream targeting applications.

More recently, pIX has developed as a practical platform for the presentation of targeting ligands. pIX is a small protein that plays the role of a cement protein, helping stabilize hexon interactions. Found in 240 copies within each virion, pIX provides drastically increased ligand presentation over fiber modifications. Structural studies and the observation that the C-terminus of pIX may be solvent exposed lead to several groups, exploring the concept of pIX presented ligands. The first reported targeting ligand incorporated into the terminus of pIX was presented by Dmitriev, Kashentseva, and Curiel (2002). By incorporating polylysine or FLAG motifs, they showed CAR-independent transduction via interactions with heparin sulfate chains on the target cell surface (Dmitriev et al., 2002). Furthering development of pIX as a targeting local, Vellinga et al. (2004) fused varying sized α-helical linkers to the terminus of pIX and used these linkers to present integrin-binding RGD motifs. Of note, longer linker length corresponded with increased gene delivery in CAR negative endothelial cells. Relatively large proteins fused to pIX have also been explored. Incorporation of hyperstable scFv against β-galactosidase fused to pIX showed that the scFv retained its binding affinity to β-galactosidase (Vellinga et al., 2007). However, this antibody has no targeting applications and the availability of hyperstable scFvs is limiting. Poulin et al. attempted to incorporate an scFv against a mutant form of the EGF receptor (EGFRvIII) fused to pIX, but the scFv failed to fold properly, resulting in a lack of targeting (Poulin et al., 2010). As a consequence, they attempted to route the pIX-scFv through the ER, but biological incompatibility between the cytoplasmically assembled Ad and the ER routed pIX resulted in low levels of incorporation and thus a lack of targeting. However, they were able to incorporate a single domain antibody (AFAI) against CD66c (CEA-related cell adhesion molecule family 6). This ligand provided CD66c-specific binding and transduction of A549 non-small-cell lung carcinoma cell line. de Vrij et al. (2008) showed that large single-chain T-cell receptors could also be attached to pIX. They fused pIX to a single-chain T-cell receptor against the CT antigen melanoma-associated antigen A1 (MAGE-A1). This vector specifically transduced melanoma cell lines. In addition, transduction was shown to correlate with the levels of MAGE-A1 peptide within the cells. However, lack of or downregulation of HLA-A1 molecules can drastically reduce the transduction efficiency of this vector. As a whole, studies have shown that pIX is a flexible platform for the display of both small and large targeting ligands.

## 17. NOVEL TRANSDUCTIONAL STRATEGIES

Although Ad-based therapeutics has progressed greatly, the vast majority of targeting strategies have relied upon a single cellular surface receptor for the target. This strategy has been shown to work remarkably well for homogenous cell line populations. However, some target cells such as cancer do not comprise a homogenous population. As such, targeting a single cellular surface receptor may lead to a selected population that is resistant to further therapy. Thus, any molecular-based therapy should take this into account when designing a targeting strategy. Several groups have begun developing Ad-targeting strategies that utilize multiple ligands within the same virion. One of the first vectors reported was an Ad5 containing both an RGD motif and a polylysine ligand (Wu et al., 2002). This vector could thus target both cell surface integrins and heparin sulfate proteoglycans. Following that, Borovjagin et al. (2005) inserted an integrin-binding RGD motif into the C-terminus of a chimeric Ad fiber composed of the shaft domain of Ad5 and the knob domain of Ad3. This vector showed 55-fold increase in gene transduction of bladder cancer cell lines. Utilizing adapter-based retargeting, Grill et al. (2001) combined an adapter composed of an scFv against EGF receptor with genetically incorporated RGD motif in the HI loop of fiber knob. This virus was shown to be able to target both EGF receptor and cell surface integrins on primary glioma cells and spheroids.

Although these strategies provide insight into the efficacy of dual targeting over single, they are limited in their ability to insert multiple ligands within the same fiber. Using multiple fibers within the same Ad virion would bypass this restriction and provides a platform for the utilization of multiple complex targeting ligands. Pereboeva, Komarova, Mahasreshti, and Curiel (2004) first showed that this strategy was feasible by generating an Ad vector incorporating both the wild-type fiber and a knob-less fiber fibritin presenting a 6-His motif. Utilizing this mosaic vector, they showed both CAR and artificial 6-His receptor-specific gene transduction, though in high-CAR-expressing cells no additional gain was seen from the 6-His containing fiber fibritin. The combination of an Ad5/3 chimeric fiber with a fiber containing the reovirus σ-1 protein in the same Ad virion has also shown to provide an increase in infectivity enhancement of ovarian cancer cell lines and primary ovarian cancer tissue slices (Tsuruta et al., 2007). Murakami, Ugai, Wang, et al. (2010) additionally provided evidence that with the correct genetic construct, equal expression, and incorporation of the two distinct fibers can be accomplished. This vector, containing both Ad5 and Ad3 fiber, provided both CAR- and CD46-specific gene transduction. Of interest, they showed that this vector could target two distinct cells, PC-3 cells expressing CD46 and Cho-CAR cells expressing CAR, in a mixed culture experiment.

## 18. AD TARGETING: RECENT CLINICAL DEVELOPMENTS

Although Ad-based therapeutics have shown great promise in preclinical studies, in the clinical setting, therapeutic efficacy of Ad vectors has not followed. Although ample evidence concludes that lack of CAR expression upon cancer cells drastically limits Ad-based therapeutic efficacy, the vast majority of clinical trials utilizing Ad rely upon native CAR-based transduction. One hurdle limiting the translation of Ad-based targeting strategies has been the additional complexity of adapter/Ad conjugates. Since these strategies are two-component systems, they entail additional production complexity and scrutiny in regard to safety before being approved for clinical use. As such, the few clinical trials to date utilizing targeting have relied upon genetically inserted targeting ligands.

Recently, clinical studies involving Ad-based therapies utilizing genetically incorporated integrin-binding RGD motifs in the fiber knob have been reported. Kimball et al. recently finished a phase I clinical trial examining the therapeutic efficacy and maximum tolerated dose of a tropism-modified, infectivity enhanced conditionally replicative adenovirus (CRAd), Ad5-Δ24-RGD, in patients with malignant gynecologic diseases (Kimball et al., 2010; Page et al., 2007). Following treatment, of the 21 patients, 71% had stable disease, while 29% still showed disease progression after 1 month of follow-up. Of note, seven patients did show a decrease in CA-121 levels, with four of these being a greater than 20% decrease. Although no patients showed regression of disease in this study, toxicity associated with therapy was limited to grade 1/2 fever, fatigue, and abdominal pain. A similarly sized trial reported by Nokisalmi et al. (2010) analyzed an integrin-targeted CRAd, ICOVIR-7, in patients with a variety of solid tumors. Similarly, this trial saw only mild to moderate treatment-related side effects. Of note, 9 of 17 evaluable patients showed evidence of antitumor activity, with 1 patient showing partial response and 2 patients with minor responses. This strategy has also shown similar safety and therapeutic outcomes when combined with expression of granulocyte-macrophage colony-stimulating factor, an immune-stimulatory molecule (Pesonen et al., 2011). Progression of targeted Ad-based cancer therapies in the clinical setting has lead to new studies utilizing chimeric Ad-based vectors replacing the Ad5 knob domain with that of the Ad3 knob (Ad5/3) (Kim et al., 2011). Pesonen et al. (2010) showed similar safety in a trial of 18 patients with varying solid tumors. Of these patients, 61% showed evidence of antitumor activity.

In addition to these CRAd-based trials, Matthews et al. (2009) reported plans for and are currently finishing another phase I clinical trial involving Ad5.SSTR/TK.RGD. This Ad-based therapeutic utilizes a genetically incorporated RGD ligand to target the expression of a therapeutic suicide gene, herpes simplex virus thymidine kinase (TK), and an imaging motif, somatostatin receptor type 2 (SSTR), for viral tracking via nuclear imaging.

In all, translation of targeted Ad-based therapeutics to the clinical setting has shown a solid safety record similar to their untargeted counterparts. Although some levels of antitumor activity and therapeutic response have been noted, the response rate is still far below that required for therapeutic usage. This emphasizes the need for further targeting trials and the utilization of additional strategies involving Ad virion/host interactions such as liver detargeting and immune evasion.

## 19. CONCLUDING REMARKS

Ad-based vectors are a widely used therapeutic platform for gene delivery. They are especially prominent in the field of cancer gene therapy where shorter gene expression times are not an issue. However, biological hurdles stand between native Ad-based vectors and their full utilization as a therapeutically effective cancer treatment platform. Of these hurdles, effective gene transduction of cancer cells drastically limits potential of Ad-based vectors. Early clinical trials highlighted this issue by reporting Ad vectors safe but therapeutically nonefficacious. In this regard, studies have clearly illustrated the case for increased transduction of target cells leading to increased therapeutic efficacy. To this end, Ad-based vectors utilizing cancer-specific targeting should continue to be progressed and examined in stringent models of cancer with the goal of full therapeutic efficacy in the clinical setting.

## ACKNOWLEDGMENT

We would like to acknowledge NIH Pancreatic Cancer SPORE Grant 2P50CA101955.

## REFERENCES

Alba, R., Bradshaw, A. C., Coughlan, L., Denby, L., McDonald, R. A., Waddington, S. N., et al. (2010). Biodistribution and retargeting of FX-binding ablated adenovirus serotype 5 vectors. *Blood*, *116*, 2656–2664.

Alba, R., Bradshaw, A. C., Parker, A. L., Bhella, D., Waddington, S. N., Nicklin, S. A., et al. (2009). Identification of coagulation factor (F)X binding sites on the adenovirus serotype 5 hexon: Effect of mutagenesis on FX interactions and gene transfer. *Blood*, *114*, 965–971.

Alemany, R., & Curiel, D. T. (2001). CAR-binding ablation does not change biodistribution and toxicity of adenoviral vectors. *Gene Therapy*, *8*, 1347–1353.

Araki, K., Yamashita, T., Reddy, N., Wang, H., Abuzeid, W. M., Khan, K., et al. (2010). Molecular disruption of NBS1 with targeted gene delivery enhances chemosensitisation in head and neck cancer. *British Journal of Cancer*, *103*, 1822–1830.

Arnberg, N., Edlund, K., Kidd, A. H., & Wadell, G. (2000). Adenovirus type 37 uses sialic acid as a cellular receptor. *Journal of Virology*, *74*, 42–48.

Bangari, D. S., & Mittal, S. K. (2005). Porcine adenovirus serotype 3 internalization is independent of CAR and alphavbeta3 or alphavbeta5 integrin. *Virology*, *332*, 157–166.

Bangari, D. S., Shukla, S., & Mittal, S. K. (2005). Comparative transduction efficiencies of human and nonhuman adenoviral vectors in human, murine, bovine, and porcine cells in culture. *Biochemical and Biophysical Research Communications*, *327*, 960–966.

Barker, S. D., Dmitriev, I. P., Nettelbeck, D. M., Liu, B., Rivera, A. A., Alvarez, R. D., et al. (2003). Combined transcriptional and transductional targeting improves the specificity and efficacy of adenoviral gene delivery to ovarian carcinoma. *Gene Therapy*, *10*, 1198–1204.

Belousova, N., Krendelchtchikova, V., Curiel, D. T., & Krasnykh, V. (2002). Modulation of adenovirus vector tropism via incorporation of polypeptide ligands into the fiber protein. *Journal of Virology*, *76*, 8621–8631.

Belousova, N., Mikheeva, G., Gelovani, J., & Krasnykh, V. (2008). Modification of adenovirus capsid with a designed protein ligand yields a gene vector targeted to a major molecular marker of cancer. *Journal of Virology*, *82*, 630–637.

Bockmann, M., Drosten, M., & Putzer, B. M. (2005). Discovery of targeting peptides for selective therapy of medullary thyroid carcinoma. *The Journal of Gene Medicine*, *7*, 179–188.

Borovjagin, A. V., Krendelchtchikov, A., Ramesh, N., Yu, D. C., Douglas, J. T., & Curiel, D. T. (2005). Complex mosaicism is a novel approach to infectivity enhancement of adenovirus type 5-based vectors. *Cancer Gene Therapy*, *12*, 475–486.

Breidenbach, M., Rein, D. T., Everts, M., Glasgow, J. N., Wang, M., Passineau, M. J., et al. (2005). Mesothelin-mediated targeting of adenoviral vectors for ovarian cancer gene therapy. *Gene Therapy*, *12*, 187–193.

Breidenbach, M., Rein, D. T., Wang, M., Nettelbeck, D. M., Hemminki, A., Ulasov, I., et al. (2004). Genetic replacement of the adenovirus shaft fiber reduces liver tropism in ovarian cancer gene therapy. *Human Gene Therapy*, *15*, 509–518.

Burnett, R. M. (1985). The structure of the adenovirus capsid. II. The packing symmetry of hexon and its implications for viral architecture. *Journal of Molecular Biology*, *185*, 125–143.

Campos, S. K., Parrott, M. B., & Barry, M. A. (2004). Avidin-based targeting and purification of a protein IX-modified, metabolically biotinylated adenoviral vector. *Molecular Therapy*, *9*, 942–954.

Carette, J. E., Graat, H. C., Schagen, F. H., Mastenbroek, D. C., Rots, M. G., Haisma, H. J., et al. (2007). A conditionally replicating adenovirus with strict selectivity in killing cells expressing epidermal growth factor receptor. *Virology*, *361*, 56–67.

Chen, C. Y., May, S. M., & Barry, M. A. (2010). Targeting adenoviruses with factor x-single-chain antibody fusion proteins. *Human Gene Therapy*, *21*, 739–749.

Coyne, C. B., & Bergelson, J. M. (2006). Virus-induced Abl and Fyn kinase signals permit coxsackievirus entry through epithelial tight junctions. *Cell*, *124*, 119–131.

Davison, A. J., Benko, M., & Harrach, B. (2003). Genetic content and evolution of adenoviruses. *The Journal of General Virology*, *84*, 2895–2908.

Davison, E., Diaz, R. M., Hart, I. R., Santis, G., & Marshall, J. F. (1997). Integrin alpha5beta1-mediated adenovirus infection is enhanced by the integrin-activating antibody TS2/16. *Journal of Virology*, *71*, 6204–6207.

de Vrij, J., Uil, T. G., van den Hengel, S. K., Cramer, S. J., Koppers-Lalic, D., Verweij, M. C., et al. (2008). Adenovirus targeting to HLA-A1/MAGE-A1-positive tumor cells by fusing a single-chain T-cell receptor with minor capsid protein IX. *Gene Therapy*, *15*, 978–989.

Diaconu, I., Denby, L., Pesonen, S., Cerullo, V., Bauerschmitz, G. J., Guse, K., et al. (2009). Serotype chimeric and fiber-mutated adenovirus Ad5/19p-HIT for targeting renal cancer and untargeting the liver. *Human Gene Therapy*, *20*, 611–620.

Dmitriev, I. P., Kashentseva, E. A., & Curiel, D. T. (2002). Engineering of adenovirus vectors containing heterologous peptide sequences in the C terminus of capsid protein IX. *Journal of Virology*, *76*, 6893–6899.

Dmitriev, I., Kashentseva, E., Rogers, B. E., Krasnykh, V., & Curiel, D. T. (2000). Ectodomain of coxsackievirus and adenovirus receptor genetically fused to epidermal growth factor mediates adenovirus targeting to epidermal growth factor receptor-positive cells. *Journal of Virology*, *74*, 6875–6884.

Dmitriev, I., Krasnykh, V., Miller, C. R., Wang, M., Kashentseva, E., Mikheeva, G., et al. (1998). An adenovirus vector with genetically modified fibers demonstrates expanded tropism via utilization of a coxsackievirus and adenovirus receptor-independent cell entry mechanism. *Journal of Virology*, *72*, 9706–9713.

Douglas, J. T., Rogers, B. E., Rosenfeld, M. E., Michael, S. I., Feng, M., & Curiel, D. T. (1996). Targeted gene delivery by tropism-modified adenoviral vectors. *Nature Biotechnology*, *14*, 1574–1578.

Ebbinghaus, C., Al-Jaibaji, A., Operschall, E., Schoffel, A., Peter, I., Greber, U. F., et al. (2001). Functional and selective targeting of adenovirus to high-affinity Fcgamma receptor I-positive cells by using a bispecific hybrid adapter. *Journal of Virology*, *75*, 480–489.

Everts, M., Kim-Park, S. A., Preuss, M. A., Passineau, M. J., Glasgow, J. N., Pereboev, A. V., et al. (2005). Selective induction of tumor-associated antigens in murine pulmonary vasculature using double-targeted adenoviral vectors. *Gene Therapy*, *12*, 1042–1048.

Fabry, C. M., Rosa-Calatrava, M., Conway, J. F., Zubieta, C., Cusack, S., Ruigrok, R. W., et al. (2005). A quasi-atomic model of human adenovirus type 5 capsid. *The EMBO Journal*, *24*, 1645–1654.

Fechner, H., Haack, A., Wang, H., Wang, X., Eizema, K., Pauschinger, M., et al. (1999). Expression of coxsackie adenovirus receptor and alphav-integrin does not correlate with adenovector targeting in vivo indicating anatomical vector barriers. *Gene Therapy*, *6*, 1520–1535.

Figures, M. R., Wobb, J., Araki, K., Liu, T., Xu, L., Zhu, H., et al. (2009). Head and neck squamous cell carcinoma targeted chemosensitization. *Otolaryngology and Head and Neck Surgery*, *141*, 177–183.

Gaggar, A., Shayakhmetov, D. M., & Lieber, A. (2003). CD46 is a cellular receptor for group B adenoviruses. *Nature Medicine*, *9*, 1408–1412.

Glasgow, J. N., Kremer, E. J., Hemminki, A., Siegal, G. P., Douglas, J. T., & Curiel, D. T. (2004). An adenovirus vector with a chimeric fiber derived from canine adenovirus type 2 displays novel tropism. *Virology*, *324*, 103–116.

Glasgow, J. N., Mikheeva, G., Krasnykh, V., & Curiel, D. T. (2009). A strategy for adenovirus vector targeting with a secreted single chain antibody. *PLoS One*, *4*, e8355.
Goldman, C. K., Rogers, B. E., Douglas, J. T., Sosnowski, B. A., Ying, W., Siegal, G. P., et al. (1997). Targeted gene delivery to Kaposi's sarcoma cells via the fibroblast growth factor receptor. *Cancer Research*, *57*, 1447–1451.
Grellier, E., Lecolle, K., Rogee, S., Couturier, C., D'Halluin, J. C., Hong, S. S., et al. (2011). A fiber-modified adenoviral vector interacts with immunoevasion molecules of the B7 family at the surface of murine leukemia cells derived from dormant tumors. *Molecular Cancer*, *10*, 105.
Grill, J., Van Beusechem, V. W., Van Der Valk, P., Dirven, C. M., Leonhart, A., Pherai, D. S., et al. (2001). Combined targeting of adenoviruses to integrins and epidermal growth factor receptors increases gene transfer into primary glioma cells and spheroids. *Clinical Cancer Research*, 7, 641–650.
Gu, D. L., Gonzalez, A. M., Printz, M. A., Doukas, J., Ying, W., D'Andrea, M., et al. (1999). Fibroblast growth factor 2 retargeted adenovirus has redirected cellular tropism: Evidence for reduced toxicity and enhanced antitumor activity in mice. *Cancer Research*, *59*, 2608–2614.
Guardado-Calvo, P., Munoz, E. M., Llamas-Saiz, A. L., Fox, G. C., Kahn, R., Curiel, D. T., et al. (2010). Crystallographic structure of porcine adenovirus type 4 fiber head and galectin domains. *Journal of Virology*, *84*, 10558–10568.
Haisma, H. J., Grill, J., Curiel, D. T., Hoogeland, S., van Beusechem, V. W., Pinedo, H. M., et al. (2000). Targeting of adenoviral vectors through a bispecific single-chain antibody. *Cancer Gene Therapy*, 7, 901–904.
Haisma, H. J., Kamps, G. K., Bouma, A., Geel, T. M., Rots, M. G., Kariath, A., et al. (2010). Selective targeting of adenovirus to alphavbeta3 integrins, VEGFR2 and Tie2 endothelial receptors by angio-adenobodies. *International Journal of Pharmaceutics*, *391*, 155–161.
Haisma, H. J., Pinedo, H. M., Rijswijk, A., der Meulen-Muileman, I., Sosnowski, B. A., Ying, W., et al. (1999). Tumor-specific gene transfer via an adenoviral vector targeted to the pan-carcinoma antigen EpCAM. *Gene Therapy*, *6*, 1469–1474.
Hakkarainen, T., Hemminki, A., Pereboev, A. V., Barker, S. D., Asiedu, C. K., Strong, T. V., et al. (2003). CD40 is expressed on ovarian cancer cells and can be utilized for targeting adenoviruses. *Clinical Cancer Research*, *9*, 619–624.
Harvey, T. J., Burdon, D., Steele, L., Ingram, N., Hall, G. D., Selby, P. J., et al. (2010). Retargeted adenoviral cancer gene therapy for tumour cells overexpressing epidermal growth factor receptor or urokinase-type plasminogen activator receptor. *Gene Therapy*, *17*, 1000–1010.
Heideman, D. A., van Beusechem, V. W., Offerhaus, G. J., Wickham, T. J., Roelvink, P. W., Craanen, M. E., et al. (2002). Selective gene transfer into primary human gastric tumors using epithelial cell adhesion molecule-targeted adenoviral vectors with ablated native tropism. *Human Gene Therapy*, *13*, 1677–1685.
Hemminki, A., Belousova, N., Zinn, K. R., Liu, B., Wang, M., Chaudhuri, T. R., et al. (2001). An adenovirus with enhanced infectivity mediates molecular chemotherapy of ovarian cancer cells and allows imaging of gene expression. *Molecular Therapy*, *4*, 223–231.
Henning, P., Andersson, K. M., Frykholm, K., Ali, A., Magnusson, M. K., Nygren, P. A., et al. (2005). Tumor cell targeted gene delivery by adenovirus 5 vectors carrying knobless fibers with antibody-binding domains. *Gene Therapy*, *12*, 211–224.
Hoffmann, D., Meyer, B., & Wildner, O. (2007). Improved glioblastoma treatment with Ad5/35 fiber chimeric conditionally replicating adenoviruses. *The Journal of Gene Medicine*, *9*, 764–778.
Huch, M., Abate-Daga, D., Roig, J. M., Gonzalez, J. R., Fabregat, J., Sosnowski, B., et al. (2006). Targeting the CYP2B 1/cyclophosphamide suicide system to fibroblast growth

factor receptors results in a potent antitumoral response in pancreatic cancer models. *Human Gene Therapy, 17*, 1187–1200.

Izumi, M., Kawakami, Y., Glasgow, J. N., Belousova, N., Everts, M., Kim-Park, S., et al. (2005). In vivo analysis of a genetically modified adenoviral vector targeted to human CD40 using a novel transient transgenic model. *The Journal of Gene Medicine, 7*, 1517–1525.

Kalyuzhniy, O., Di Paolo, N. C., Silvestry, M., Hofherr, S. E., Barry, M. A., Stewart, P. L., et al. (2008). Adenovirus serotype 5 hexon is critical for virus infection of hepatocytes in vivo. *Proceedings of the National Academy of Sciences of the United States of America, 105*, 5483–5488.

Kashentseva, E. A., Seki, T., Curiel, D. T., & Dmitriev, I. P. (2002). Adenovirus targeting to c-erbB-2 oncoprotein by single-chain antibody fused to trimeric form of adenovirus receptor ectodomain. *Cancer Research, 62*, 609–616.

Kelly, F. J., Miller, C. R., Buchsbaum, D. J., Gomez-Navarro, J., Barnes, M. N., Alvarez, R. D., et al. (2000). Selectivity of TAG-72-targeted adenovirus gene transfer to primary ovarian carcinoma cells versus autologous mesothelial cells in vitro. *Clinical Cancer Research, 6*, 4323–4333.

Kim, K. H., Ryan, M. J., Estep, J. E., Miniard, B. M., Rudge, T. L., Peggins, J. O., et al. (2011). A new generation of serotype chimeric infectivity-enhanced conditionally replicative adenovirals: The safety profile of ad5/3-Delta24 in advance of a phase I clinical trial in ovarian cancer patients. *Human Gene Therapy, 22*, 821–828.

Kim, J., Smith, T., Idamakanti, N., Mulgrew, K., Kaloss, M., Kylefjord, H., et al. (2002). Targeting adenoviral vectors by using the extracellular domain of the coxsackie-adenovirus receptor: Improved potency via trimerization. *Journal of Virology, 76*, 1892–1903.

Kimball, K. J., Preuss, M. A., Barnes, M. N., Wang, M., Siegal, G. P., Wan, W., et al. (2010). A phase I study of a tropism-modified conditionally replicative adenovirus for recurrent malignant gynecologic diseases. *Clinical Cancer Research, 16*, 5277–5287.

Korokhov, N., de Gruijl, T. D., Aldrich, W. A., Triozzi, P. L., Banerjee, P. T., Gillies, S. D., et al. (2005). High efficiency transduction of dendritic cells by adenoviral vectors targeted to DC-SIGN. *Cancer Biology & Therapy, 4*, 289–294.

Kraaij, R., van Rijswijk, A. L., Oomen, M. H., Haisma, H. J., & Bangma, C. H. (2005). Prostate specific membrane antigen (PSMA) is a tissue-specific target for adenoviral transduction of prostate cancer in vitro. *The Prostate, 62*, 253–259.

Krasnykh, V., Belousova, N., Korokhov, N., Mikheeva, G., & Curiel, D. T. (2001). Genetic targeting of an adenovirus vector via replacement of the fiber protein with the phage T4 fibritin. *Journal of Virology, 75*, 4176–4183.

Krasnykh, V., Dmitriev, I., Mikheeva, G., Miller, C. R., Belousova, N., & Curiel, D. T. (1998). Characterization of an adenovirus vector containing a heterologous peptide epitope in the HI loop of the fiber knob. *Journal of Virology, 72*, 1844–1852.

Lavilla-Alonso, S., Bauerschmitz, G., Abo-Ramadan, U., Halavaara, J., Escutenaire, S., Diaconu, I., et al. (2010). Adenoviruses with an alphavbeta integrin targeting moiety in the fiber shaft or the HI-loop increase tumor specificity without compromising antitumor efficacy in magnetic resonance imaging of colorectal cancer metastases. *Journal of Translational Medicine, 8*, 80.

Li, E., Brown, S. L., Stupack, D. G., Puente, X. S., Cheresh, D. A., & Nemerow, G. R. (2001). Integrin alpha(v)beta1 is an adenovirus coreceptor. *Journal of Virology, 75*, 5405–5409.

Li, E., Brown, S. L., Von Seggern, D. J., Brown, G. B., & Nemerow, G. R. (2000). Signaling antibodies complexed with adenovirus circumvent CAR and integrin interactions and improve gene delivery. *Gene Therapy, 7*, 1593–1599.

Li, H. J., Everts, M., Yamamoto, M., Curiel, D. T., & Herschman, H. R. (2009). Combined transductional untargeting/retargeting and transcriptional restriction enhances adenovirus gene targeting and therapy for hepatic colorectal cancer tumors. *Cancer Research, 69*, 554–564.

Li, D., Guang, W., Abuzeid, W. M., Roy, S., Gao, G. P., Sauk, J. J., et al. (2008). Novel adenoviral gene delivery system targeted against head and neck cancer. *The Laryngoscope, 118*, 650–658.

Li, E., Stupack, D., Bokoch, G. M., & Nemerow, G. R. (1998). Adenovirus endocytosis requires actin cytoskeleton reorganization mediated by Rho family GTPases. *Journal of Virology, 72*, 8806–8812.

Li, E., Stupack, D., Klemke, R., Cheresh, D. A., & Nemerow, G. R. (1998). Adenovirus endocytosis via alpha(v) integrins requires phosphoinositide-3-OH kinase. *Journal of Virology, 72*, 2055–2061.

Magnusson, M. K., Hong, S. S., Boulanger, P., & Lindholm, L. (2001). Genetic retargeting of adenovirus: Novel strategy employing "deknobbing" of the fiber. *Journal of Virology, 75*, 7280–7289.

Magnusson, M. K., Kraaij, R., Leadley, R. M., De Ridder, C. M., van Weerden, W. M., Van Schie, K. A., et al. (2011). A transductionally retargeted adenoviral vector for virotherapy of Her2/neu-expressing prostate cancer. *Human Gene Therapy, 23*, 70–82.

Marsh, M. P., Campos, S. K., Baker, M. L., Chen, C. Y., Chiu, W., & Barry, M. A. (2006). Cryoelectron microscopy of protein IX-modified adenoviruses suggests a new position for the C terminus of protein IX. *Journal of Virology, 80*, 11881–11886.

Matthews, K., Noker, P. E., Tian, B., Grimes, S. D., Fulton, R., Schweikart, K., et al. (2009). Identifying the safety profile of Ad5.SSTR/TK.RGD, a novel infectivity-enhanced bicistronic adenovirus, in anticipation of a phase I clinical trial in patients with recurrent ovarian cancer. *Clinical Cancer Research, 15*, 4131–4137.

Meier, O., Boucke, K., Hammer, S. V., Keller, S., Stidwill, R. P., Hemmi, S., et al. (2002). Adenovirus triggers macropinocytosis and endosomal leakage together with its clathrin-mediated uptake. *The Journal of Cell Biology, 158*, 1119–1131.

Miller, C. R., Buchsbaum, D. J., Reynolds, P. N., Douglas, J. T., Gillespie, G. Y., Mayo, M. S., et al. (1998). Differential susceptibility of primary and established human glioma cells to adenovirus infection: Targeting via the epidermal growth factor receptor achieves fiber receptor-independent gene transfer. *Cancer Research, 58*, 5738–5748.

Miura, Y., Yoshida, K., Nishimoto, T., Hatanaka, K., Ohnami, S., Asaka, M., et al. (2007). Direct selection of targeted adenovirus vectors by random peptide display on the fiber knob. *Gene Therapy, 14*, 1448–1460.

Murakami, M., Ugai, H., Belousova, N., Pereboev, A., Dent, P., Fisher, P. B., et al. (2010). Chimeric adenoviral vectors incorporating a fiber of human adenovirus 3 efficiently mediate gene transfer into prostate cancer cells. *The Prostate, 70*, 362–376.

Murakami, M., Ugai, H., Wang, M., Belousova, N., Dent, P., Fisher, P. B., et al. (2010). An adenoviral vector expressing human adenovirus 5 and 3 fiber proteins for targeting heterogeneous cell populations. *Virology, 407*, 196–205.

Murugesan, S. R., Akiyama, M., Einfeld, D. A., Wickham, T. J., & King, C. R. (2007). Experimental treatment of ovarian cancers by adenovirus vectors combining receptor targeting and selective expression of tumor necrosis factor. *International Journal of Oncology, 31*, 813–822.

Myhre, S., Henning, P., Friedman, M., Stahl, S., Lindholm, L., & Magnusson, M. K. (2009). Re-targeted adenovirus vectors with dual specificity; binding specificities conferred by two different Affibody molecules in the fiber. *Gene Therapy, 16*, 252–261.

Nakayama, M., Both, G. W., Banizs, B., Tsuruta, Y., Yamamoto, S., Kawakami, Y., et al. (2006). An adenovirus serotype 5 vector with fibers derived from ovine atadenovirus

demonstrates CAR-independent tropism and unique biodistribution in mice. *Virology, 350*, 103–115.
Nettelbeck, D. M., Miller, D. W., Jerome, V., Zuzarte, M., Watkins, S. J., Hawkins, R. E., et al. (2001). Targeting of adenovirus to endothelial cells by a bispecific single-chain diabody directed against the adenovirus fiber knob domain and human endoglin (CD105). *Molecular Therapy, 3*, 882–891.
Nettelbeck, D. M., Rivera, A. A., Kupsch, J., Dieckmann, D., Douglas, J. T., Kontermann, R. E., et al. (2004). Retargeting of adenoviral infection to melanoma: Combining genetic ablation of native tropism with a recombinant bispecific single-chain diabody (scDb) adapter that binds to fiber knob and HMWMAA. *International Journal of Cancer, 108*, 136–145.
Nicklin, S. A., Dishart, K. L., Buening, H., Reynolds, P. N., Hallek, M., Nemerow, G. R., et al. (2003). Transductional and transcriptional targeting of cancer cells using genetically engineered viral vectors. *Cancer Letters, 201*, 165–173.
Nishimoto, T., Yoshida, K., Miura, Y., Kobayashi, A., Hara, H., Ohnami, S., et al. (2009). Oncolytic virus therapy for pancreatic cancer using the adenovirus library displaying random peptides on the fiber knob. *Gene Therapy, 16*, 669–680.
Nokisalmi, P., Pesonen, S., Escutenaire, S., Sarkioja, M., Raki, M., Cerullo, V., et al. (2010). Oncolytic adenovirus ICOVIR-7 in patients with advanced and refractory solid tumors. *Clinical Cancer Research, 16*, 3035–3043.
Okada, Y., Okada, N., Mizuguchi, H., Hayakawa, T., Nakagawa, S., & Mayumi, T. (2005). Transcriptional targeting of RGD fiber-mutant adenovirus vectors can improve the safety of suicide gene therapy for murine melanoma. *Cancer Gene Therapy, 12*, 608–616.
Okegawa, T., Pong, R. C., Li, Y., Bergelson, J. M., Sagalowsky, A. I., & Hsieh, J. T. (2001). The mechanism of the growth-inhibitory effect of coxsackie and adenovirus receptor (CAR) on human bladder cancer: A functional analysis of car protein structure. *Cancer Research, 61*, 6592–6600.
Page, J. G., Tian, B., Schweikart, K., Tomaszewski, J., Harris, R., Broadt, T., et al. (2007). Identifying the safety profile of a novel infectivity-enhanced conditionally replicative adenovirus, Ad5-delta24-RGD, in anticipation of a phase I trial for recurrent ovarian cancer. *American Journal of Obstetrics and Gynecology, 196*, 389.e9–389.e10; discussion 389 e9–389.e10.
Papanikolopoulou, K., Forge, V., Goeltz, P., & Mitraki, A. (2004). Formation of highly stable chimeric trimers by fusion of an adenovirus fiber shaft fragment with the foldon domain of bacteriophage t4 fibritin. *The Journal of Biological Chemistry, 279*, 8991–8998.
Parrott, M. B., Adams, K. E., Mercier, G. T., Mok, H., Campos, S. K., & Barry, M. A. (2003). Metabolically biotinylated adenovirus for cell targeting, ligand screening, and vector purification. *Molecular Therapy, 8*, 688–700.
Pereboeva, L., Komarova, S., Mahasreshti, P. J., & Curiel, D. T. (2004). Fiber-mosaic adenovirus as a novel approach to design genetically modified adenoviral vectors. *Virus Research, 105*, 35–46.
Pereboeva, L., Komarova, S., Roth, J., Ponnazhagan, S., & Curiel, D. T. (2007). Targeting EGFR with metabolically biotinylated fiber-mosaic adenovirus. *Gene Therapy, 14*, 627–637.
Pesonen, S., Diaconu, I., Cerullo, V., Escutenaire, S., Raki, M., Kangasniemi, L., et al. (2011). Integrin targeted oncolytic adenoviruses Ad5-D24-RGD and Ad5-RGD-D24-GMCSF for treatment of patients with advanced chemotherapy refractory solid tumors. *International Journal of Cancer, 130*, 1937–1947.
Pesonen, S., Nokisalmi, P., Escutenaire, S., Sarkioja, M., Raki, M., Cerullo, V., et al. (2010). Prolonged systemic circulation of chimeric oncolytic adenovirus Ad5/3-Cox2L-D24 in patients with metastatic and refractory solid tumors. *Gene Therapy, 17*, 892–904.

Philipson, L., & Pettersson, R. F. (2004). The coxsackie-adenovirus receptor—A new receptor in the immunoglobulin family involved in cell adhesion. *Current Topics in Microbiology and Immunology*, *273*, 87–111.

Piao, Y., Jiang, H., Alemany, R., Krasnykh, V., Marini, F. C., Xu, J., et al. (2009). Oncolytic adenovirus retargeted to Delta-EGFR induces selective antiglioma activity. *Cancer Gene Therapy*, *16*, 256–265.

Poulin, K. L., Lanthier, R. M., Smith, A. C., Christou, C., Risco Quiroz, M., Powell, K. L., et al. (2010). Retargeting of adenovirus vectors through genetic fusion of a single-chain or single-domain antibody to capsid protein IX. *Journal of Virology*, *84*, 10074–10086.

Printz, M. A., Gonzalez, A. M., Cunningham, M., Gu, D. L., Ong, M., Pierce, G. F., et al. (2000). Fibroblast growth factor 2-retargeted adenoviral vectors exhibit a modified biolocalization pattern and display reduced toxicity relative to native adenoviral vectors. *Human Gene Therapy*, *11*, 191–204.

Reddy, V. S., Natchiar, S. K., Stewart, P. L., & Nemerow, G. R. (2010). Crystal structure of human adenovirus at 3.5 Å resolution. *Science*, *329*, 1071–1075.

Rein, D. T., Breidenbach, M., Wu, H., Han, T., Haviv, Y. S., Wang, M., et al. (2004). Gene transfer to cervical cancer with fiber-modified adenoviruses. *International Journal of Cancer*, *111*, 698–704.

Rein, D. T., Volkmer, A., Beyer, I. M., Curiel, D. T., Janni, W., Dragoi, A., et al. (2011). Treatment of chemotherapy resistant ovarian cancer with a MDR1 targeted oncolytic adenovirus. *Gynecologic Oncology*, *123*, 138–146.

Renaut, L., Colin, M., Leite, J. P., Benko, M., & D'Halluin, J. C. (2004). Abolition of hCAR-dependent cell tropism using fiber knobs of Atadenovirus serotypes. *Virology*, *321*, 189–204.

Reynolds, P. N., Zinn, K. R., Gavrilyuk, V. D., Balyasnikova, I. V., Rogers, B. E., Buchsbaum, D. J., et al. (2000). A targetable, injectable adenoviral vector for selective gene delivery to pulmonary endothelium in vivo. *Molecular Therapy*, *2*, 562–578.

Rocconi, R. P., Zhu, Z. B., Stoff-Khalili, M., Rivera, A. A., Lu, B., Wang, M., et al. (2007). Treatment of ovarian cancer with a novel dual targeted conditionally replicative adenovirus (CRAd). *Gynecologic Oncology*, *105*, 113–121.

Saban, S. D., Silvestry, M., Nemerow, G. R., & Stewart, P. L. (2006). Visualization of alpha-helices in a 6-angstrom resolution cryoelectron microscopy structure of adenovirus allows refinement of capsid protein assignments. *Journal of Virology*, *80*, 12049–12059.

San Martin, C., Glasgow, J. N., Borovjagin, A., Beatty, M. S., Kashentseva, E. A., Curiel, D. T., et al. (2008). Localization of the N-terminus of minor coat protein IIIa in the adenovirus capsid. *Journal of Molecular Biology*, *383*, 923–934.

Schmidt, A., Eipel, C., Furst, K., Sommer, N., Pahnke, J., & Putzer, B. M. (2011). Evaluation of systemic targeting of RET oncogene-based MTC with tumor-selective peptide-tagged Ad vectors in clinical mouse models. *Gene Therapy*, *18*, 418–423.

Shayakhmetov, D. M., Gaggar, A., Ni, S., Li, Z. Y., & Lieber, A. (2005). Adenovirus binding to blood factors results in liver cell infection and hepatotoxicity. *Journal of Virology*, *79*, 7478–7491.

Short, J. J., Pereboev, A. V., Kawakami, Y., Vasu, C., Holterman, M. J., & Curiel, D. T. (2004). Adenovirus serotype 3 utilizes CD80 (B7.1) and CD86 (B7.2) as cellular attachment receptors. *Virology*, *322*, 349–359.

Short, J. J., Rivera, A. A., Wu, H., Walter, M. R., Yamamoto, M., Mathis, J. M., et al. (2010). Substitution of adenovirus serotype 3 hexon onto a serotype 5 oncolytic adenovirus reduces factor X binding, decreases liver tropism, and improves antitumor efficacy. *Molecular Cancer Therapeutics*, *9*, 2536–2544.

Silver, J., & Mei, Y. F. (2011). Transduction and oncolytic profile of a potent replication-competent adenovirus 11p vector (RCAd11pGFP) in colon carcinoma cells. *PLoS One*, *6*, e17532.

Smith, T., Idamakanti, N., Kylefjord, H., Rollence, M., King, L., Kaloss, M., et al. (2002). In vivo hepatic adenoviral gene delivery occurs independently of the coxsackievirus-adenovirus receptor. *Molecular Therapy*, *5*, 770–779.
Smith, T. A., Idamakanti, N., Rollence, M. L., Marshall-Neff, J., Kim, J., Mulgrew, K., et al. (2003). Adenovirus serotype 5 fiber shaft influences in vivo gene transfer in mice. *Human Gene Therapy*, *14*, 777–787.
Stoff-Khalili, M. A., Rivera, A. A., Stoff, A., Michael Mathis, J., Rocconi, R. P., Matthews, Q. L., et al. (2007). Combining high selectivity of replication via CXCR4 promoter with fiber chimerism for effective adenoviral oncolysis in breast cancer. *International Journal of Cancer*, *120*, 935–941.
Takahashi, S., Kato, K., Nakamura, K., Nakano, R., Kubota, K., & Hamada, H. (2011). Neural cell adhesion molecule 2 as a target molecule for prostate and breast cancer gene therapy. *Cancer Science*, *102*, 808–814.
Tao, N., Gao, G. P., Parr, M., Johnston, J., Baradet, T., Wilson, J. M., et al. (2001). Sequestration of adenoviral vector by Kupffer cells leads to a nonlinear dose response of transduction in liver. *Molecular Therapy*, *3*, 28–35.
Tillman, B. W., de Gruijl, T. D., Luykx-de Bakker, S. A., Scheper, R. J., Pinedo, H. M., Curiel, T. J., et al. (1999). Maturation of dendritic cells accompanies high-efficiency gene transfer by a CD40-targeted adenoviral vector. *Journal of Immunology*, *162*, 6378–6383.
Tsuruta, Y., Pereboeva, L., Glasgow, J. N., Luongo, C. L., Komarova, S., Kawakami, Y., et al. (2005). Reovirus sigma1 fiber incorporated into adenovirus serotype 5 enhances infectivity via a CAR-independent pathway. *Biochemical and Biophysical Research Communications*, *335*, 205–214.
Tsuruta, Y., Pereboeva, L., Glasgow, J. N., Rein, D. T., Kawakami, Y., Alvarez, R. D., et al. (2007). A mosaic fiber adenovirus serotype 5 vector containing reovirus sigma 1 and adenovirus serotype 3 knob fibers increases transduction in an ovarian cancer ex vivo system via a coxsackie and adenovirus receptor-independent pathway. *Clinical Cancer Research*, *13*, 2777–2783.
Vellinga, J., de Vrij, J., Myhre, S., Uil, T., Martineau, P., Lindholm, L., et al. (2007). Efficient incorporation of a functional hyper-stable single-chain antibody fragment protein-IX fusion in the adenovirus capsid. *Gene Therapy*, *14*, 664–670.
Vellinga, J., Rabelink, M. J., Cramer, S. J., van den Wollenberg, D. J., Van der Meulen, H., Leppard, K. N., et al. (2004). Spacers increase the accessibility of peptide ligands linked to the carboxyl terminus of adenovirus minor capsid protein IX. *Journal of Virology*, *78*, 3470–3479.
Vellinga, J., Van der Heijdt, S., & Hoeben, R. C. (2005). The adenovirus capsid: Major progress in minor proteins. *The Journal of General Virology*, *86*, 1581–1588.
Vigne, E., Dedieu, J. F., Brie, A., Gillardeaux, A., Briot, D., Benihoud, K., et al. (2003). Genetic manipulations of adenovirus type 5 fiber resulting in liver tropism attenuation. *Gene Therapy*, *10*, 153–162.
Vigne, E., Mahfouz, I., Dedieu, J. F., Brie, A., Perricaudet, M., & Yeh, P. (1999). RGD inclusion in the hexon monomer provides adenovirus type 5-based vectors with a fiber knob-independent pathway for infection. *Journal of Virology*, *73*, 5156–5161.
Waddington, S. N., McVey, J. H., Bhella, D., Parker, A. L., Barker, K., Atoda, H., et al. (2008). Adenovirus serotype 5 hexon mediates liver gene transfer. *Cell*, *132*, 397–409.
Wickham, T. J., Mathias, P., Cheresh, D. A., & Nemerow, G. R. (1993). Integrins alpha v beta 3 and alpha v beta 5 promote adenovirus internalization but not virus attachment. *Cell*, *73*, 309–319.
Wickham, T. J., Tzeng, E., Shears, L. L., 2nd, Roelvink, P. W., Li, Y., Lee, G. M., et al. (1997). Increased in vitro and in vivo gene transfer by adenovirus vectors containing chimeric fiber proteins. *Journal of Virology*, *71*, 8221–8229.

Wood, M., Perrotte, P., Onishi, E., Harper, M. E., Dinney, C., Pagliaro, L., et al. (1999). Biodistribution of an adenoviral vector carrying the luciferase reporter gene following intravesical or intravenous administration to a mouse. *Cancer Gene Therapy*, *6*, 367–372.
Wu, H., Han, T., Belousova, N., Krasnykh, V., Kashentseva, E., Dmitriev, I., et al. (2005). Identification of sites in adenovirus hexon for foreign peptide incorporation. *Journal of Virology*, *79*, 3382–3390.
Wu, H., Seki, T., Dmitriev, I., Uil, T., Kashentseva, E., Han, T., et al. (2002). Double modification of adenovirus fiber with RGD and polylysine motifs improves coxsackievirus-adenovirus receptor-independent gene transfer efficiency. *Human Gene Therapy*, *13*, 1647–1653.
Zubieta, C., Schoehn, G., Chroboczek, J., & Cusack, S. (2005). The structure of the human adenovirus 2 penton. *Molecular Cell*, *17*, 121–135.

CHAPTER THREE

# Syrian Hamster as an Animal Model to Study Oncolytic Adenoviruses and to Evaluate the Efficacy of Antiviral Compounds

**William S.M. Wold[1], Karoly Toth**
Department of Molecular Microbiology and Immunology, Saint Louis University School of Medicine, St. Louis, Missouri, USA
[1]Corresponding author: e-mail address: woldws@slu.edu

## Contents



## Abstract

The Syrian (golden) hamster (*Mesocricetus auratus*) has served as a useful model for different aspects of biology for at least 50 years, and its use has been expanding recently. In earlier years, among other things, it was a model for cancer development. More recently, it has become a model for many different infectious diseases. It has also become an alternative model for the study of oncolytic adenovirus vectors for cancer gene therapy. Among several other human pathogens, the hamster is permissive for the replication of human species C adenoviruses, which are the parental virus for the majority of adenovirus vectors in use today. These vectors replicate in some of the established hamster tumor cell lines that can be used to generate tumors *in vivo*, that is, one can study oncolytic (replication competent) adenoviruses in a permissive, immunocompetent model. This has afforded the opportunity to study the effect of the host immune system on the vector-infected tumor and has allowed the use of a more relevant animal model to determine the safety and biodistribution of

*Advances in Cancer Research*, Volume 115
ISSN 0065-230X
http://dx.doi.org/10.1016/B978-0-12-398342-8.00003-3

replication-competent adenoviruses. The hamster has also been used to evaluate antiviral compounds and vaccines against many viruses, including adenoviruses, flaviviruses, alphaviruses, arenaviruses, bunyaviruses, and paramyxoviruses.

# 1. INTRODUCTION

Despite recent advances in the diagnosis and therapy of cancer, it is still the second leading cause of mortality in the developed world. Oncolytic viruses represent an important new modality in the fight against cancer (Eager & Nemunaitis, 2011; Hernandez-Alcoceba, 2011). These viruses replicate in and kill cancer cells and produce progeny virus in the process. This progeny virus is then able to infect surrounding tumor cells and in theory perpetuate the process until all viable tumor cells are killed.

Adenoviruses (Ads), in particular human type 5 (Ad5), a member of human species C Ads, are one of the most popular virus species that is used to design replication-competent (RC) anticancer vectors (Toth & Wold, 2010; Toth, Dhar, & Wold, 2010; Yamamoto & Curiel, 2010). These viruses are easy to propagate, infect a broad range of cell types, are genetically stable, infect both dividing and quiescent cells, and their genome does not integrate into the host genome. Ads have long been serving as a workhorse for scientists; many of the seminal findings of the past half century in molecular biology resulted from experiments that studied the virus or used it as a tool (Berk, 2007; Wold & Horwitz, 2007). Ads were used to model cellular processes including DNA replication, pre-mRNA processing, oncogenic transformation, and apoptosis (reviewed in Berk, 2007; Russell, 2009). Virus replication and the interaction of virus and the host cell have been investigated exhaustively. Clearly, the molecular biology of Ad is one of the best studied among viruses. The epidemiology of the virus is thoroughly described as well. Studies have demonstrated the transmission routes, the prevalence, and the diseases caused by the virus.

Unfortunately, the workings of Ad infection between the molecular and societal scale are less well understood. We do not know much about the spread of the virus in an infected individual, the way Ads cause pathology, the role of the immune response in Ad-induced diseases, and the phenomenon of latent infection/persistence. This paucity of knowledge can be traced back in part to the lack of an adequate animal model. Ads are fairly species specific; thus human Ads replicate very poorly in most commonly used laboratory rodent species. Besides hampering the characterization of adenoviral pathology, this inability to study human Ad infection *in vivo* has negatively affected the development of oncolytic Ad research as well. Until recently, the only choice for scientists to study oncolytic Ads *in vivo* has been the exceedingly artificial immunodeficient mouse–human xenograft model, in which neither the immune response nor the effect of the vector on normal tissues could be studied.

It is clear that another animal model is necessary to further the development of Ad research. In this review, we briefly discuss the existing animal models, and then give a more detailed description of the Syrian hamster model, which is, in our opinion, the model of choice for the *in vivo* evaluation of oncolytic Ad5-based vectors, for research on Ad pathology, and for testing anti-Ad drugs. The Syrian hamster is also very useful for the study of many other viruses and the evaluation of compounds to inhibit the replication of these viruses.

# 2. EXISTING ANIMAL MODELS

## 2.1. Mouse models

Immunocompetent mice have been used in the past extensively, primarily to assess the toxicity of Ad vectors used as gene therapy tools. After intravenous injection into mice, the virus distributes to most organs (Ying et al., 2009). The primary target of species C Ads is the liver. Viral DNA in ever-decreasing quantities could be recovered from the spleen, liver, and lungs for up to a year after intravenous injection (Ying et al., 2009). The virus causes hepatocellular necrosis in mice, the severity of which depends on the injected dose (Lichtenstein et al., 2009). Mice succumb to intravenous injections of high doses; the $LD_{50}$ is approximately $10^9$ plaque-forming units (PFU)/animal of virus, but only marginal virus replication (i.e., recovery of infectious progeny virus from the liver) can be detected even at these conditions (Duncan et al., 1978; William Wold, personal communication). Other studies have seen minimal to no Ad replication in mice (Ginsberg et al., 1991; Oualikene, Gonin, & Eloit, 1994; Ying et al., 2009).

Mouse cell lines are generally regarded as nonpermissive for human Ad replication. Hallden et al. (2003) described mouse cell lines that were semipermissive for Ad replication and demonstrated that intratumoral injection of Ad5 into tumors generated by injecting these cells into immunocompetent mice increased the survival of the animals. However, it was unclear if virus replication was necessary for this effect. Unfortunately, no further development of this model was reported. We have tried to replicate these data without success (Ying et al., 2009).

Up to now, the most frequently used animal model to study oncolytic Ads has been the immunodefective mouse–human tumor xenograft model. To study oncolytic Ad vectors *in vivo*, investigators have been using the human tumor tissue grafted into immunodeficient (nude, SCID, etc.) mice. With this animal model, the tumor, which is of human origin, is permissive for virus replication, but as discussed above, the normal organs of the mouse do not support human Ad replication. Also, as these animals have to be immunodeficient to accept human tissue, the model does not allow for the study of the effect of the host immune system on the vector and the virus-infected tumor. Despite these disadvantages, until recently, scientists have used the immunodefective mouse–human tumor xenograft model almost exclusively. This resulted in the accumulation of a vast amount of data, which are available for comparison for ongoing and future studies. This, and the availability of a plethora of reagents specific to mice, has been a strong incentive for the continued use of this model.

One way to take advantage of the available mouse-specific reagents in a less artificial model would be to use oncolytic vectors based on mouse Ads and treat mouse tumors generated by transplanting mouse cell lines into syngeneic mice. Unfortunately, the biology of mouse Ads is very different from that of their human counterparts (Lenaerts, Verbeken, De Clercq, & Naesens, 2005). Thus, no oncolytic vectors that are analogous in structure and tropism to the human ones can be constructed using mouse Ads.

## 2.2. Pig

Human species C Ads were reported to replicate in the lungs of pigs after intravenous injection of $1 \times 10^{10}$ virus particles (vp)/animal of Ad5 (Jogler et al., 2006). Viral hexon gene expression was also detected by quantitative PCR in the liver, kidneys, and blood of the infected animals. The virus caused severe interstitial pneumonia but did not cause any disease in the liver or kidneys, two important sites for disseminated Ad infection in humans. Besides this discrepancy of target organs, the size of these animals would make them an impractical laboratory animal.

## 2.3. Cotton rat

Species C Ads replicated and caused pathological damage in the lungs of cotton rats (*Sygmodon hyspidus*) following intranasal inoculation (Prince et al., 1993). The intranasal $LD_{50}$ was $\sim 2.5 \times 10^9$ PFU/animal. The pathology resulted mostly from the immune reaction to the virus, and it was proportional to the infecting dose.

A tumor model, in which RC Ad vectors can be investigated, was also developed using cotton rats (Steel et al., 2007; Toth et al., 2005). In these studies, intratumoral injection of RC Ad vectors delayed the growth of subcutaneous cotton rat tumors. Virus replication was detected in the injected tumors, but not in other organs.

While species C Ads replicated in the lung of cotton rats and caused pathology, the model has several complications arising from the difficult nature of the species. These animals have a very strong fight-or-flight instinct and are exceedingly problematic to handle. This is a drawback for the experimenters but, more importantly, severely limits the animals' availability and increases its price due to the commercial breeders' inability to produce animals in sufficient numbers.

### 2.4. New Zealand white rabbit ocular infection model

Adenoviral epidemic keratoconjunctivitis (EKC), caused by Ad species D types 8 and 37, is a serious disease that can cause permanent damage to the eye and result in loss of vision (Gordon, Aoki, & Kinchington, 1996). To study the pathogenesis of the disease, an animal model based on the New Zealand white rabbit was developed (Gordon, Romanowski, & Araullo-Cruz, 1992). In this model, Ad5 is administered to the cornea of the animal via scarification, and this causes lesions similar to those seen in human cases of EKC. The model has been used extensively to assess the efficacy of topical antiviral agents (Clement et al., 2011; Romanowski & Gordon, 2008).

## 3. THE SYRIAN HAMSTER MODEL

### 3.1. Syrian hamsters as laboratory animals

Syrian (golden) hamsters (*Mesocricetus auratus*) belong to the Cricetidae family of the Rodentia order. The dorsal fur of the animal is reddish brown (hence the name "golden"), while its ventral area is light gray. They are medium-sized (150–170 cm long) rodents; young adults weigh approximately 100 g. They have short legs and an approximately 10 mm long tail. Because of this lack of a tail, intravenous injections are best performed through the jugular vein (Thomas, Spencer, & Wold, 2007). The hamsters' skin is very loose; the animal can easily "turn around" in its own skin if scruffed like one would a mouse. This needs to be taken into consideration when injecting the animals subcutaneously or intratumorally; it is not enough to fix the skin, rather the whole animal needs to be immobilized. This can be achieved with a little practice. Another distinctive anatomic feature of hamsters is the cheek pouch. These muscular sacs cover the area of both sides of the neck from the mouth to the shoulder and can extend to about 20 × 40 mm, allowing the animal to store considerable amount of food, bedding, etc. The cheek pouch of hamsters has been used to study chemical-induced carcinogenesis (Vairaktaris et al., 2008). Interestingly, the cheek pouch is an immunoprivileged site in the hamster; human tumors could be transplanted there with reasonable efficacy (Blumenthal, Kashi, Stephens, Sharkey, & Goldenberg, 1991).

Today, hamster colonies are maintained by Charles River and Harlan Sprague–Dawley. Hamsters are not considered truly inbred, although all of today's laboratory hamsters are thought of being the descendants of three or four individual animals captured in 1930. Distinct laboratory strains of Syrian hamsters were available from Charles River Laboratories until the 1980s, when they were discontinued. At present, there is no transgenic technique developed for hamsters, although hamster embryonic stem cell lines have been established (Doetschman, Williams, & Maeda, 1988).

While there is a general lack of hamster-specific reagents, there are several antibodies that can be used to detect hamster leukocyte subsets. With a few exceptions, antibodies generated against human, mouse, or rat proteins do not cross-react with their hamster counterparts. However, there is an anti-mouse CD3ε polyclonal goat serum (sc-1127) available from Santa Cruz that reacts with hamster CD3. This antibody was generated against the cytoplasmic tail of CD3; therefore, it works only with permeabilized cells. We have successfully used it for both flow cytometry and immunohistochemistry. The anti-mouse CD4 monoclonal antibody GK4.5 cross-reacts with hamster CD4 and can be used for flow cytometry. To detect tissue macrophages by immunohistochemistry in frozen sections, one can use the PAL1 monoclonal antibody (Palecanda et al., 1999). There are ongoing efforts in several laboratories to generate antibodies against various targets in hamsters; no doubt these will be made available for the hamster research community. A molecular biological approach was utilized to detect hamster cytokines; two laboratories published primer–probe sets that can be used in quantitative reverse transcriptase PCR assays (Espitia et al., 2010; Zivcec, Safronetz, Haddock, Feldmann, & Ebihara, 2011). These assays can be used to characterize the immune response to an infectious agent in hamsters (Espitia et al., 2010; Rockx et al., 2011). This approach will be extended as more of the hamster genome will be sequenced. At present, the sequences of approximately 800 hamster ESTs have been published; there is an ongoing effort to sequence both the hamster genome and transcriptome.

## 3.2. Syrian hamster model for oncolytic Ads

### *3.2.1 Hamsters are permissive for species C Ad replication*

Recently, others and we proposed the Syrian hamster as a model to test the efficacy and safety of oncolytic Ad vectors, to study Ad pathology, and to study the efficacy of antiviral drugs. There are many arguments for the practicality of this model. Many organs of hamsters are permissive for species C human Ad replication; thus the effect of an RC vector on the normal tissues can be assessed (Toth et al., 2008). Only Ad5 was tested in detail, but we assume that Syrian hamsters would be equally permissive to other species C human Ads (Ad1, 2, and 6). It was known previously that when infected intranasally with Ad5, infant and adult hamsters developed bronchopneumonia and interstitial pneumonia. Infectious viral titers in the lung peaked 2–3 days after infection and then declined (Hjorth et al., 1988; Thomas et al., 2006; Fig. 3.1). When infected enterally, the infant animals shed live virus with the feces (Hjorth et al., 1988). All infected animals developed humoral immunity to Ad. Thus, these routes of infection generated a disease mimicking the respiratory and enteric infections of immunocompetent humans. Further, it was demonstrated that injection of Ad2 into subcutaneous hamster melanoma tumors in Syrian hamsters inhibited tumor growth (Khoobyarian, Barone, Sabet, El Domeiri, & Das Gupta, 1975), and Ad6 established persistent infection of the hamster brain (Yabe, Matsumoto, & Ogura, 1988).

Parenteral administration of Ad5 causes disseminated infection in hamsters. In newborn hamsters, the subcutaneous injection of Ad5 caused hepatocellular necrosis and inflammation in the liver. Viral replication could be detected in the liver, with peak titers at 4 days after injection (Zarubaev et al., 2007). In adult hamsters, intravenous injection of either Ad5 or an oncolytic Ad named VRX-007 resulted in infection of most organs. Replication could be detected in the liver, lung, adrenal gland, heart, and kidney (Ying et al., 2009). Infectious progeny virus could be recovered from the liver and lung starting at 1 day after infection up to 6 days after injection. The virus caused the most significant toxicity in the liver, which could be monitored by testing serum transaminase levels. Histopathology revealed

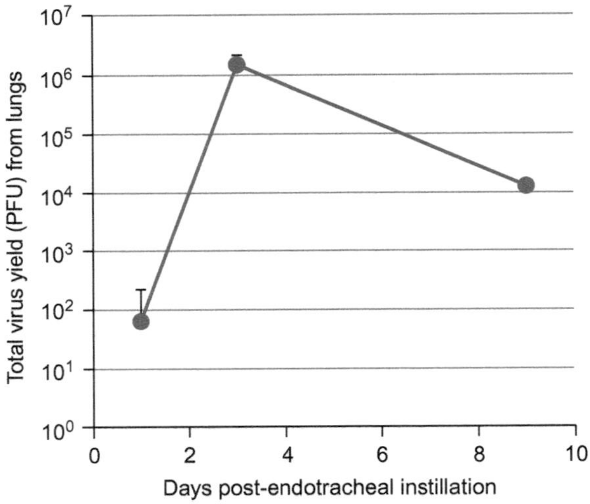


**Figure 3.1** Human type 5 adenovirus replicates in the lung of Syrian hamsters. Ad5 ($1 \times 10^7$ PFU in 100 μl total volume) was administered to Syrian hamsters by endotracheal instillation. Hamsters were sacrificed, and lungs were harvested on 1, 3, and 9 days after instillation ($n=2$ for 1 and 3 days after instillation, $n=1$ for 9 days after instillation). Lungs were homogenized and titered by plaque assay on A549 cells, and the yields are reported as the total virus yield (PFU) per animal (Thomas et al., 2006). (For color version of this figure, the reader is referred to the online version of this chapter.)

hepatocellular necrosis with mixed mononuclear infiltration, concentrated to the periportal areas (Lichtenstein et al., 2009). The intravenous maximum tolerated dose of Ad5 is $\sim 3 \times 10^{10}$ PFU/animal. The Syrian hamster model was used to determine the safety and biodistribution of several oncolytic Ad before entering clinical trials (Kim et al., 2011; Lichtenstein et al., 2009; Matthews et al., 2009; Ying et al., 2009). Intracranial injection of Ad5 or an oncolytic Ad into immunocompetent hamsters resulted in limited local replication. The viruses entered the bloodstream, and virus replication could be detected at distant organs like the lung and liver (Sonabend et al., 2009).

### *3.2.2 Syrian hamster tumor models*

Various tumor cell lines of hamster origin are available, and these cell lines form tumors in hamsters upon injection (Table 3.1). Further, a collection of naturally arisen transplantable hamster tumors (Kirkman, 1964; Kirkman & Chesterman, 1972) is maintained by the Biological Testing Branch of the Developmental Therapeutics Program at the National Cancer Institute (Table 3.2; DTP, 2011), which can be used for direct implantation into hamsters or for

**Table 3.1** Established Syrian hamster cell lines

| Cell line | Origin | Source |
|---|---|---|
| *Cell lines available from cell banks* | | |
| BHK-21 | Kidney | ATCC,[a] JCRB,[b] DSMZ[c] |
| ts13 | Kidney | ATCC |
| tk-ts13 | Kidney | ATCC |
| BHK21-pcDNA3.1-HC (BHK-21-secreting human erythropoietin) | Kidney | ATCC |
| HaK | Kidney | ATCC |
| BHK570 | Kidney | ATCC |
| HKT-1097 | Kidney | DSMZ, HPACC[d] |
| HIT-T15 | Pancreas | ATCC, DSMZ |
| HAP-T1 | Pancreas | DSMZ |
| HPD-1NR | Pancreas | DSMZ |
| HPD-2NR | Pancreas | DSMZ |
| RPMI-1846 | Melanoma | ATCC, JCRB |
| DDT MF2 | Smooth muscle | ATCC, HPACC |
| M3E3/C3 | Lung | DSMZ |
| HT100 | Lung | JCRB |
| HCT-2 | Lymphocyte | JCRB |
| HM-1 | Monocyte | JCRB |
| MCB3901 | Fibroblast | ATCC |
| DSPT200 | Fibroblast | JCRB |
| NIL | Fibroblast | JCRB |
| NIL-SF2 | Fibroblast | JCRB |
| SHOK | Fibroblast | JCRB |
| *Cells not deposited in cell banks* | | |
| SHPC6 | Pancreas | Spencer, Sagartz, Wold, and Toth (2009) |

*Continued*

**Table 3.1** Established Syrian hamster cell lines—cont'd

| Cell line | Origin | Source |
|---|---|---|
| SHRC1 | Kidney | Spencer et al. (2009) |
| SHLC1 | Lung | Spencer et al. (2009) |
| SHHC1 | Liver | Spencer et al. (2009) |
| PC1 | Pancreas | Hirota et al. (1993) |
| PC1.0 | Pancreas | Hirota et al. (1993) |
| HMAM4A | Mammary tumor | Coburn et al. (2011) |
| HMAM4B | Mammary tumor | Coburn et al. (2011) |
| HMAM5 | Mammary tumor | Coburn et al. (2011) |

[a]American Type Culture Collection.
[b]Japanese Collection of Research Bioresources.
[c]German Collection of Microorgansims and Cell Cultures.
[d]Health Protection Agency Culture Collections.

establishing permanent cell lines (Dhar, Toth, & Wold, 2011; Spencer et al., 2009). All the cell lines and transplantable tumors we have tried so far formed subcutaneous tumors in hamsters purchased from HSD without the need for immunosuppression. Thus, the vector can be studied in an immunocompetent model, allowing researchers to study the effect of the host immune response. Hamsters are an overly docile species that can be easily handled and can be multihoused (as opposed to cotton rats). Syrian hamsters are cheaper than immunodeficient mice, and their larger size allows for larger tumor sizes that reflect the clinical situation better and make certain surgical procedures easier. The one major drawback of the model is the relative lack of reagents, which makes thorough mechanistic studies difficult.

Several tumor cell lines were reported to be permissive for the replication of species C Ads and oncolytic Ad vectors based on species C Ads (Bortolanza, Bunuales, Alzuguren, et al., 2009; Spencer et al., 2009; Thomas et al., 2006; Ying et al., 2009). Virus yields with these cells were about 10- to 100-fold less than yields with human A549 cells, which are one of the most permissive human cells. When injected subcutaneously into immunocompetent hamsters, these hamster cell lines formed tumors. The subcutaneous tumors formed by injecting the HaK renal cells were highly metastatic, forming

**Table 3.2** Transplantable tumors

| **Tumor** | **Histologic type** |
|---|---|
| Fibrosarcoma | Fibrosarcoma |
| H-12 | Mesothelioma |
| H-75 | Mesothelioma |
| TG1-4 | Mesothelioma |
| 10–24 | Mesothelioma |
| Islet Cell | Pancreatic adenocarcinoma |
| Pan#1 | Pancreatic duct adenocarcinoma |
| 2309V | Pancreatic islet B-cell adenocarcinoma |
| 4671 | Pancreatic duct adenocarcinoma |
| 1382J | Liver carcinoma |
| SB#1 | Small bowel adenocarcinoma |
| 8721R | Renal carcinoma |
| 11348P | Pulmonary squamous cell carcinoma |
| Melanoma | Melanotic melanoma |
| TS1-4 | Epidermoid carcinoma |
| 11963V | Leiomyosarcoma |
| 6973P | Leiomyosarcoma |
| 8746Q | Uterine adenocarcinoma |
| 9242 | Parotid acinar cell adenocarcinoma |
| 10838 | Seminoma |
| Lymphosarcoma | Lymphosarcoma |
| 22047 | Adenocarcinoma |

metastases in the draining lymph nodes, the kidneys, the liver, and most prominently in the lung (Thomas et al., 2006). Upon intravenous injection, HaK cells formed multiple tumor nodules in the lung (Wold, personal communication). The pancreatic ductal carcinoma cell line SHPC6 can be injected intraperitoneally to form disseminated tumor nodules in the peritoneal cavity and nodules attaching to the pancreas (Spencer et al., 2009).

### 3.2.3 Oncolytic Ads in Syrian hamsters

Multiple oncolytic Ad vectors were tested in tumor-bearing Syrian hamsters. The most data were generated using VRX-007, a vector with enhanced cell-to-cell spread (Doronin et al., 2003; Toth, Kuppuswamy, Shashkova, Spencer, & Wold, 2010). When injected intratumorally, VRX-007 significantly inhibited the growth of subcutaneous renal, pancreatic, and leiomyosarcoma tumors (Spencer et al., 2009; Thomas et al., 2006; Fig. 3.2). In these experiments, virus replication could be demonstrated in the tumors by recovering infectious virions and by detecting infected cells via immunohistochemistry. Intraperitoneal injection of VRX-007 completely eliminated disseminated intraperitoneal pancreatic tumors and prevented mortality induced by the tumor burden (Spencer et al., 2009).

### 3.2.4 The effect of immune response on the efficacy of oncolytic ads

Interesting data were generated while investigating the effect of host immune response on the antitumor efficacy of VRX-007. Treatment with cyclophosphamide, an alkylating agent that is known to suppress the immune

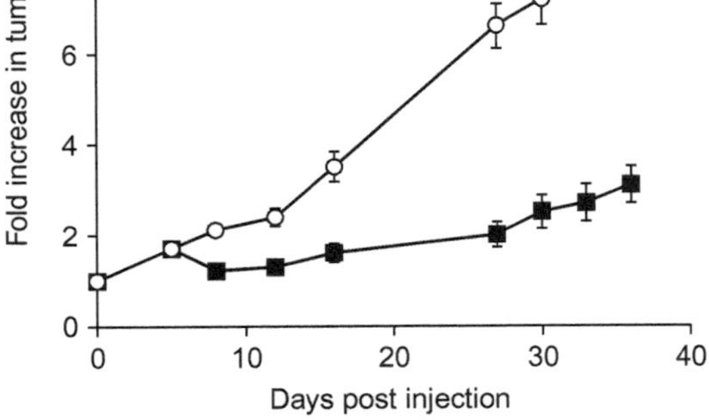


**Figure 3.2** Established HaK (hamster kidney) tumors with a mean initial volume of 490 μl were injected with $1 \times 10^{10}$ PFU of VRX-007 or vehicle ($n = 18$ for both groups) for 6 consecutive days. A second round of treatment (three injections of $1 \times 10^{10}$ PFU/dose) was initiated 16 days after the start of the first round of treatment. The graph shows the mean fold increase in tumor volume; the error bars represent the standard error of the mean. v, vehicle. VRX-007 $P < 0.02$ (Student's *t*-test) (Thomas et al., 2006).

system by preventing cell proliferation (i.e., suppressing the bone marrow and clonal expansion of lymphocytes), potentiated the suppression of tumor growth by VRX-007 (Thomas et al., 2008; Fig. 3.3). As the treatment also increased the persistence of the vector in the tumor, we hypothesized that the increase in efficacy is a result of the inability of the host to mount an immune response to the vector. Thus, the oncolytic Ad is not eliminated prematurely and has more time to exert its effect on the tumor. While this hypothesis has not been validated by using other types of immunosuppressants (cyclophosphamide has a pleiotropic effect on the body), this seems to be a logical scenario. To further investigate whether the antivector immune response has a significant negative effect on vector efficacy, an experiment was carried out in which the tumor-bearing animals were immunized with human type 5 Ad (Ad5). As a large proportion of adults have circulating antibodies against Ad5, these animals closely resemble the situation what can be expected for many tumor patients. Interestingly, preexisting immunity did not have a significant effect on the antitumor efficacy of VRX-007 in this

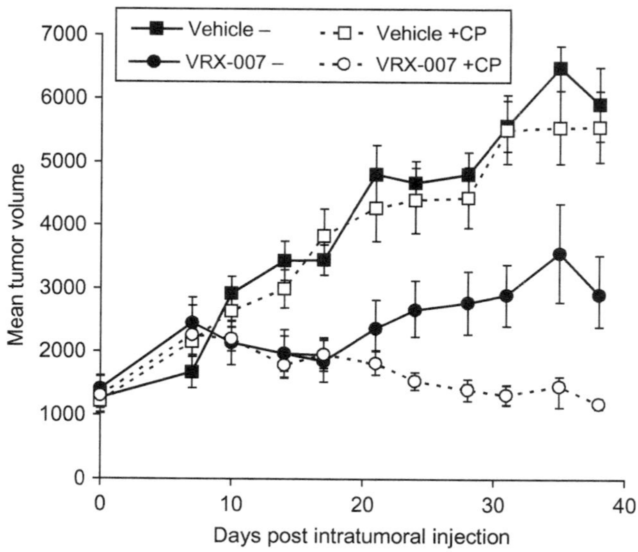


**Figure 3.3** Immunosuppression with cyclophosphamide (CP) resulted in significantly enhanced tumor control with oncolytic adenoviruses VRX-007. Animals with established HaK tumors (mean tumor volume of ca.1300 μl) received daily intratumoral injections of VRX-007 or vehicle ($n=12$ for each group) on days 0–6. Median tumor volumes are shown for each group. Virus treatment alone (VRX-007 –) versus virus plus CP (VRX-007 + CP; $P \leq 0.001$). No significant difference was detected between "Vehicle–" and "Vehicle + CP" ($P=0.692$) (Thomas et al., 2008).

experiment (Dhar, Spencer, Toth, & Wold, 2009a). However, the vector was less efficacious in immunosuppressed animals that were immunized previously (i.e., had circulating anti-Ad antibodies) than in naïve immunosuppressed animals.

Data from another laboratory speak to the importance to the host immune response in vector clearance. Rather than general immunosuppression, another approach to blunt the host immune response to the vector is to make use of the virus's own defense. Ad encodes a variety of proteins that function to modulate the host's immune system. Bortolanza et al. demonstrated that the deletion of two such genes, *6.7K* and *gp19K*, results in accelerated clearance of the virus in immunocompetent hamsters but not in immunodeficient mice (Bortolanza, Bunuales, Alzuguren, et al., 2009).

To recapitulate these results, the host rapidly mounts an antivector immune response (neutralizing antibodies can be detected at 7 days postinfection; Dhar et al., 2009a), which is deleterious to vector efficacy. Because of the rapid nature of this response, preexisting immunity does not significantly add to this decrease of efficacy. The inhibitory effect of the immune response is at least partially mediated by neutralizing antibodies, inasmuch as circulating neutralizing antibodies can inhibit vector activity in otherwise immunosuppressed animals.

Neutralizing antibodies can have another, beneficial effect as well. It was demonstrated that even after intratumoral injection, a significant amount of the vector is dispersed into the liver and other organs (Thomas et al., 2006). This is especially pronounced with immunosuppressed animals (Thomas et al., 2008). In immunosuppressed hamsters, circulating antibodies either induced by immunization before cyclophosphamide treatment or passively introduced could fend off a lethal challenge with intravenous Ad5 injection (Dhar, Spencer, Toth, & Wold, 2009b). Thus, circulating antibodies, even in severely immunosuppressed patients, can protect against unwanted disseminated infection with the vector.

### 3.2.5 The effect of immunostimulatory transgenes on the efficacy of oncolytic Ads in hamsters

The Syrian hamster model enabled researchers to explore another aspect of the immune response to the vector. One of the strategies to increase the efficacy of oncolytic Ad vectors is to incorporate immunostimulatory genes in the vector (Toth & Wold, 2010). The goal of this approach is to try to boost the host immune response to the vector and channel it to attack the tumor.

Granulocyte–macrophage colony-stimulating factor (GM-CSF) is one such immunostimulatory gene that induces antitumor immunity by stimulating tumor-specific cytotoxic T-lymphocytes and natural killer cells (Arellano & Lonial, 2008). Interestingly, human GM-CSF functions in hamsters as well (Cho et al., 2006). The Hemminki lab injected GM-CSF-expressing oncolytic Ads into subcutaneous pancreatic tumors and demonstrated that they were more effective than the control oncolytic vector with no transgene (Koski et al., 2010). The authors showed a significant local increase of GM-CSF in the tumor. They also showed that animals that cleared tumors due to injection with the GM-CSF-expressing vector cannot be rechallenged with the same tumor cells (Cerullo et al., 2010). This indicates that a tumor-specific immune response was evoked in these animals, which may have been responsible for the antitumor effects.

Another cytokine used to stimulate the host's antitumor responses is interleukin 12 (IL-12) (Del Vecchio et al., 2007). Bortolanza et al. demonstrated that an oncolytic Ad expressing single-chain IL-12 (IL-12 is a heterodimer) exhibited greater efficacy than its parental vector when injected into intrahepatic pancreatic tumors (Bortolanza, Bunuales, Otano, et al., 2009). They also showed that interferon gamma expression was elevated in tumors that were injected with the IL-12-expressing vector, pointing to a probable mechanism.

Taken together, the Syrian hamster model for oncolytic Ads presents multiple opportunities to investigate facets of the vector–tumor–host interaction that were not available with the most widely used immunodeficient mouse–human tumor xenograft model.

## 3.3. Syrian hamster as a test system to evaluate anti-adenoviral drugs

### *3.3.1 Adenovirus is an important human pathogen*

While human Ads generally cause mild, self-limiting infections in immunocompetent adults, the virus can present serious problems for immunocompromised patients. Most importantly, systemic Ad infection causes significant mortality in hematopoietic stem cell transplant recipients, but Ad infection can have grave consequences in any patient with a compromised immune system (Ison, 2006; Ison & Green, 2009). Presently, there are no drugs approved by the FDA specifically to treat Ad infection. Physicians use a variety of off-label drugs on a "compassionate use" basis, but there are no placebo-controlled studies to confirm the efficacy of these treatments. One major reason for the lack of an approved drug to treat Ad infection is the unmet need for an animal model in which these drugs could be tested before clinical trials. Others and we have been proposing that the Syrian hamster would be a suitable model for this purpose. As mentioned above, the organs and tissues of the hamster are permissive for species C Ad replication, and the infection causes a pathology resembling that observed in humans (Lichtenstein et al., 2009; Ying et al., 2009). Immunosuppression markedly increased virus replication in various organs (mainly in the liver; Fig. 3.4) and the resulting viral pathology ($LD_{50}$~$1\text{–}3 \times 10^{10}$ PFU/animal). The ongoing infection and the liver damage can be followed by titering the virus in the blood and checking serum transaminase levels, respectively (Toth et al., 2008). Hence, intravenous administration of Ads, especially to immunosuppressed hamsters, can be used as a model for disseminate Ad infection in immunocompromised humans.

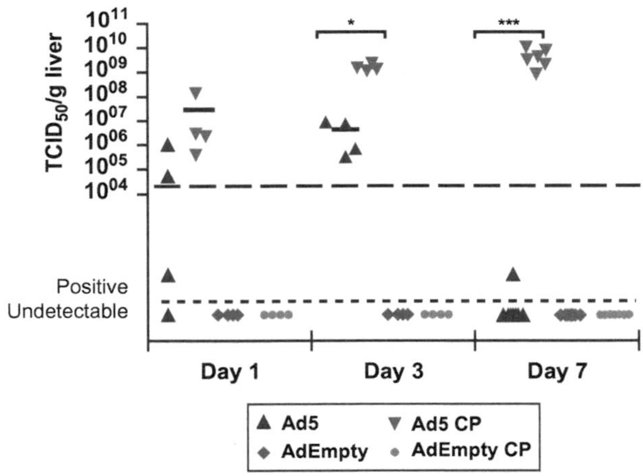


**Figure 3.4** Immunosuppression increases the replication of Ad5 in the livers of Syrian hamsters. Immunocompetent hamsters and hamsters immunosuppressed with cyclophosphamide were injected intravenously with $1.9 \times 10^{12}$ vp/kg of Ad5 or AdEmpty, a replication-defective Ad vector with no transgene. Animals were sacrificed at the indicated times postinfection ($n = 4$ for the day 1 and day 3 time points, $n = 6$ or 8 for the day 7 time point), and the infectious virus titer in the livers was determined by tissue culture infective dose ($TCID_{50}$) assay on HEK-293 cells. The symbols represent values from individual animals, the dashed line with the longer dashes signifies the limit of quantifiability, and symbols under the line with the shorter dashes represent samples in which no virus was detected. The horizontal bars correspond to means. CP, cyclophosphamide; $*P < 0.05$ (Toth et al., 2008). (For color version of this figure, the reader is referred to the online version of this chapter.)

### 3.3.2 *The immunosuppressed Syrian hamster is a valid model to test anti-Ad drugs*

We have tested the efficacy of one antiviral drug candidate named CMX001, an acyclic nucleoside phosphonate with a covalently linked lipid moiety to facilitate cellular uptake, against lethal intravenous challenge with Ad5 in hamsters immunosuppressed by use of cyclophosphamide. After intravenous administration of Ad5, treatment with CMX001 dramatically reduced Ad5 replication in the liver of immunosuppressed hamsters and greatly inhibited the resulting liver toxicity (Fig. 3.5). The drug completely prevented mortality when applied preventatively, and significantly reduced both mortality and morbidity even when administered 2 days post-virus challenge (Toth et al., 2008). This latter result is especially promising in the light that a considerable amount of progeny virus can be recovered from the liver 24 h postinfection; that is, a large part of viral replication and the resulting organ

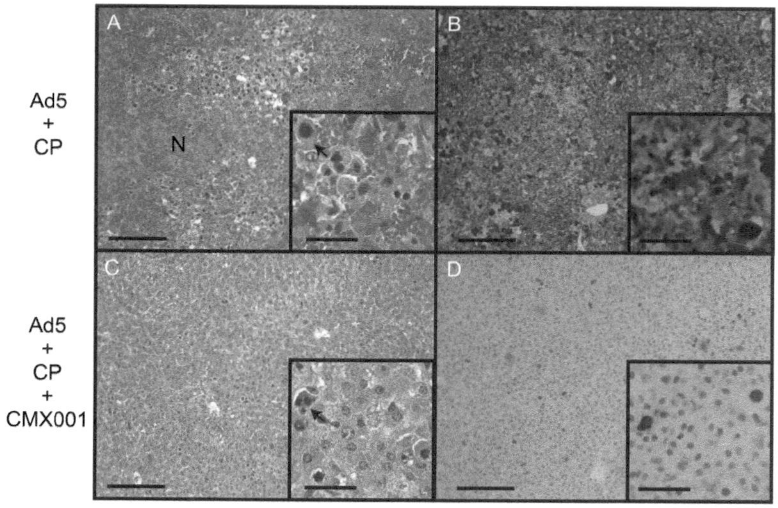


**Figure 3.5** The antiviral drug CMX001 decreases replication of Ad5 and Ad5-induced lesions in the liver. Livers of immunosuppressed hamsters sacrificed 6 days after challenge with the $LD_{50}$ of Ad5 were subjected to histopathological and immunohistochemistry (IHC) evaluation. Animals infected with Ad5 and not treated with CMX001 exhibited extensive coagulation necrosis throughout the liver (A) and widespread replication of Ad5, demonstrated by staining for fiber (B). Treatment of Ad5-infected hamsters with CMX001 resulted in a significant reduction in hepatocellular injury (C) and greatly reduced IHC staining for fiber (D). The arrows indicate intranuclear inclusion bodies (scale bars: 200 μm for the larger images and 50 μm for insets). N, necrosis (Toth et al., 2008). (See Page 3 in Color Section at the back of book.)

damage had happened by the time the drug was applied. This drug is now in a Phase II clinical trial and produces similar results in Ad-infected human patients (Paolino et al., 2011). The success of this drug in humans lends support to the idea that data obtained using the hamster model have biological relevance in humans.

### 3.3.3 Immunocompetent hamster models

Other laboratories have utilized the hamster model to test anti-adenoviral drugs as well. Diaconu et al. (2010) treated tumor-bearing hamsters with cidofovir (the parental drug for CMX001) or chlorpromazine. They injected the animals intratumorally with Ad5 and showed that treatment with either drug significantly attenuated the replication of Ad5 in both the tumor and the liver.

Another research group used subcutaneous injection of newborn hamsters as their model. They demonstrated that treatment with 6-azacytidine decreased virus burden and sped up the clearance of the virus in the liver, heart, kidney, and the lung (Zarubaev et al., 2007). This research group tested another drug, Ingavirin, and recorded a modest inhibition of Ad5 replication in the liver and lung of infected hamsters (Zarubaev et al., 2011).

## 3.4. Syrian hamster as a model for other infectious diseases

For reasons that are not completely understood, hamsters are permissive for replication of many different microorganisms regardless of their natural host. Infection with these microorganisms often induces disease in hamsters that is similar to that in humans. These features have made the Syrian hamster an attractive model for several diseases.

Hamsters are especially sensitive to arthropod-borne viruses, among them viruses of important clinical significance such as the flaviviruses West Nile virus (WNV) and Yellow fever virus (YFV). Hamsters are used to study the pathology of WNV (Morrey et al., 2004b). The neurological symptoms of WNV infection were studied in great detail in hamsters (Samuel, Wang, Siddharthan, Morrey, & Diamond, 2007; Siddharthan et al., 2009; Xiao, Guzman, Zhang, Travassos Da Rosa, & Tesh, 2001). The hamster model was also used to test anti-WNV vaccines and drugs (Morrey et al., 2004a, 2007, 2008; Tesh et al., 2002; Watts et al., 2007). The course of YFV infection, the pathology caused, and the immune response to the infection were examined extensively (Li et al., 2008; Sbrana, Xiao, Popov, Newman, & Tesh, 2006; Tesh et al., 2001; Xiao, Zhang, Guzman, & Tesh, 2001), as were the effects of various antivirals and vaccines (Julander et al., 2010;

Julander, Ennis, Turner, & Morrey, 2011; Mateo et al., 2007; Monath et al., 2010). In addition to WNV and YFV, hamsters served as models for two other flaviviruses of equal importance, St. Louis encephalitis virus and Japanese encephalitis virus (Burke & Monath, 2001; Siirin et al., 2007).

Other human pathogen viruses, namely the alphaviruses Eastern (Paessler et al., 2004), Western (Julander et al., 2007), and Venezuelan (Turell & Parker, 2008) equine encephalitis viruses, the paramyxoviruses Nipah and Hendra virus (Rockx et al., 2011), are also studied in Syrian hamsters. Hamsters are important model organisms for human bunyavirus and arenavirus infection. The study of these viruses requires high-containment laboratories, so researchers use viruses that do not infect or are less pathogenic for humans but cause human-like pathology in hamsters to describe the pathology of the disease and test antiviral drugs (Gowen et al., 2007, 2008, 2010). Besides viruses, protozoan and bacterial infections were also studied in hamsters. The cutaneous leishmaniasis of hamsters is described in great detail (Espitia et al., 2010; Osorio, Bonilla, Peniche, Melby, & Travi, 2008).

In summary, the Syrian hamster is used as an animal model for various human diseases. In certain cases, such as for human infection with Ads, it is the most practical model in which to study the pathogenesis of a human disease and to test antivirals. In other cases, such as for the study of oncolytic Ads, it allows us to study aspects of the mechanism of the vectors that is not possible with other models. In yet other cases, such as for certain highly pathogenic human virus infections, Syrian hamsters provide a surrogate model of infection by using a similar virus of like pathogenesis. Certainly, more work is needed to obtain the necessary reagents to utilize the model to its full potential. We are confident that more and more antibodies, nucleic acid sequences, and methods will be accessible as research groups working with hamsters publish their results.

## REFERENCES

Arellano, M., & Lonial, S. (2008). Clinical uses of GM-CSF, a critical appraisal and update. *Biologics*, *2*, 13–27.

Berk, A. J. (2007). Adenoviridae: The viruses and their replication. In D. M. Knipe & P. M. Howley (Eds.), *Field's virology*. Philadelphia, USA: Lippincott Williams & Wilkins.

Blumenthal, R. D., Kashi, R., Stephens, R., Sharkey, R. M., & Goldenberg, D. M. (1991). Improved radioimmunotherapy of colorectal cancer xenografts using antibody mixtures against carcinoembryonic antigen and colon-specific antigen-p. *Cancer Immunology, Immunotherapy*, *32*, 303–310.

Bortolanza, S., Bunuales, M., Alzuguren, P., Lamas, O., Aldabe, R., Prieto, J., et al. (2009). Deletion of the E3-6.7K/gp19K region reduces the persistence of wild-type adenovirus in a permissive tumor model in Syrian hamsters. *Cancer Gene Therapy, 16*, 703–712.

Bortolanza, S., Bunuales, M., Otano, I., Gonzalez-Aseguinolaza, G., Ortiz-de-Solorzano, C., Perez, D., et al. (2009). Treatment of pancreatic cancer with an oncolytic adenovirus expressing interleukin-12 in Syrian hamsters. *Molecular Therapy, 17*, 614–622.

Burke, D. S., & Monath, T. P. (2001). Flaviviruses. In D. M. Knipe & P. M. Howley (Eds.), *Field's virology*. Philadelphia: Lippincott Williams & Wilkins.

Cerullo, V., Pesonen, S., Diaconu, I., Escutenaire, S., Arstila, P. T., Ugolini, M., et al. (2010). Oncolytic adenovirus coding for granulocyte macrophage colony-stimulating factor induces antitumoral immunity in cancer patients. *Cancer Research, 70*, 4297–4309.

Cho, S. A., Park, J. H., Seok, S. H., Juhn, J. H., Kim, S. J., Ji, H. J., et al. (2006). Effect of granulocyte macrophage-colony stimulating factor (GM-CSF) on 5-FU-induced ulcerative mucositis in hamster buccal pouches. *Experimental and Toxicologic Pathology, 57*, 321–328.

Clement, C., Capriotti, J. A., Kumar, M., Hobden, J. A., Foster, T. P., Bhattacharjee, P. S., et al. (2011). Clinical and antiviral efficacy of an ophthalmic formulation of dexamethasone povidone-iodine in a rabbit model of adenoviral keratoconjunctivitis. *Investigative Ophthalmology & Visual Science, 52*, 339–344.

Coburn, M. A., Brueggemann, S., Bhatia, S., Cheng, B., Li, B. D., Li, X. L., et al. (2011). Establishment of a mammary carcinoma cell line from Syrian hamsters treated with N-methyl-N-nitrosourea. *Cancer Letters, 312*, 82–90.

Del Vecchio, M. F., Bajetta, E., Canova, S., Lotze, M. T., Wesa, A., Parmiani, G., et al. (2007). Interleukin-12: Biological properties and clinical application. *Clinical Cancer Research, 13*, 4677–4685.

Dhar, D., Spencer, J. F., Toth, K., & Wold, W. S. M. (2009a). Effect of preexisting immunity on oncolytic adenovirus vector INGN 007 antitumor efficacy in immunocompetent and immunosuppressed Syrian hamsters. *Journal of Virology, 83*, 2130–2139.

Dhar, D., Spencer, J. F., Toth, K., & Wold, W. S. M. (2009b). Pre-existing immunity and passive immunity to adenovirus 5 prevents toxicity caused by an oncolytic adenovirus vector in the syrian hamster model. *Molecular Therapy, 17*, 1724–1732.

Dhar, D., Toth, K., & Wold, W. S. M. (2011). Syrian hamster tumor model to study oncolytic Ad5 based vectors. In D. Kirn, T. C. Liu & S. Thorne (Eds.), *Oncolytic viruses: Methods and protocols*. New York: Humana Press.

Diaconu, I., Cerullo, V., Escutenaire, S., Kanerva, A., Bauerschmitz, G. J., Hernandez-Alcoceba, R., et al. (2010). Human adenovirus replication in immunocompetent Syrian hamsters can be attenuated with chlorpromazine or cidofovir. *The Journal of Gene Medicine, 12*, 435–445.

Doetschman, T., Williams, P., & Maeda, N. (1988). Establishment of hamster blastocyst-derived embryonic stem (ES) cells. *Developmental Biology, 127*, 224–227.

Doronin, K., Toth, K., Kuppuswamy, M., Krajcsi, P., Tollefson, A. E., & Wold, W. S. M. (2003). Overexpression of the ADP (E3-11.6K) protein increases cell lysis and spread of adenovirus. *Virology, 305*, 378–387.

DTP. (2011). DCTD tumor repository: A catalog of in vitro cell lines, transplantable animal and human tumors and microarrays. http://dtp.nci.nih.gov/index.html. Retrieved on May 25th, 2012.

Duncan, S. J., Gordon, F. C., Gregory, D. W., McPhie, J. L., Postlethwaite, R., White, R., et al. (1978). Infection of mouse liver by human adenovirus type 5. *The Journal of General Virology, 40*, 45–61.

Eager, R. M., & Nemunaitis, J. (2011). Clinical development directions in oncolytic viral therapy. *Cancer Gene Therapy, 18*, 305–317.

Espitia, C. M., Zhao, W., Saldarriaga, O., Osorio, Y., Harrison, L. M., Cappello, M., et al. (2010). Duplex real-time reverse transcriptase PCR to determine cytokine mRNA expression in a hamster model of New World cutaneous leishmaniasis. *BMC Immunology*, *11*, 31.

Ginsberg, H. S., Moldawer, L. L., Sehgal, P. B., Redington, M., Kilian, P. L., Chanock, R. M., et al. (1991). A mouse model for investigating the molecular pathogenesis of adenovirus pneumonia. *Proceedings of the National Academy of Sciences of the United States of America*, *88*, 1651–1655.

Gordon, Y. J., Aoki, K., & Kinchington, P. R. (1996). Adenovirus keratoconjunctivitis. In J. S. Pepose, G. N. Holland & K. R. Wilhelmus (Eds.), *Ocular infectin and immunity*. St. Louis: Mosby.

Gordon, Y. J., Romanowski, E., & Araullo-Cruz, T. (1992). An ocular model of adenovirus type 5 infection in the NZ rabbit. *Investigative Ophthalmology & Visual Science*, *33*, 574–580.

Gowen, B. B., Judge, J. W., Wong, M. H., Jung, K. H., Aylsworth, C. F., Melby, P. C., et al. (2008). Immunoprophylaxis of Punta Toro virus (Phlebovirus, Bunyaviridae) infection in hamsters with recombinant Eimeria profilin-like antigen. *International Immunopharmacology*, *8*, 1089–1094.

Gowen, B. B., Julander, J. G., London, N. R., Wong, M. H., Larson, D., Morrey, J. D., et al. (2010). Assessing changes in vascular permeability in a hamster model of viral hemorrhagic fever. *Virology Journal*, 7, 240.

Gowen, B. B., Wong, M. H., Jung, K. H., Sanders, A. B., Mendenhall, M., Bailey, K. W., et al. (2007). In vitro and in vivo activities of T-705 against arenavirus and bunyavirus infections. *Antimicrobial Agents and Chemotherapy*, *51*, 3168–3176.

Hallden, G., Hill, R., Wang, Y., Anand, A., Liu, T. C., Lemoine, N. R., et al. (2003). Novel immunocompetent murine tumor models for the assessment of replication-competent oncolytic adenovirus efficacy. *Molecular Therapy*, *8*, 412–424.

Hernandez-Alcoceba, R. (2011). Recent advances in oncolytic virus design. *Clinical & Translational Oncology*, *13*, 229–239.

Hirota, M., Egami, H., Corra, S., Fujii, H., Chaney, W. G., Rizzino, A., et al. (1993). Production of scatter factor-like activity by a nitrosamine-induced pancreatic cancer cell line. *Carcinogenesis*, *14*, 259–264.

Hjorth, R. N., Bonde, G. M., Pierzchala, W. A., Vernon, S. K., Wiener, F. P., Levner, M. H., et al. (1988). A new hamster model for adenoviral vaccination. *Archives of Virology*, *100*, 279–283.

Ison, M. G. (2006). Adenovirus infections in transplant recipients. *Clinical Infectious Diseases*, *43*, 331–339.

Ison, M. G., & Green, M. (2009). Adenovirus in solid organ transplant recipients. *American Journal of Transplantation*, *9*, S161–S165.

Jogler, C., Hoffmann, D., Theegarten, D., Grunwald, T., Uberla, K., & Wildner, O. (2006). Replication properties of human adenovirus in vivo and in cultures of primary cells from different animal species. *Journal of Virology*, *80*, 3549–3558.

Julander, J. G., Ennis, J., Turner, J., & Morrey, J. D. (2011). Treatment of yellow fever virus with an adenovirus-vectored interferon, DEF201, in a hamster model. *Antimicrobial Agents and Chemotherapy*, *55*, 2067–2073.

Julander, J. G., Jha, A. K., Choi, J. A., Jung, K. H., Smee, D. F., Morrey, J. D., et al. (2010). Efficacy of 2'-C-methylcytidine against yellow fever virus in cell culture and in a hamster model. *Antiviral Research*, *86*, 261–267.

Julander, J. G., Siddharthan, V., Blatt, L. M., Schafer, K., Sidwell, R. W., & Morrey, J. D. (2007). Effect of exogenous interferon and an interferon inducer on western equine encephalitis virus disease in a hamster model. *Virology*, *360*, 454–460.

Khoobyarian, N., Barone, F., Sabet, T., El Domeiri, A. A., & Das Gupta, T. K. (1975). Inhibition of melanoma growth in hamsters by type-2 adenovirus. *Journal of Surgical Oncology*, 7, 421–425.

Kim, K. H., Ryan, M. J., Estep, J. E., Miniard, B. M., Rudge, T. L., Peggins, J. O., et al. (2011). A new generation of serotype chimeric infectivity-enhanced conditionally replicative adenovirals: The safety profile of ad5/3-delta24 in advance of a phase I clinical trial in ovarian cancer patients. *Human Gene Therapy*, *22*, 821–828.

Kirkman, H. (1964). A preliminary report concerning tumors observed in Syrian hamsters. *Stanford Medical Bulletin*, 163–166.

Kirkman, H., & Chesterman, F. C. (1972). Additional data on transplanted tumours of the golden hamster. *Progress in Experimental Tumor Research*, *16*, 580–621.

Koski, A., Kangasniemi, L., Escutenaire, S., Pesonen, S., Cerullo, V., Diaconu, I., et al. (2010). Treatment of cancer patients with a serotype 5/3 chimeric oncolytic adenovirus expressing GMCSF. *Molecular Therapy*, *18*, 1874–1884.

Lenaerts, L., Verbeken, E., De Clercq, E., & Naesens, L. (2005). Mouse adenovirus type 1 infection in SCID mice: An experimental model for antiviral therapy of systemic adenovirus infections. *Antimicrobial Agents and Chemotherapy*, *49*, 4689–4699.

Li, G., Duan, T., Wu, X., Tesh, R. B., Soong, L., & Xiao, S. Y. (2008). Yellow fever virus infection in Syrian golden hamsters: Relationship between cytokine expression and pathologic changes. *International Journal of Clinical and Experimental Pathology*, *1*, 169–179.

Lichtenstein, D. L., Spencer, J. F., Doronin, K., Patra, D., Meyer, J., Shashkova, E. V., et al. (2009). An acute toxicology study with INGN 007, an oncolytic adenovirus vector, in mice and permissive Syrian hamsters; comparisons with wild-type Ad5 and a replication-defective adenovirus vector. *Cancer Gene Therapy*, *16*, 644–654.

Mateo, R. I., Xiao, S. Y., Travassos Da Rosa, A. P., Lei, H., Guzman, H., Lu, L., et al. (2007). Yellow fever 17-D vaccine is neurotropic and produces encephalitis in immunosuppressed hamsters. *The American Journal of Tropical Medicine and Hygiene*, *77*, 919–924.

Matthews, K., Noker, P. E., Tian, B., Grimes, S. D., Fulton, R., Schweikart, K., et al. (2009). Identifying the safety profile of Ad5.SSTR/TK.RGD, a novel infectivity-enhanced bicistronic adenovirus, in anticipation of a phase I clinical trial in patients with recurrent ovarian cancer. *Clinical Cancer Research*, *15*, 4131–4137.

Monath, T. P., Lee, C. K., Julander, J. G., Brown, A., Beasley, D. W., Watts, D. M., et al. (2010). Inactivated yellow fever 17D vaccine: Development and nonclinical safety, immunogenicity and protective activity. *Vaccine*, *28*, 3827–3840.

Morrey, J. D., Day, C. W., Julander, J. G., Blatt, L. M., Smee, D. F., & Sidwell, R. W. (2004). Effect of interferon-alpha and interferon-inducers on West Nile virus in mouse and hamster animal models. *Antiviral Chemistry & Chemotherapy*, *15*, 101–109.

Morrey, J. D., Day, C. W., Julander, J. G., Olsen, A. L., Sidwell, R. W., Cheney, C. D., et al. (2004). Modeling hamsters for evaluating West Nile virus therapies. *Antiviral Research*, *63*, 41–50.

Morrey, J. D., Siddharthan, V., Olsen, A. L., Wang, H., Julander, J. G., Hall, J. O., et al. (2007). Defining limits of treatment with humanized neutralizing monoclonal antibody for West Nile virus neurological infection in a hamster model. *Antimicrobial Agents and Chemotherapy*, *51*, 2396–2402.

Morrey, J. D., Taro, B. S., Siddharthan, V., Wang, H., Smee, D. F., Christensen, A. J., et al. (2008). Efficacy of orally administered T-705 pyrazine analog on lethal West Nile virus infection in rodents. *Antiviral Research*, *80*, 377–379.

Osorio, Y., Bonilla, D. L., Peniche, A. G., Melby, P. C., & Travi, B. L. (2008). Pregnancy enhances the innate immune response in experimental cutaneous leishmaniasis through hormone-modulated nitric oxide production. *Journal of Leukocyte Biology*, *83*, 1413–1422.

Oualikene, W., Gonin, P., & Eloit, M. (1994). Short and long term dissemination of deletion mutants of adenovirus in permissive (cotton rat) and non-permissive (mouse) species. *The Journal of General Virology*, *75*, 2765–2768.

Paessler, S., Aguilar, P., Anishchenko, M., Wang, H. Q., Aronson, J., Campbell, G., et al. (2004). The hamster as an animal model for eastern equine encephalitis—And its use in studies of virus entrance into the brain. *The Journal of Infectious Diseases, 189*, 2072–2076.
Palecanda, A., Paulauskis, J., Al-Mutairi, E., Imrich, A., Qin, G., Suzuki, H., et al. (1999). Role of the scavenger receptor MARCO in alveolar macrophage binding of unopsonized environmental particles. *Journal of Experimental Medical Sciences, 189*, 1497–1506.
Paolino, K., Sande, J., Perez, E., Loechelt, B., Jantausch, B., Painter, W., et al. (2011). Eradication of disseminated adenovirus infection in a pediatric hematopoietic stem cell transplantation recipient using the novel antiviral agent CMX001. *Journal of Clinical Virology, 50*, 167–170.
Prince, G. A., Porter, D. D., Jenson, A. B., Horswood, R. L., Chanock, R. M., & Ginsberg, H. S. (1993). Pathogenesis of adenovirus type 5 pneumonia in cotton rats (Sigmodon hispidus). *Journal of Virology, 67*, 101–111.
Rockx, B., Brining, D., Kramer, J., Callison, J., Ebihara, H., Mansfield, K., et al. (2011). Clinical outcome of henipavirus infection in hamsters is determined by the route and dose of infection. *Journal of Virology, 85*, 7658–7671.
Romanowski, E. G., & Gordon, Y. J. (2008). Update on antiviral treatment of adenoviral ocular infections. *American Journal of Ophthalmology, 146*, 635–637.
Russell, W. C. (2009). Adenoviruses: Update on structure and function. *The Journal of General Virology, 90*, 1–20.
Samuel, M. A., Wang, H., Siddharthan, V., Morrey, J. D., & Diamond, M. S. (2007). Axonal transport mediates West Nile virus entry into the central nervous system and induces acute flaccid paralysis. *Proceedings of the National Academy of Sciences of the United States of America, 104*, 17140–17145.
Sbrana, E., Xiao, S. Y., Popov, V. L., Newman, P. C., & Tesh, R. B. (2006). Experimental yellow fever virus infection in the golden hamster (Mesocricetus auratus). III. Clinical laboratory values. *The American Journal of Tropical Medicine and Hygiene, 74*, 1084–1089.
Siddharthan, V., Wang, H., Motter, N. E., Hall, J. O., Skinner, R. D., Skirpstunas, R. T., et al. (2009). Persistent West Nile virus associated with a neurological sequela in hamsters identified by motor unit number estimation. *Journal of Virology, 83*, 4251–4261.
Siirin, M. T., Duan, T., Lei, H., Guzman, H., Da Rosa, A. P., Watts, D. M., et al. (2007). Chronic St. Louis encephalitis virus infection in the golden hamster (Mesocricetus auratus). *The American Journal of Tropical Medicine and Hygiene, 76*, 299–306.
Sonabend, A. M., Ulasov, I. V., Han, Y., Rolle, C. E., Nandi, S., Cao, D., et al. (2009). Biodistribution of an oncolytic adenovirus after intracranial injection in permissive animals: A comparative study of Syrian hamsters and cotton rats. *Cancer Gene Therapy, 16*, 362–372.
Spencer, J. F., Sagartz, J. E., Wold, W. S. M., & Toth, K. (2009). New pancreatic carcinoma model for studying oncolytic adenoviruses in the permissive Syrian hamster. *Cancer Gene Therapy, 16*, 912–922.
Steel, J. C., Morrison, B. J., Mannan, P., Abu-Asab, M. S., Wildner, O., Miles, B. K., et al. (2007). Immunocompetent syngeneic cotton rat tumor models for the assessment of replication-competent oncolytic adenovirus. *Virology, 369*, 131–142.
Tesh, R. B., Arroyo, J., Travassos Da Rosa, A. P., Guzman, H., Xiao, S. Y., & Monath, T. P. (2002). Efficacy of killed virus vaccine, live attenuated chimeric virus vaccine, and passive immunization for prevention of West Nile virus encephalitis in hamster model. *Emerging Infectious Diseases, 8*, 1392–1397.
Tesh, R. B., Guzman, H., Da Rosa, A. P., Vasconcelos, P. F., Dias, L. B., Bunnell, J. E., et al. (2001). Experimental yellow fever virus infection in the Golden Hamster (Mesocricetus auratus). I. Virologic, biochemical, and immunologic studies. *The Journal of Infectious Diseases, 183*, 1431–1436.
Thomas, M. A., Spencer, J. F., La Regina, M. C., Dhar, D., Tollefson, A. E., Toth, K., et al. (2006). Syrian hamster as a permissive immunocompetent animal model for the study of oncolytic adenovirus vectors. *Cancer Research, 66*, 1270–1276.

Thomas, M. A., Spencer, J. F., Toth, K., Sagartz, J. E., Phillips, N., & Wold, W. S. M. (2008). Immunosuppression enhances oncolytic adenovirus replication and anti tumor efficacy in the Syrian hamster model. *Molecular Therapy, 16*, 1665–1673.
Thomas, M. A., Spencer, J. F., & Wold, W. S. M. (2007). Use of the Syrian hamster as an animal model for oncolytic adenovirus vectors. *Methods in Molecular Medicine, 130*, 169–183.
Toth, K., Dhar, D., & Wold, W. S. M. (2010). Oncolytic (replication-competent) adenoviruses as anticancer agents. *Expert Opinion on Biological Therapy, 10*, 353–368.
Toth, K., Kuppuswamy, M., Shashkova, E. V., Spencer, J. F., & Wold, W. S. M. (2010). A fully replication-competent adenovirus vector with enhanced oncolytic properties. *Cancer Gene Therapy, 17*, 761–770.
Toth, K., Spencer, J. F., Dhar, D., Sagartz, J. E., Buller, R. M., Painter, G. R., et al. (2008). Hexadecyloxypropyl-cidofovir, CMX001, prevents adenovirus-induced mortality in a permissive, immunosuppressed animal model. *Proceedings of the National Academy of Sciences of the United States of America, 105*, 7293–7297.
Toth, K., Spencer, J. F., Tollefson, A. E., Kuppuswamy, M., Doronin, K., Lichtenstein, D. L., et al. (2005). Cotton rat tumor model for the evaluation of oncolytic adenoviruses. *Human Gene Therapy, 16*, 139–146.
Toth, K., & Wold, W. S. M. (2010). Increasing the efficacy of oncolytic adenovirus vectors. *Viruses, 2*, 1844–1866.
Turell, M. J., & Parker, M. D. (2008). Protection of hamsters by Venezuelan equine encephalitis virus candidate vaccine V3526 against lethal challenge by mosquito bite and intraperitoneal injection. *The American Journal of Tropical Medicine and Hygiene, 78*, 328–332.
Vairaktaris, E., Spyridonidou, S., Papakosta, V., Vylliotis, A., Lazaris, A., Perrea, D., et al. (2008). The hamster model of sequential oral oncogenesis. *Oral Oncology, 44*, 315–324.
Watts, D. M., Tesh, R. B., Siirin, M., Rosa, A. T., Newman, P. C., Clements, D. E., et al. (2007). Efficacy and durability of a recombinant subunit West Nile vaccine candidate in protecting hamsters from West Nile encephalitis. *Vaccine, 25*, 2913–2918.
Wold, W. S. M., & Horwitz, M. S. (2007). Adenoviruses. In D. M. Knipe & P. M. Howley (Eds.), *Field's virology*. (5th ed.). Philadelphia, USA: Lippincott, Williams & Wilkins.
Xiao, S. Y., Guzman, H., Zhang, H., Travassos Da Rosa, A. P., & Tesh, R. B. (2001). West Nile virus infection in the golden hamster (Mesocricetus auratus): A model for West Nile encephalitis. *Emerging Infectious Diseases, 7*, 714–721.
Xiao, S. Y., Zhang, H., Guzman, H., & Tesh, R. B. (2001). Experimental yellow fever virus infection in the Golden hamster (Mesocricetus auratus) II. Pathology. *The Journal of Infectious Diseases, 183*, 1437–1444.
Yabe, Y., Matsumoto, K., & Ogura, H. (1988). Lifelong persistent infection of hamster brain by human adenovirus type 6. *Acta Medica Okayama, 42*, 45–47.
Yamamoto, M., & Curiel, D. T. (2010). Current issues and future directions of oncolytic adenoviruses. *Molecular Therapy, 18*, 243–250.
Ying, B., Toth, K., Spencer, J. F., Meyer, J., Tollefson, A. E., Patra, D., et al. (2009). INGN 007, an oncolytic adenovirus vector, replicates in Syrian hamsters but not mice: Comparison of biodistribution studies. *Cancer Gene Therapy, 16*, 625–637.
Zarubaev, V. V., Garshinina, A. V., Kalinina, N. A., Shtro, A. A., Belyaevskaya, S. V., Slita, A. V., et al. (2011). Activity of ingavirin (6-[2-(1H-imidazol-4-yl)ethylamino]-5-oxo-hexanoic acid) against human respiratory viruses in in vivo experiments. *Pharmaceuticals, 4*, 1518–1534.
Zarubaev, V. V., Slita, A. V., Sukhinin, V. P., Nosach, L. N., Dyachenko, N. S., Povnitsa, O. Y., et al. (2007). Effect of 6-azacytidine on the course of experimental adenoviral infection in newborn Syrian hamsters. *Journal of Chemotherapy, 19*, 44–51.
Zivcec, M., Safronetz, D., Haddock, E., Feldmann, H., & Ebihara, H. (2011). Validation of assays to monitor immune responses in the Syrian golden hamster (Mesocricetus auratus). *Journal of Immunological Methods, 368*, 24–35.

CHAPTER FOUR

# Design of Improved Oncolytic Adenoviruses

**Ramon Alemany**[1]
Translational Research Laboratory, Institut Català d'Oncologia-IDIBELL, L'Hospitalet de Llobregat, Barcelona, Spain
[1]Corresponding author: e-mail address: ralemany@iconcologia.net

## Contents



## Abstract

During the last decade adenovirus has lost its appeal in gene therapy due to a high immunogenicity that leads to a transient gene expression. However, adenovirus has gained attention as replication-competent vector to treat cancer. Designed for virotherapy, adenovirus has been successfully modified to replicate selectively in tumor cells. After the initial clinical trials with tumor-selective adenoviruses, it has become clear that further improvements on tumor targeting, intratumoral dissemination, and modulation of antiviral and antitumor immune responses are needed to effectively treat cancer. The non-viral delivery of infectious DNA encoding an oncolytic adenovirus armed with extracellular matrix-degrading genes and with genes that regulate the immune system to favor antitumor instead of antiviral immunity are key in the design oncolytic adenovirus.

## 1. INTRODUCTION

Adenovirus has been a major player in the revival of virotherapy. Despite Martuza and collaborators already designed oncolytic herpes viruses with antitumor selectivity traits (Martuza, Malick, Markert, Ruffner, & Coen, 1991), it is the paper on oncolytic adenovirus published from Onyx in Science in 1996 that boosts the design of viruses for virotherapy (Bischoff et al., 1996). This is partly due to the large amount of groups working with

*Advances in Cancer Research*, Volume 115
ISSN 0065-230X
http://dx.doi.org/10.1016/B978-0-12-398342-8.00004-5

adenovirus for cancer gene therapy, using p53, TK, and many other genes, that were facing the enormous barriers of efficient gene transfer that adenovectors had to overcome in order to impact clinically against cancer. The results of clinical trials with adenovirus vectors against cancer were not encouraging and the concept of achieving transduction beyond the needle track by means of the virus replication felt in a fertilized field. After some years of effervescent activity proving concepts at a preclinical level, trials with oncolytic adenoviruses arrived, initially via intratumoral administration and later intravenously (for a recent review, see Pesonen, Kangasniemi, & Hemminki, 2011). In the aftermath, the efficacy observed was almost negligible, in particular when administered intravenously. This highlighted the major barriers for efficacy and fostered new research to address them. It also fostered the research in other viruses that might show better oncolytic properties taking into account such barriers and that had been less covered with patents, but this is not the focus of this chapter. The focus here is on the barriers that oncolytic adenoviruses have to face for efficacy and the designs that can help to overcome them. Unfortunately, limitations for adenovirus virotherapy have been found at all levels: delivery, intratumoral spread, and immunity. Fortunately, many different groups still believe in the oncolytic potential of adenovirus and work to understand and solve these limitations (for a recent review, see Choi, Lee, Kim, & Yun, 2012). It is very difficult to compare one oncolytic virus against another in preclinical tumor models and has seldom been done. These results are biased by infectivity, permissiveness to replication, tumor structures, and immunogenicity of the models selected. Adenovirus has been discredited in view to the many challenges to overcome, but whether adenovirus is really a bad choice or an informed choice is still an open question. It is possible that the more we know about the limitations of a particular oncolytic virus, the closer we are to the design of a good virus for virotherapy. This chapter only describes the challenges and solutions that may help to overcome them bearing in mind that there is a long path until success.

## 2. ONCOLYTIC ADENOVIRUS TUMOR TARGETING

Proving the consistent tumor-targeting capability of oncolytic vaccinia virus in humans has been regarded as a major breakthrough (Breitbach et al., 2011). This is in part due to the previous failure with other viruses, mostly adenovirus, in similar trials. After a dozen trials with systemic oncolytic adenoviruses, the evidence of tumor targeting and antitumor activity is scant (Table 4.1). One could say that the most commonly tested virus (Onyx-015) was too crippled to burst in tumors, but more potent

**Table 4.1** Clinical trials using oncolytic adenoviruses injected systemically

| Virus | Dosing schedule | *N* | Adverse events | Tumor type | Response | References |
|---|---|---|---|---|---|---|
| Onyx-015 Phase I | $2 \times 10^{12}$ vp/d at d1 (cycle 1), 8 (cycle 2), 22 (cycle 3), 50 (cycle 4), 78 + hepatic artery + Cx cycles 3 and later | 11 | Flu-like symptoms (fiber, nausea,). No hepatic or clotting tox | CRC | 1/11 PR at highest dose | Reid et al. Gene Ther. 8, 1618–1626 (2001) |
| Onyx-015 Phase I | As above | 27 | Flu-like. One systemic inflammatory response | CRC | 3/27 PR | Reid et al. Cancer Res. 62, 6070–6079 (2002) |
| Onyx-015 Phase I | $2 \times 10^{13}$ d1 + $2 \times 10^{12}$ later 11 times (each 21 d-cycle: d1, d8, and d15) (total $4.2 \times 10^{13}$) + Cx | 10 | Flu-like. Transaminitis transient at $> 2 \times 10^{12}$ IL1, IL6, TNFa, IFNg | Lung mets | 1/10 PR | Neumunaitis et al. Gene Ther. 8, 746–759 (2001) |
| Onyx-015 Phase I/II | $1 \times 10^{13}$ total (in 5 consecutive days). Hepatic artery + Cx | 5 | No hepatic or clotting tox | HCC | 3/6 Lower CEA | Habib et al. Human Gene Ther. 12, 219–226 (2001) |
| Onyx-015 Phase II | Up to $2 \times 10^{13}$ ($2 \times 10^{12}$ at d1, d15 of each 28 d-cycle) | 18 | Flu-like. Biodistr after death: spleen and liver | CRC liver mets | 4/18 Lower CEA All progression | Hamid et al. J. Clin. Oncol. 21, 1498–1504 (2003) |

*Continued*

**Table 4.1** Clinical trials using oncolytic adenoviruses injected systemically—cont'd

| Virus | Dosing schedule | *N* | Adverse events | Tumor type | Response | References |
|---|---|---|---|---|---|---|
| Onyx-015 Phase I/II | As Reid (2001). Hepatic artery | 35 | Very low tox (CAR is not exposed) | CRC mets | Few PR | Au et al. Cancer Gene Ther. 14, 139–150 (2007) |
| CG7870 Phase I | Up to $6 \times 10^{12}$ vp in a single administration | 23 | Flu-like. Transaminitis transient at $>1 \times 10^{12}$ D dimmer rise at $>1 \times 10^{12}$ IL6, IL10 | Prostate | 5/23 Lower PSA | Small et al. Mol. Ther. 14, 107–117 (2006) |
| Onyx-015 + Enbrel Phase I | $(1 \times 10^{10}, 1 \times 10^{11}$, or $1 \times 10^{12}$ vp weekly $\times 3) = 1 \times 10^{12}$ vp weekly $\times$ 3) = one cycle $\times$ then up to three cycles + so enbrel during cycle 1 | 9 | None | Several: CCR, breast, ovarian | 4SD | Nemunaitis et al. Cancer Gene Ther. 14, 885–893 (2007) |

vp, Virus particles; Cx, chemotherapy; d, days; CRC, colorectar carcinoma; HCC, hepatocellular carcinona; mets, metastases; PR, partial responses; SD, stable disease.

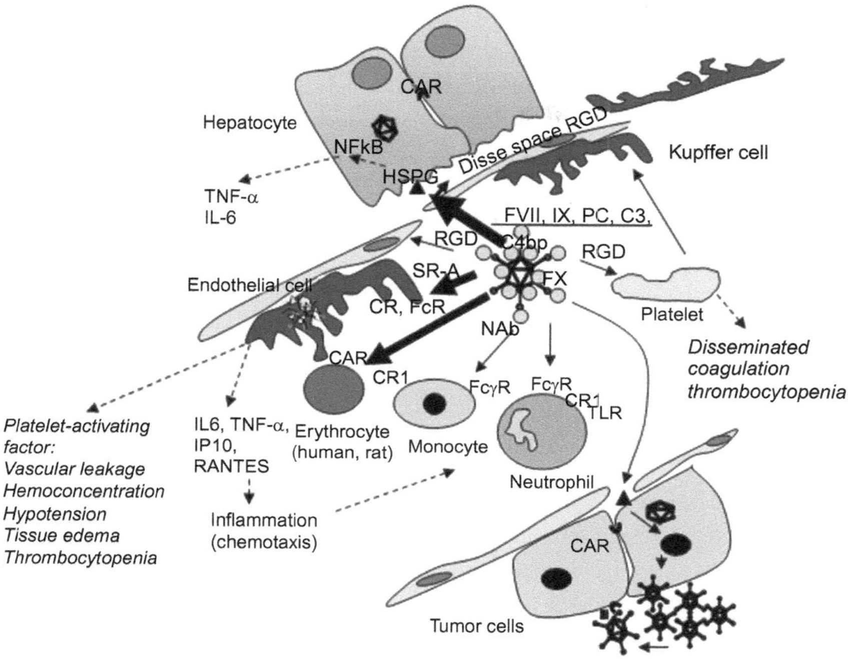


**Figure 4.1** Interactions of Ad5 in blood. The arrow size width is intended to be proportional to the contribution to clearance from blood. Passive diffusion into the Disse space is the major trapping mechanism in liver. The contribution of antibodies by direct neutralization or binding to Fc receptors will depend on the level of antibodies (seropositivity). Full arrows indicate binding. Dashed arrows indicate toxicity mechanisms. (See Page 4 in Color Section at the back of book.)

viruses have been tested with similar results. Taking into account the emerging picture of multiple neutralizing interactions that adenovirus suffers when injected in human blood, the poor tumor targeting is not surprising (Fig. 4.1) (for a recent review, see Coughlan et al., 2011). Adenovirus type 5 binds directly to CAR in human erythrocytes, to integrins in platelets, and to scavenger receptors in Kupffer cells, and then indirectly through any adenovirus-binding antibody, binds to complement receptor 1 in erythrocytes and to Fc receptor in monocytes and neutrophils. Antibodies also can neutralize adenovirus infectivity directly by binding to fiber or penton base. Adenovirus is cleared also passively from blood by extravasation in liver and spleen, where in addition to the passive extravasation, an active uptake is driven by the interaction of hexon-bound FVIII with HSGP of hepatocytes and other cells. Infection of hepatocytes leads to gene expression (transduction) which is often measured by researchers, but such liver transduction represents a minor part of the fast blood clearance of low bioavailability problem for oncolytic adenoviruses. Except for the erythrocyte CR1 and CAR binding, the rest of the clearance mechanisms seem preserved in mice. The result is a half-life of infectious-competent adenovirus in mouse blood of2 min and in human blood of12 min (Alemany, Suzuki, & Curiel, 2000 Reid et al., 2002). Recent studies point to liver sinusoidal endothelial cells as the major target for initial adenovirus type 5 binding in liver (Ganesan et al., 2011). However, the clearance mediated by these cells relative to Kupffer cells remains to be demonstrated, and it could depend on variable factors such as virus aggregation or the presence of pre-existing antibodies to the virus. The phagocytic capacity of Kupffer cells may dominate over the pinocytic capacity of endothelial cells in the absence of antibodies or virus aggregation as the scavenger receptor of the endothelial cell seems much less efficient at adenovirus type 5 binding than the scavenger receptor of the Kupffer cell (Khare, Reddy, Nemerow, & Barry, 2012).

There are different strategies to inhibit adenovirus blood clearance and promote tumor targeting. Genetic modification of the capsid would be definitive and would preserve the single-product simplicity of an oncolytic adenovirus. The interactions with CAR and integrins were the first ones to be eliminated genetically but without satisfactory results (Alemany & Curiel, 2001 Martin et al., 2003). However, the true bioavailability improvement related to these modifications needs to be reevaluated in the context of a FVIII-nonbinding capsid. An adenovirus unable to bind FVIII, CAR, and integrins is now easy to construct as all these mutations have been well identified. The major caveat of this design if anyone tries to demonstrate tumor targeting is that tumor cells may not get infected with such an ideally detargeted virus. The FVIII-HSGP entry pathway has been demonstrated in tumor cells and the inhibition of this pathway results in less tumor transduction (Gimenez-Alejandre, Cascallo, Bayo-Puxan, & Alemany, 2008). Therefore, *in vivo*, the FVIII pathway is important for tumor transduction even in the context of a virus that can mediate cell entry using CAR and integrins as primary receptors. We also have noted that FVIII-nonbinding CRAds have less antitumor activity even when injected directly in tumors, so the FVIII entry pathways may also at the stage of intratumoral spread (Ramon Alemany and Marta Gimenez-Alejandre unpublished results). Therefore, the design of FVIII-nonbinding CRAds, with or without additional CAR and integrin-binding ablations, or even the use of adenovirus serotypes that do not bind FVIII, such as Ad48, may require compensatory entry pathways that work efficiently *in vivo*. This may require high-affinity interactions of the modified fiber with internalizing cell receptors. Insertion of hyperstable scFv, affibodies, or darpins could be an option, as much as these avoid disulfide bridges noncompatible with the cytoplasmic synthesis of the fiber. However, the progress of several groups in this direction has been very unfruitful beyond some *in vitro* proof of concept experiments, despite the different tricks used to promote the trimerization of the modified fibers (reviewed in Coughlan et al., 2011). On the other hand, sequence repeats are very unstable in the adenovirus genome, and structures based on repeats such as darpins may induce rearrangements.

Masking the capsid with polymers is an alternative option to improve tumor targeting. Polymers such as PEG or PHMA have been used (reviewed in Coughlan et al., 2011). This strategy has the advantage that it not only ablates FVIII, CAR, and integrin binding but also may contribute to neutralize the negative charge of the capsid to reduce the binding to scavenger receptors on Kupffer cells, may mask antibody-binding sites of the capsid, and may slightly increase the 100-nm diameter of the capsid to reduce the extravasation in liver and spleen. With such a modified capsid, it can operate the enhanced retention and permeability effect (Fang, Nakamura, & Maeda, 2011). While liver detargeting and improved blood persistence have been proven *in vivo*, the impact of these strategies on tumor targeting has been elusive. The broad non-specific masking of the capsid with such polymers decreases infectivity and the liver detargeting achieved with such extensive masking strategies is likely to be a consequence of the impaired infectivity of the capsid. Recently, targeted PEGylation has allowed to identify hexon hypervariable loops 1, 2, 5, and 7 of adenovirus type 5 as the binding sites for the Kupffer cell scavenger receptor (Khare et al., 2012). When Kupffer cell binding is ablated by targeted PEGylation, liver transduction increases 40-fold compared to wild-type adenovirus 5 capsid. With more available virus in circulation, liver detargeting will require ablation of clotting factors binding, CAR, and complement. As mentioned above, high-affinity tumor-binding ligands may be needed to harvest the benefits of these liver-detargeting strategies. In addition, a major caveat of polymer masking strategies is the commercialization challenge associated with the GMP production of a complexed virus.

A more advanced solution to the tumor-targeting problem is the nonviral delivery of the virus or the oncolytic virus genome. Different types of cells have been used to deliver oncolytic viruses to tumors (for a review, see Willmon et al., 2009). Adenovirus has been carried with tumor cells (Alcayaga-Miranda, Cascallo, Rojas, Pastor, & Alemany, 2010 Garcia-Castro et al., 2005 Hamada et al., 2007) and mesenchymal stem cells (MSCs) (Komarova, Kawakami, Stoff-Khalili, Curiel, & Pereboeva, 2006

Garcia-Castro et al., 2010). For intracranial administration in the proximity of intracranial glioblastoma tumor models, neural stem cells were superior than MSCs (Ahmed et al., 2011); however, the later may be used also systemically to reach intracranial glioblastomas (Yong et al., 2009). From the perspective of product development, infected carrier cells are quite difficult to characterize to comply with good manufacturing procedures as other cellular therapy products. It is difficult to expect a commercial development of a product based on autologous or heterologous infected cells unless the efficacy is outstanding and there is no other way to get a similar delivery efficiency. Synthetic vectors or nanoparticles is an emerging area for oncolytic virus delivery. For example, the whole adenovirus has been encapsulated in liposomes and injected directly in tumors to achieve transduction (Van den Bossche et al., 2011). Even more promising is to use nonviral vectors to deliver the oncolytic virus genome. This idea has been proven by direct tumor injection of infectious RNA of an oncolytic coxsackievirus (Hadac, Kelly, & Russell, 2011). With oncolytic adenovirus, Kwon and collaborators have demonstrated that liposome-enveloped genomes improve tumor targeting and have more antitumor activity upon systemic delivery than naked adenovirus (Kwon, Kang, Kim, & Yun, 2011).

## 3. THE STROMA BARRIER TO ONCOLYTIC ADENOVIRUS SPREAD

Once the oncolytic virus reaches the tumor, it has to spread efficiently in order to eliminate it. Under an immunocentric view of oncolysis, this intratumoral spread efficiency may not be as important as if no antitumor immunity is achieved. If antitumor immunity is elicited, then the immune response may reach far beyond the infected cells. However, even in this immunocentric scenario, the virus mediates tumor debulking and transforms the local tumor immune inhibitory environment (caused by TGFB, IDO/tryptophan deprivation, IL10, and other mechanisms) in an inflammatory site (caused by TLR -toll-like receptors- and other innate immune receptors that trigger proinflammatory interleukins in response to infection) that allows the proper function of the effector phase of the immune system (for a review, see Alemany & Cascallo, 2009). If this is true, efficient tumor targeting and spread will also correlate with antitumor activity. Some information about the spreading problem can be deduced from clinical results obtained when the virus is directly injected in tumors. In a review of different trials performed with adenovirus by intratumoral injection (Table 4.2),

**Table 4.2** Clinical trials using oncolytic adenoviruses injected directly into tumors

| Virus | Dosing schedule | *N* | Adverse events | Tumor type | Response | Refs. |
|---|---|---|---|---|---|---|
| Onyx-015 Phase I | Dose prepared in 30% of tumor volume, then injected/distributed into equal 1 $cm^2$ areas.<br>Up to $1 \times 10^{11}$ pfu.<br>Retreatment at 4 w if no progression, up to five cycles. | 22 | Flu-like symptoms | SCCHN (refractory) | No response | Ganly et al. Clin. Cancer Res. 6, 798–806 (2000) |
| Onyx-015 + Cx Phase II | $1 \times 10^{10}$ pfu/d $\times$ 5 consecutive d (a different quadrant every day) + Cisplatin + FU<br>Retreatment at 3 w if no progression<br>Later $1 \times 10^{10}$ pfu twice/d $\times$ 5 consecutive d + repeated after 1 w rest. | 30 | | SCCHN (refractory) | 8 CR, 11 PR | Khuri et al. Nat. Med. 6, 879–885 (2000) |
| Onyx-015 + Cx Phase II | As above. | 37 | ND | SCCHN | 21% PR: 2 CR, 3 PR. | Nemunaitis et al. J. Clin. Oncol. 19, 289–298 (2001) |
| Onyx-015 Phase I | Three cohorts determined that the maximum tolerated injected volume was 20% of the tumor volume. Then a fixed dose up to $2\times 10 \times 10^{12}$ vp ($1 \times 10^{11}$ pfu) distributed using CT guidance or at laparotomy.<br>Number of injections adjusted to tumor diameter. All at day 1.<br>Retreatment cycles every 4 w | 23 | Flu-like symptoms (fiber, nausea) 1 pancreatitis | Pancreas (first line in several pts) | No response | Mulvihill et al. Gene Ther. 8, 308–315 (2001) |

Continued

**Table 4.2** Clinical trials using oncolytic adenoviruses injected directly into tumors—cont'd

| Virus | Dosing schedule | *N* | Adverse events | Tumor type | Response | Refs. |
|---|---|---|---|---|---|---|
| Onyx-015 + Cx Phase I/II | To allow for repeated injection change from CT to endoscopic ultrasound guiding via gastric route. Then up to $2 \times 10 \times 10^{11}$ vp/session (repeated in days 1, 5, 8, 15, 36, 43, 50, and 57) in 10% of tumor volume.<br>+ Gemcitabine iv at days 36, 43, 50, and 57. | 21: 6(I) 15 (II) | Flu-like symptoms 1PTT, leucopenia | Pancreas (first line in many pts) | 2 PR only after Gemcitabine | Hetch et al. Clin Cancer Res. 9, 555–561 (2003) |
| Onyx-015 + Cx Phase I–II | $10 \times 10^{9}$ pfu 1 patient, $3 \times 10 \times 10^{9}$ another one, and $10 \times 10^{10}$ four pt.<br>Up to six cycles.<br>+Mytomycin, doxorubicin, cisplatin. | 6 | Fever | Soft tissue sarcomas | 1 PR | Galanis et al. Gene Ther. 12, 437–445 (2005) |
| H101 + Cx Phase II | ($5 \times 10^{11}$ vp/d × 5 consecutive days) normalized volume injected in one section each day.<br>3 w-cycles<br>+Cisplatin or adriamycin + 5FU | 45 | Fever (30%) and pain at injection site (27%) | Several:<br>15 SCCHN<br>8 Esoph. | 30% RR (3 CR, 11 PR) In HN1CR and 3 PR out of 15. | Lu et al. World J. Gastroenterol. 10, 3634–3638 (2004) |
| Ad5y CD/mT Krep A + Rx Phase I | $10 \times 10^{12}$ at d1 × 2, repeated at d22.<br>Intraprostatic<br>+GCV | 9 | Transaminitis | Prostate Ca | Negative biopsy (78%) better than expected (60%) | Freytag et al. Mol Ther. 15, 1016–1023 (2007) |

| | | | | | | |
|---|---|---|---|---|---|---|
| H103 Phase I | la: Single injection on d1 (up to $1.5 \times 10^{12}$ vp)<br>lb: $3 \times 10 \times 10^{12}$ vp divided by six injections (d1, 3, 5, 8, 10, and 12)<br>Dose adjusted to 30% tumor volume. | 27 | Fever<br>Thrombocytopenia<br>MTD not reached | Several:<br>11 Mel<br>5 RCC<br>5 Lung<br>6 Other | 3 PR<br>18 SD<br>DR in 2 Mel patients | Li et al. Gene Ther. 16, 376–382 (2009) |
| KH901 (TelE1a/ GM CSF in E3) Phase I | Single dose group ($3 \times 10^{11}$ vp/ $\times 10^{12}$ vp/$3 \times 10^{12}$/$\times 10^{13}$) versus multiple (two injections/ w × 3 w) ($1 \times 10^{12}$ or $3 \times 10 \times 10^{12}$ each time) | 23 | Fever | SCCHN | No OR. Injected tumors necrosed. | Chang et al. Cancer Biol Ther 8, 676–682 (2009) |
| Telomelysin hTERT-E1aIRESE1b Phase I | Up to $10 \times 10^{12}$ vp/tumor single injection but distributed.<br>Target tumor periphery. | 16 | Fever | Several:<br>4 Mel<br>8 SCC<br>1 NSCLC | 1 PR<br>7 SD | Nemunaitis et al. Mol Ther. 18, 429–434 (2010) |

vp, virus particles; pfu, plaque-forming units; Cx, chemotherapy; Rx, radiotherapy; d, days; w, weeks; PTT, protrombin time delay, CRC, colorectar carcinoma; HCC, hepatocellular carcinoma; SCC, squamous cell carcinoma; NSCLC, non-small cell lung carcinoma; Mel, melanoma; mets, metastases; PR, partial responses; CR, complete responses; OR, objective responses; SD, stable disease; DR, distal responses.

there is a clear evidence of partial responses, but it is clear that the virus seldom spreads enough to eliminate the tumor. The virus will be halted by stroma barriers and by a neutralizing immune system. Fibrosis or desmoplasia is a typical hallmark of solid tumors that grow beyond a few centimeters in diameter (Whatcott, Han, Posner, Hostetter, & Von Hoff, 2011). Tumor cells promote an environment rich in connective tissue formed by extracellular matrix proteins (collagen, fibronectin, laminin, fibrin, and sparc/osteonectin), polysaccharides (proteoglycan glycosaminoglycans such as protein-attached heparan, chondroitin and keratan sulfates, and nonproteoglycan glycosaminoglycans such as hyaluronan), and cells such as fibroblasts and inflammatory cells. This stroma blocks the transit of chemotherapy drugs from the blood vessels to tumor cells contributing to chemoresistance. For an oncolytic adenovirus of 100 nm diameter, this barrier is even harder to cross. Smaller viruses, such as picornaviruses like Seneca Valley Virus or coxsackievirus, may have clear advantage when compared to adenovirus. In addition, these viruses are much faster in terms of virus life cycle, and they spread faster than adenovirus.

Several strategies are proposed to solve the spreading limitation of adenovirus. The bioselection of faster adenoviruses both in cell lines *in vitro* and in tumors *in vivo* has led to the identification of mutants that spread faster in tumors (Gros et al., 2008 Subramanian, Vijayalingam, & Chinnadurai, 2006 Yan et al., 2003). The identified mutations resulted in a nonfunctional E1b-19k (an antiapoptotic gene), the c-terminal truncation of a functional iLeader protein (of unknown function), and the C-terminal truncation of a functional E3-19K protein (a MHC-I-binding protein). The common phenotype of such mutations was the faster release of virus from the infected cells and the subsequent generation of large plaques in an infected monolayer. This converging phenotype seems to indicate that adenovirus exit from the cell is probably the least efficient step of the infectious cycle in terms of cellular propagation speed. Adenovirus release is not an active process of lysis. An array of virus particles in the form of an inclusion body accumulates in the nucleus until an attrition of the cell allows the leakage of virus from the infected cell. The functional loss of E1b-19k promotes early apoptosis of the infected cell and then the faster release of virus. E1b-19k mutants have been proposed as improved oncolytic viruses, a trait that can be enhanced with proapoptotic drugs (Oberg et al., 2010). The mechanism that underlies the faster release associated with the c-terminus truncation of the iLeader protein is unknown as it is the function of this small viral protein.

The truncation of the c-terminus of E3-19K results in the loss of the endoplasmic reticulum retention signal of this protein. The mutant protein is no longer entrapped in the reticulum and traffics to the plasma membrane. The main function of E3-19K is to retain the MHC-I molecules in the ER, but the earlier release of adenovirus caused by E3-19K at the plasmatic membrane is not related to this function (Gros et al., 2008). Another less-explored function of E3-19K is the mobilization of calcium pools from the ER. This led us to explore the importance of calcium channels in the adenovirus release, and we could demonstrate that the fast-release phenotype was associated to ionic permeabilization and intracellular calcium levels, reminiscent of the virus release action of viroporins used by other viruses. We also found out that inhibitors of calcium channels such as verapamil could mimic the fast-release and large-plaque phenotype and that verapamil could be used to enhance oncolytic activity of adenovirus. Recently, this has been translated into patients with an improved viremia that may indicate faster virus replication (Koski et al., 2011). The main caveat with faster-release oncolytic adenoviruses is that they have been selected *in vitro* or in immune-deficient animal models. A fast virus capsid release in the context of a functional immune system may lead to a faster or more efficient rejection of the oncolytic virus. Immune-competent hamster tumor models permissive to adenovirus replication have been presented (Khoobyarian, Barone, Sabet, El-Domeiri, & Das Gupta, 1975 Thomas et al., 2006) and can be used to answer this question; however, the T-cell immune response in hamster may be quite unique (Witte & Streilein, 1986). Some mouse tumor models may allow a certain human adenovirus replication, but their permissiveness is still 100-fold lower than in human cells and the impact of the fast-release mutants in those cell lines is not clear.

Kim and collaborators demonstrated another strategy to facilitate intratumoral spread in the digestion of the stroma that an oncolytic adenovirus-expressing relaxin, an hormone that induces metalloprotease expression with degradation of collagen and other stromal proteins, showed enhanced spread and antitumor activity (Kim et al., 2006), and the potency of this relaxin-armed adenovirus can be further enhanced with Ad35 fibers (Ganesh et al., 2007). A similar enhanced spread and antitumor activity result has been obtained with the combination of an oncolytic virus and a metalloproteinase-8-expressing adenovirus vector (Cheng et al., 2007) and with an oncolytic adenovirus armed with decorin, a collagen-binding protein that limits the collagen fibril size and induces metalloprotease 1 (Choi et al., 2010).

The polysaccharides of the ECM have also been targeted in order to improve the spread of oncolytic adenoviruses, both the glycosaminoglycans attached to proteins (proteoglycans) or protein-free (hyaluronan). The combination of an adenovirus vector expressing heparanase to degrade heparan sulfate proteoglycans with an oncolytic adenovirus improved tumor spread and efficacy (Watanabe et al., 2010). Hyaluronidase is being reviewed in the context of cancer therapy (Whatcott et al., 2011), and it has been coinjected with oncolytic adenoviruses (Ganesh, Gonzalez-Edick, Gibbons, Van Roey, & Jooss, 2008) or expressed from the oncolytic adenovirus (Guedan et al., 2010) in order to enhance intratumoral spread. The risk of enhancing tumor cell spread or metastasis has been discarded with collagen, proteoglycan, and hyaluronan-degrading enzymes confirming the current idea that metastasis is an early event that requires multiple complex steps (Chaffer & Weinberg, 2011).

Cells, mainly fibroblasts, represent a key obstacle for oncolytic adenovirus in desmoplastic tumors such as breast, stomach, and pancreas, in which up to 90% of the tumor mass consists of stroma. A vascularized fibroblast-rich loose connective tissue is typical of the tumor growing edge or peripheral zone and evolves around a poorly vascularized scar-like collagenous stroma typical of the inner areas of the tumor mass. The vascularized peripheral area is where initial viral infection occurs after intravascular delivery (Breitbach et al., 2007), and an oncolytic adenovirus growing in stromal fibroblasts, in addition to tumor cells, would be ideal to achieve intratumoral spread. This has been designed using E1a transcriptional regulation with the SPARC promoter (Lopez et al., 2009). Eventually, the combination of ECM-degrading enzymes with fibroblast replication may be needed to tackle the complexity of the stroma barrier. Destroying the stroma may not only improve the spread of the oncolytic virus but also allow chemotherapy drugs to penetrate (Neesse et al., 2010), and the combination of oncolytic viruses that soften the stroma with chemotherapy is very promising. Chemotherapy drugs can kill tumor in the tumor mass softened by the virus but also can transiently inhibit the immune response to the virus to improve its spread.

## 4. ANTIADENOVIRUS AND ANTITUMOR IMMUNITY IN THE OUTCOME OF VIROTHERAPY

To expect that an oncolytic virus will find every tumor cell of a given patient in an advanced stage of disease is very presumptuous. The most reasonable is to expect that, without the cooperation of the immune system, a complete clearance of the tumor is not possible. Even for chemotherapy, this seems to be the case (Zitvogel, Kepp, & Kroemer, 2011). Eventually, the immune system response to the virus infection will play the key role in virotherapy. However, one can raise the question whether delaying that response may be beneficial or not. Allowing a virus more time to reach and destroy tumor cells may be needed to finally get the proper immunity against the residual tumor. Debulking the tumor mass is a major requisite to revert the strong local immune suppression elicited by tumors. A virus is the best danger signal that one can introduce in a tumor in order to debulk and revert such a local immune suppression (Alemany & Cascallo, 2009). Our innate immune system is specially trained to detect viruses, and adenovirus is no exception. An adenovirus growing in a tumor will elicit inflammation and eventually adaptive immunity. In fact, this inflammation often confounds the antitumor effect of virotherapy when measured by tumor size or nonspecific imaging techniques such as RMN, a phenomenon described as pseudoprogression (Koski et al., 2011). There is a clear trend toward arming adenoviruses and other oncolytic viruses with immunostimulatory genes. Oncolytic adenovirus has been armed with immunostimulatory genes such as IL2 (Bortolanza et al., 2009), IL4 (Post et al., 2007), IL12 (Bortolanza et al., 2009), IL18 (Zheng et al., 2009), IL24 (Luo et al., 2008), GM CSF (Cerullo et al., 2010), IFN alpha (Shashkova, Kuppuswamy, Wold, & Doronin, 2008), RANTES (Lapteva et al., 2009), HSP-70 (Li et al., 2009), CD40L (Fernandes et al., 2009), and anti-CTLA4 antibody (Dias et al., 2011). However, the contrary direction, that of immune suppressing the response in order to allow for more extensive virus replication and propagation during a longer period of time, is also being explored for other oncolytic adenoviruses (Le Boeuf et al., 2011). Of course, the level of virus selectivity and genomic stability in this immune-evasion approach has to be very high in order to rebate any argument on safety for the patient and for the population. Immune-evasion genes are virulence genes, and enhancing the virulence of a nonpathogenic virus needs to be carefully evaluated. In any case, a complete immune evasion will not be desirable and the eventual neutralization by the immune system is always warranted. Assuming then an antiviral immune response in any scenario, the key question is how to obtain a concomitant antitumor immune response. A major drawback with the use of oncolytic viruses is their immunodominance. Even a nonreplicating adenovirus vector when used as a vaccine for another strong antigenic protein such as the surface antigen of hepatitis B virus precludes a response against the vectored antigen (Schirmbeck, Reimann, Kochanek, & Kreppel, 2008). If we now think in a replication-competent virus, which is much more immunogenic than an E1-deleted vector with regard to antiadenovirus-specific responses, and in a tumor-associated antigen that is much less immunogenic than the surface antigen of hepatitis B virus, then we cannot be surprised about finding immune responses always against the oncolytic adenovirus in patients but never against the injected and infected tumor. In the future, the oncolytic adenoviruses will be designed to modulate immune responses but not to stimulate or inhibit them. How can we favor the response against the tumor antigens in contrast to the virus antigens? Can we redirect the antivirus response toward the tumor cells? A strategy to favor immune responses against tumor antigens could be to display the tumor antigen itself in the capsid of the oncolytic adenovirus. Antigenic epitopes have been inserted into the hypervariable loops of the hexon, fiber, and pIX (Crompton, Toogood, Wallis, & Hay, 1994 Krause et al., 2006; Shiratsuchi, Rai, Krause, Worgall, & Tsuji, 2011; Worgall et al., 2007), and recently, whole antigens have been displayed as pIX fusion proteins (Bayer et al., 2010). Even when the tumor antigen is expressed by an oncolytic adenovirus or displayed in its capsid, the virus proteins are likely to be immunodominant (Schirmbeck et al., 2008). The immunodominance of the virus will be even worse in case the patient has been preexposed to the virus. For adenovirus type 5, as we are commonly exposed during childhood, the chances of dominant responses are higher in adults than in children. Seeking serotypes that are less common may help to achieve less viral immunodominance and favor tumor responses. This is a major issue in the vaccination field with adenoviruses, and the use of antigenically different (heterologous) vector systems to prime and boost the response against the vaccine antigen is considered fundamental. With the same rationale, using different oncolytic vectors will boost only the response against the tumor antigens. This strategy has already been explored alternating oncolytic adenovirus and vaccinia viruses (Tysome et al., 2011). Another strategy could be a virus-selective transient tolerization prior to the virotherapy treatment to foster responses against tumor antigens or epitopes. MSCs are potent transient immune suppressors (Bassi, Aita, & Camara, 2011) and could be used in this direction. It is tempting to speculate that our success in treating advanced neuroblastoma in children using MSC infected with an oncolytic virus is associated with the specific immunosuppressive activity of infected MSC against adenovirus antigens (Garcia-Castro et al., 2010). In this limited experience (currently two complete responses out of nine children; Manuel Ramirez Orellana, personal communication), the evidence of MSC as carrier cells to the tumor is scarce, and the low level of virus in blood and in tumor of responding children suggests that complete tumor clearance is mediated by

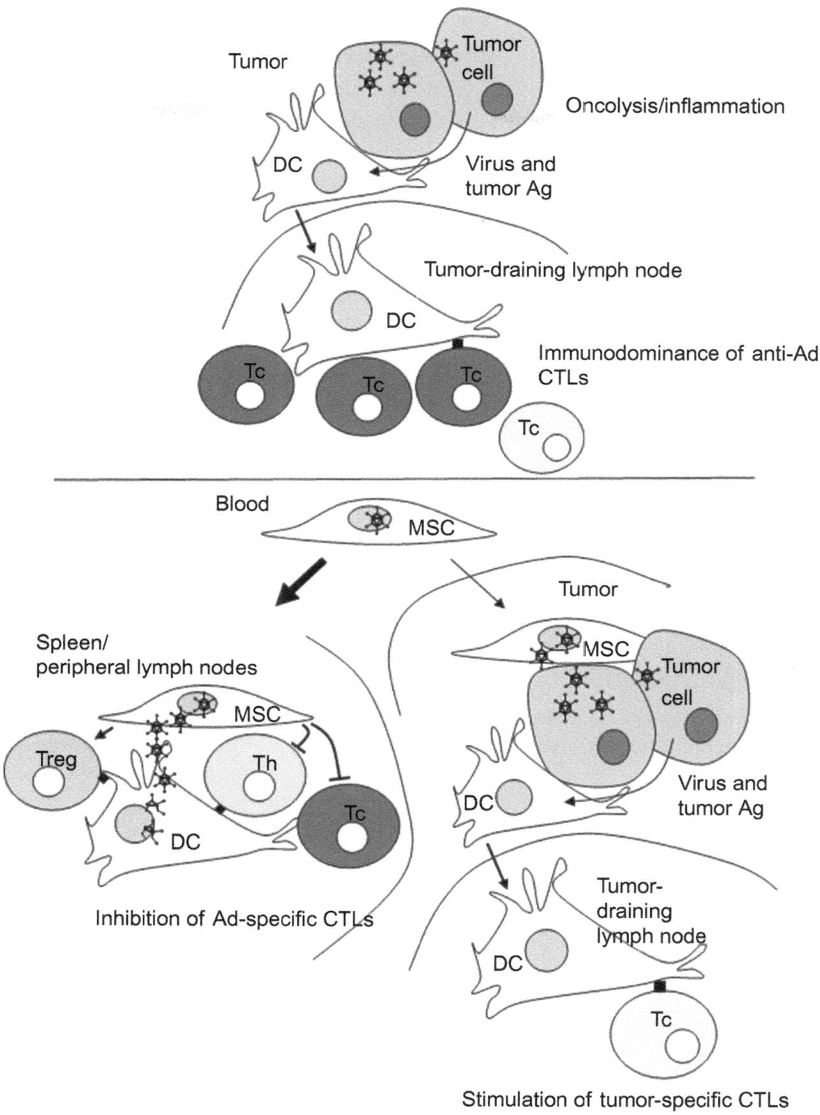


**Figure 4.2** The carrier-immune suppressor hypothesis to explain the activation of anti-tumor immunity mediated by mesenchymal stem cells (MSCs) infected with an oncolytic virus. In the absence of MSC (above), the immune dominance of adenovirus-derived epitopes mask the response against tumor-derived epitopes. Only virus-infected cells will be eliminated. When MSCs are infected *ex vivo* and injected systemically (below), they inhibit the response to virus epitopes and the proliferation of virus-specific CTLs through multiple mechanisms. In the tumor-draining lymph nodes, the absence of antivirus immune-dominant CTLs allows the response to epitopes derived from tumor antigens. (See Page 5 in Color Section at the back of book.)

antitumor immune response. The oncolytic virus-infected MSC injected intravenously may both carry virus to tumors and inhibit adenovirus-specific T cells. Thus, the infected MSC would provide few virus particles to tumors which in turn lyse tumor cells and induce danger signals at the tumor site to spark the immune system. Simultaneously, the peripheral virus-infected MSC (putatively in lymph nodes in the absence of tumor antigens) would suppress the virus-specific CTLs and therefore the antiviral dominance of the immune elicited by the oncolysis in the tumor. Based on this hypothesis, the combination of systemic administration of virus-infected MSC, or any other treatment leading to transient virus tolerization, along with intratumoral virus injection of the same virus modified to display tumor antigens should maximize chances of eliciting immunity against tumor antigens (Fig. 4.2).

In summary, the main barriers for adenovirus oncolytic efficacy have been identified and are being addressed in novel designs. In the past decade, there has been a major advance in the design of promoters and mutations that yield tumor-selective replication. In the future, the advance is expected from designs that favor tumor targeting, intratumoral spread, and immunity against tumor antigens.

## ACKNOWLEDGMENTS

I would like to acknowledge the contribution of all the members of the "Gene and Virus Therapy Group" at the Catalan Institute of Oncology to discussions and results, and funding granted to our group from the "Plan Nacional of the Ministerio de Ciencia e Innovación," the "Departament d'Universitats, Recerca i Societat de la Informació, Generalitat de Catalunya," and the Network of Cooperative Research on Cancer of the "Instituto de Salud Carlos III, Ministerio de Sanidad y Consumo." I also acknowledge Cristina Balagué for editorial help.

## REFERENCES

Ahmed, A. U., Tyler, M. A., Thaci, B., Alexiades, N. G., Han, Y., Ulasov, I. V., et al. (2011). A comparative study of neural and mesenchymal stem cell-based carriers for oncolytic adenovirus in a model of malignant glioma. *Molecular Pharmaceutics*, *8*, 1559–1572.

Alcayaga-Miranda, F., Cascallo, M., Rojas, J. J., Pastor, J., & Alemany, R. (2010). Osteosarcoma cells as carriers to allow antitumor activity of canine oncolytic adenovirus in the presence of neutralizing antibodies. *Cancer Gene Therapy*, *17*, 792–802.

Alemany, R., & Cascallo, M. (2009). Oncolytic adenoviruses from the perspective of the immune system. *Future Microbiology*, *4*, 527–536.

Alemany, R., & Curiel, D. T. (2001). CAR-binding ablation does not change biodistribution and toxicity of adenoviral vectors. *Gene Therapy*, *8*, 1347–1353.

Alemany, R., Suzuki, K., & Curiel, D. T. (2000). Blood clearance rates of adenovirus type 5 in mice. *Journal of General Virology*, *81*(Pt 11), 2605–2609.

Bassi, E. J., Aita, C. A., & Camara, N. O. (2011). Immune regulatory properties of multipotent mesenchymal stromal cells: Where do we stand? *World Journal of Stem Cells*, *3*, 1–8.

Bayer, W., Tenbusch, M., Lietz, R., Johrden, L., Schimmer, S., Uberla, K., et al. (2010). Vaccination with an adenoviral vector that encodes and displays a retroviral antigen induces improved neutralizing antibody and CD4 + T-cell responses and confers enhanced protection. *Journal of Virology*, *84*, 1967–1976.

Bischoff, J. R., Kirn, D. H., Williams, A., Heise, C., Horn, S., Muna, M., et al. (1996). An adenovirus mutant that replicates selectively in p53-deficient human tumor cells. *Science*, *274*, 373–376.

Bortolanza, S., Bunuales, M., Otano, I., Gonzalez-Aseguinolaza, G., Ortiz-de-Solorzano, C., Perez, D., et al. (2009). Treatment of pancreatic cancer with an oncolytic adenovirus expressing interleukin-12 in Syrian hamsters. *Molecular Therapy*, *17*, 614–622.

Breitbach, C. J., Burke, J., Jonker, D., Stephenson, J., Haas, A. R., Chow, L. Q., et al. (2011). Intravenous delivery of a multi-mechanistic cancer-targeted oncolytic poxvirus in humans. *Nature*, *477*, 99–102.

Breitbach, C. J., Paterson, J. M., Lemay, C. G., Falls, T. J., McGuire, A., Parato, K. A., et al. (2007). Targeted inflammation during oncolytic virus therapy severely compromises tumor blood flow. *Molecular Therapy*, *15*, 1686–1693.

Cerullo, V., Pesonen, S., Diaconu, I., Escutenaire, S., Arstila, P. T., Ugolini, M., et al. (2010). Oncolytic adenovirus coding for granulocyte macrophage colony-stimulating factor induces antitumoral immunity in cancer patients. *Cancer Research*, *70*, 4297–4309.

Chaffer, C. L., & Weinberg, R. A. (2011). A perspective on cancer cell metastasis. *Science*, *331*, 1559–1564.

Cheng, J., Sauthoff, H., Huang, Y., Kutler, D. I., Bajwa, S., Rom, W. N., et al. (2007). Human matrix metalloproteinase-8 gene delivery increases the oncolytic activity of a replicating adenovirus. *Molecular Therapy*, *15*, 1982–1990.

Choi, I. K., Lee, Y. S., Yoo, J. Y., Yoon, A. R., Kim, H., Kim, D. S., et al. (2010). Effect of decorin on overcoming the extracellular matrix barrier for oncolytic virotherapy. *Gene Therapy*, *17*, 190–201.

Choi, J. W., Lee, J. S., Kim, S. W., & Yun, C. O. (2012). Evolution of oncolytic adenovirus for cancer treatment. *Advanced Drug Delivery Reviews*, *64*, 720–729.

Coughlan, L., Alba, R., Parker, A. L., Bradshaw, A. C., McNeish, I. A., Nicklin, S. A., et al. (2011). Tropism-modification strategies for targeted gene delivery using adenoviral vectors. *Viruses*, *2*, 2290–2355.

Crompton, J., Toogood, C. I., Wallis, N., & Hay, R. T. (1994). Expression of a foreign epitope on the surface of the adenovirus hexon. *Journal of General Virology*, *75*, 133–139.

Dias, J. D., Hemminki, O., Diaconu, I., Hirvinen, M., Bonetti, A., Guse, K., et al. (2011). Targeted cancer immunotherapy with oncolytic adenovirus coding for a fully human monoclonal antibody specific for CTLA-4. *Gene Therapy*, http://dx.doi.org/10.1038/gt.2011.176 (10 November, 2011).

Fang, J., Nakamura, H., & Maeda, H. (2011). The EPR effect: Unique features of tumor blood vessels for drug delivery, factors involved, and limitations and augmentation of the effect. *Advanced Drug Delivery Reviews*, *63*, 136–151.

Fernandes, M. S., Gomes, E. M., Butcher, L. D., Hernandez-Alcoceba, R., Chang, D., Kansopon, J., et al. (2009). Growth inhibition of human multiple myeloma cells by an oncolytic adenovirus carrying the CD40 ligand transgene. *Clinical Cancer Research*, *15*, 4847–4856.

Ganesan, L. P., Mohanty, S., Kim, J., Clark, K. R., Robinson, J. M., & Anderson, C. L. (2011). Rapid and efficient clearance of blood-borne virus by liver sinusoidal endothelium. *PLoS Pathogens*, *7*, e1002281.

Ganesh, S., Gonzalez Edick, M., Idamakanti, N., Abramova, M., Vanroey, M., Robinson, M., et al. (2007). Relaxin-expressing, fiber chimeric oncolytic adenovirus prolongs survival of tumor-bearing mice. *Cancer Research*, *67*, 4399–4407.

Ganesh, S., Gonzalez-Edick, M., Gibbons, D., Van Roey, M., & Jooss, K. (2008). Intratumoral coadministration of hyaluronidase enzyme and oncolytic adenoviruses enhances virus potency in metastatic tumor models. *Clinical Cancer Research, 14*, 3933–3941.

Garcia-Castro, J., Alemany, R., Cascallo, M., Martinez-Quintanilla, J., Arriero Mdel, M., Lassaletta, A., et al. (2010). Treatment of metastatic neuroblastoma with systemic oncolytic virotherapy delivered by autologous mesenchymal stem cells: An exploratory study. *Cancer Gene Therapy, 17*, 476–483.

Garcia-Castro, J., Martinez-Palacio, J., Lillo, R., Garcia-Sanchez, F., Alemany, R., Madero, L., et al. (2005). Tumor cells as cellular vehicles to deliver gene therapies to metastatic tumors. *Cancer Gene Therapy, 12*, 341–349.

Gimenez-Alejandre, M., Cascallo, M., Bayo-Puxan, N., & Alemany, R. (2008). Coagulation factors determine tumor transduction in vivo. *Human Gene Therapy, 19*, 1415–1419.

Gros, A., Martinez-Quintanilla, J., Puig, C., Guedan, S., Mollevi, D. G., Alemany, R., et al. (2008). Bioselection of a gain of function mutation that enhances adenovirus 5 release and improves its antitumoral potency. *Cancer Research, 68*, 8928–8937.

Guedan, S., Rojas, J. J., Gros, A., Mercade, E., Cascallo, M., & Alemany, R. (2010). Hyaluronidase expression by an oncolytic adenovirus enhances its intratumoral spread and suppresses tumor growth. *Molecular Therapy, 18*, 1275–1283.

Hadac, E. M., Kelly, E. J., & Russell, S. J. (2011). Myeloma xenograft destruction by a nonviral vector delivering oncolytic infectious nucleic acid. *Molecular Therapy, 19*, 1041–1047.

Hamada, K., Desaki, J., Nakagawa, K., Zhang, T., Shirakawa, T., Gotoh, A., et al. (2007). Carrier cell-mediated delivery of a replication-competent adenovirus for cancer gene therapy. *Molecular Therapy, 15*, 1121–1128.

Khare, R., Reddy, V. S., Nemerow, G. R., & Barry, M. A. (2012). Identification of adenovirus serotype 5 hexon regions that interact with scavenger receptors. *Journal of Virology, 86*, 2293–2301.

Khoobyarian, N., Barone, F., Sabet, T., El-Domeiri, A. A., & Das Gupta, T. K. (1975). Inhibition of melanoma growth in hamsters by type-2 adenovirus. *Journal of Surgical Oncology, 7*, 421–425.

Kim, J. H., Lee, Y. S., Kim, H., Huang, J. H., Yoon, A. R., & Yun, C. O. (2006). Relaxin expression from tumor-targeting adenoviruses and its intratumoral spread, apoptosis induction, and efficacy. *Journal of the National Cancer Institute, 98*, 1482–1493.

Komarova, S., Kawakami, Y., Stoff-Khalili, M. A., Curiel, D. T., & Pereboeva, L. (2006). Mesenchymal progenitor cells as cellular vehicles for delivery of oncolytic adenoviruses. *Molecular Cancer Therapeutics, 5*, 755–766.

Koski, A., Raki, M., Nokisalmi, P., Liikanen, I., Kangasniemi, L., Joensuu, T., et al. (2012). Verapamil results in increased blood levels of oncolytic adenovirus in treatment of patients with advanced cancer. *Molecular Therapy, 20*, 221–229.

Krause, A., Joh, J. H., Hackett, N. R., Roelvink, P. W., Bruder, J. T., Wickham, T. J., et al. (2006). Epitopes expressed in different adenovirus capsid proteins induce different levels of epitope-specific immunity. *Journal of Virology, 80*, 5523–5530.

Kwon, O. J., Kang, E., Kim, S., & Yun, C. O. (2011). Viral genome DNA/lipoplexes elicit in situ oncolytic viral replication and potent antitumor efficacy via systemic delivery. *Journal of Controlled Release, 155*, 317–325.

Lapteva, N., Aldrich, M., Weksberg, D., Rollins, L., Goltsova, T., Chen, S. Y., et al. (2009). Targeting the intratumoral dendritic cells by the oncolytic adenoviral vaccine expressing RANTES elicits potent antitumor immunity. *Journal of Immunotherapy, 32*, 145–156.

Le Boeuf, F., Niknejad, N., Wang, J., Auer, R., Weberpals, J. I., Bell, J. C., et al. (2011). Sensitivity of cervical carcinoma cells to vesicular stomatitis virus induced oncolysis: Potential role of human papilloma virus infection. *International Journal of Cancer, 134*, 204–215.

Li, J. L., Liu, H. L., Zhang, X. R., Xu, J. P., Hu, W. K., Liang, M., et al. (2009). A phase I trial of intratumoral administration of recombinant oncolytic adenovirus overexpressing HSP70 in advanced solid tumor patients. *Gene Therapy, 16*, 376–382.

Lopez, M. V., Viale, D. L., Cafferata, E. G., Bravo, A. I., Carbone, C., Gould, D., et al. (2009). Tumor associated stromal cells play a critical role on the outcome of the oncolytic efficacy of conditionally replicative adenoviruses. *PLoS One, 4*, e5119.

Luo, J., Xia, Q., Zhang, R., Lv, C., Zhang, W., Wang, Y., et al. (2008). Treatment of cancer with a novel dual-targeted conditionally replicative adenovirus armed with mda-7/IL-24 gene. *Clinical Cancer Research, 14*, 2450–2457.

Martin, K., Brie, A., Saulnier, P., Perricaudet, M., Yeh, P., & Vigne, E. (2003). Simultaneous CAR- and alpha V integrin-binding ablation fails to reduce Ad5 liver tropism. *Molecular Therapy, 8*, 485–494.

Martuza, R. L., Malick, A., Markert, J. M., Ruffner, K. L., & Coen, D. M. (1991). Experimental therapy of human glioma by means of a genetically engineered virus mutant. *Science, 252*, 854–856.

Neesse, A., Michl, P., Frese, K. K., Feig, C., Cook, N., Jacobetz, M. A., et al. (2010). Stromal biology and therapy in pancreatic cancer. *Gut, 60*, 861–868.

Oberg, D., Yanover, E., Adam, V., Sweeney, K., Costas, C., Lemoine, N. R., et al. (2010). Improved potency and selectivity of an oncolytic E1ACR2 and E1B19K deleted adenoviral mutant in prostate and pancreatic cancers. *Clinical Cancer Research, 16*, 541–553.

Pesonen, S., Kangasniemi, L., & Hemminki, A. (2011). Oncolytic adenoviruses for the treatment of human cancer: Focus on translational and clinical data. *Molecular Pharmaceutics, 8*, 12–28.

Post, D. E., Sandberg, E. M., Kyle, M. M., Devi, N. S., Brat, D. J., Xu, Z., et al. (2007). Targeted cancer gene therapy using a hypoxia inducible factor dependent oncolytic adenovirus armed with interleukin-4. *Cancer Research, 67*, 6872–6881.

Reid, T., Galanis, E., Abbruzzese, J., Sze, D., Wein, L. M., Andrews, J., et al. (2002). Hepatic arterial infusion of a replication-selective oncolytic adenovirus (dl1520): Phase II viral, immunologic, and clinical endpoints. *Cancer Research, 62*, 6070–6079.

Schirmbeck, R., Reimann, J., Kochanek, S., & Kreppel, F. (2008). The immunogenicity of adenovirus vectors limits the multispecificity of CD8 T-cell responses to vector-encoded transgenic antigens. *Molecular Therapy, 16*, 1609–1616.

Shashkova, E. V., Kuppuswamy, M. N., Wold, W. S., & Doronin, K. (2008). Anticancer activity of oncolytic adenovirus vector armed with IFN-alpha and ADP is enhanced by pharmacologically controlled expression of TRAIL. *Cancer Gene Therapy, 15*, 61–72.

Shiratsuchi, T., Rai, U., Krause, A., Worgall, S., & Tsuji, M. (2011). Replacing adenoviral vector HVR1 with a malaria B cell epitope improves immunogenicity and circumvents preexisting immunity to adenovirus in mice. *Journal of Clinical Investigation, 120*, 3688–3701.

Subramanian, T., Vijayalingam, S., & Chinnadurai, G. (2006). Genetic identification of adenovirus type 5 genes that influence viral spread. *Journal of Virology, 80*, 2000–2012.

Thomas, M. A., Spencer, J. F., La Regina, M. C., Dhar, D., Tollefson, A. E., Toth, K., et al. (2006). Syrian hamster as a permissive immunocompetent animal model for the study of oncolytic adenovirus vectors. *Cancer Research, 66*, 1270–1276.

Tysome, J., Li, X., Wang, S., Wang, P., Gao, D., Chen, D., et al. (2011). Sequential use of two different oncolytic viruses: A novel therapeutic regime to eradicate established tumors with an effective induction of tumour-specific immunity. *Human Gene Therapy, 22*, A124.

Van den Bossche, J., Al-Jamal, W. T., Yilmazer, A., Bizzarri, E., Tian, B., & Kostarelos, K. (2011). Intracellular trafficking and gene expression of pH-sensitive, artificially enveloped adenoviruses in vitro and in vivo. *Biomaterials, 32*, 3085–3093.

Watanabe, Y., Kojima, T., Kagawa, S., Uno, F., Hashimoto, Y., Kyo, S., et al. (2010). A novel translational approach for human malignant pleural mesothelioma: Heparanase-assisted dual virotherapy. *Oncogene, 29*, 1145–1154.

Whatcott, C. J., Han, H., Posner, R. G., Hostetter, G., & Von Hoff, M. D. (2011). Targeting the tumor microenvironment in cancer: Why hyaluronidase deserves a second look. *Cancer Discovery, 1*, 291–296.

Willmon, C., Harrington, K., Kottke, T., Prestwich, R., Melcher, A., & Vile, R. (2009). Cell carriers for oncolytic viruses: Fed Ex for cancer therapy. *Molecular Therapy, 17*, 1667–1676.

Witte, P. L., & Streilein, J. W. (1986). Development and ontogeny of hamster T cell subpopulations. *The Journal of Immunology, 137*, 45–54.

Worgall, S., Krause, A., Qiu, J., Joh, J., Hackett, N. R., & Crystal, R. G. (2007). Protective immunity to pseudomonas aeruginosa induced with a capsid-modified adenovirus expressing P. aeruginosa OprF. *Journal of Virology, 81*, 13801–13808.

Yan, W., Kitzes, G., Dormishian, F., Hawkins, L., Sampson-Johannes, A., Watanabe, J., et al. (2003). Developing novel oncolytic adenoviruses through bioselection. *Journal of Virology, 77*, 2640–2650.

Yong, R. L., Shinojima, N., Fueyo, J., Gumin, J., Vecil, G. G., Marini, F. C., et al. (2009). Human bone marrow-derived mesenchymal stem cells for intravascular delivery of oncolytic adenovirus Delta24-RGD to human gliomas. *Cancer Research, 69*, 8932–8940.

Zheng, J. N., Pei, D. S., Sun, F. H., Liu, X. Y., Mao, L. J., Zhang, B. F., et al. (2009). Potent antitumor efficacy of interleukin-18 delivered by conditionally replicative adenovirus vector in renal cell carcinoma-bearing nude mice via inhibition of angiogenesis. *Cancer Biology & Therapy, 8*, 599–606.

Zitvogel, L., Kepp, O., & Kroemer, G. (2011). Immune parameters affecting the efficacy of chemotherapeutic regimens. *Nature Reviews. Clinical Oncology, 8*, 151–160.

CHAPTER FIVE

# The Development of Transcription-Regulated Adenoviral Vectors with High Cancer-Selective Imaging Capabilities

**Ziyue Karen Jiang**[*,†], **Makoto Sato**[*,†], **Lily Wu**[*,†,‡,1]
[*]Department of Molecular and Medical Pharmacology, University of California Los Angeles, Los Angeles, California, USA
[†]Institute of Molecular Medicine, University of California Los Angeles, Los Angeles, California, USA
[‡]Department of Urology, David Geffen School of Medicine, University of California Los Angeles, Los Angeles, California, USA
[1]Corresponding author: e-mail address: lwu@mednet.ucla.edu

## Contents



## Abstract

A clear benefit of molecular imaging is to enable noninvasive, repetitive monitoring of intrinsic signals within tumor cells as a means to identify the lesions as malignant or to assess the ability of treatment to perturb key pathways within the tumor cells. Due to the promising utility of molecular imaging in oncology, preclinical research to refine molecular imaging techniques in small animals is a blossoming field. We will

*Advances in Cancer Research*, Volume 115
ISSN 0065-230X
http://dx.doi.org/10.1016/B978-0-12-398342-8.00005-7

first discuss the several imaging modalities such as fluorescent imaging, bioluminescence imaging, and positron emission tomography that are now commonly used in small animal settings. The indirect imaging approach, which can be adapted to a wide range of imaging reporter genes, is a useful platform to develop molecular imaging. In particular, reporter gene-based imaging is well suited for transcriptional-targeted imaging that can be delivered by recombinant adenoviral vectors. In this review, we will summarize transcription-regulated strategies used in adenoviral-mediated molecular imaging to visualize metastasis and monitor oncolytic therapy in preclinical models.

## 1. INTRODUCTION

Different from anatomic imaging techniques, molecular imaging can provide functional insights regarding diseases or biological processes, such as tumor proliferation, metabolic preference, as well as response toward therapeutics. There are generally two categories of approach to measure such functionality—direct and indirect imaging. In direct imaging, the target process is examined directly using its specific probe. For instance, $^{18}$F-FDG-mediated positron emission tomography (PET) imaging can be exploited to directly assess a tumor's glucose uptake, and $^{18}$F-FLT PET imaging can reveal first-hand information about the proliferative status of a tumor. On the other hand, cellular and molecular processes can also be linked to the expression of reporter genes and examined by indirect means (Iyer, Sato, Johnson, Gambhir, & Wu, 2005). For example, to measure hypoxia, a common condition in tumor environment due to poor oxygen perfusion, one can introduce a gene expression cassette with a reporter gene (e.g., luciferase) placed under the control of the HIF1$\alpha$ (hypoxia-induced factor 1 $\alpha$) responsive element (HRE). As HIF1$\alpha$ is a transcriptional factor activated under low oxygen tension, hypoxic cells will express a heightened level of such HRE-regulated luciferase gene and bioluminescent signal. In this review, we will focus on the indirect arm of imaging, namely, using reporter genes, to visualize molecular and cellular events in oncologic research.

A reporter gene usually encodes a protein that emits some form of light—visible light, fluorescence, or radiation—upon appropriate stimulation. It is delivered exogenously, and its expression can be captured by a proper imaging device. Reporter genes can be exploited to study the expression level of a linked therapeutic gene (reviewed by Iyer et al., 2005), to assess the interaction of two proteins (Pichler, Prior, Luker, & Piwnica-Worms, 2008; Ray et al., 2011), or to reflect the activation of specific inducible transcriptional factors or a signaling pathway, such as NF-κB, MAPK, TGFβ, HIF1α, or androgen receptor (AR) activation/silencing, by molecular imaging (Ilagan et al., 2006; Korpal et al., 2009; Nyati et al., 2011; Penet et al., 2009; Sato et al., 2005; Tsai et al., 2011). In all these scenarios, an efficient gene-delivery method is essential for the success of the assay. Adenovirus (Ad) has been widely used as an *in vivo* gene-delivery vehicle because of its high efficiency of transduction, ease of preparation, and low pathogenicity. The essence of molecular imaging such as Ad-mediated imaging emphasizes the possibility to repetitively query the cellular pathways in tumor in living subjects. This possibility means that growth status or treatment responses of a tumor can be monitored in real time in a noninvasive manner, without the necessity of repetitive invasive tissue samplings and histochemical analyses over time. The noninvasive nature of Ad-mediated imaging has enabled the monitoring of short-term therapeutic effect of AR blockade treatments (Ilagan et al., 2005) and tracking down disseminated, metastatic lesions of prostate cancer (Burton et al., 2008; Ilagan et al., 2006; Jiang, Sato, Wei, Kao, & Wu, 2011). In a second aspect, Ad-mediated gene expression is advantageous over cell marking by retrovirus or stable gene transfer methods as Ad transduction is transient and does not involve genome integration; hence, minimizing any permanent cellular perturbation and the chance of insertional mutagenesis in an imaging analysis. In this review, we will focus on the usage of Ad as gene-delivery vehicle for the applications in oncologic imaging. However, it is important to keep in mind that other viral, nonviral, and physical delivery methods can also be applied for this purpose, each with its own specific merit and advantage.

In this concise review, we will begin by first briefly introducing several commonly used preclinical imaging modalities (bioluminescent imaging (BLI), fluorescent imaging (FI), and PET), followed by discussion of transcriptional targeting of cancer in Ad-mediated cancer imaging.

## 2. PRECLINICAL MOLECULAR IMAGING MODALITIES

Imaging approaches can be generally categorized into anatomical imaging and functional/molecular imaging. The former category includes X-ray computed tomography (CT) and structural magnetic resonance imaging (MRI), which generates images based on the nature of a tissue's innate physical properties, for instance, size, density, or vascularity of a tumor. CT is the most widely used diagnostic imaging techniques in clinics.

It reconstructs images based on tissue's ability to transmit X-rays. CT has the spatial resolution of 50–100 μm. MRI has the highest spatial resolution among all currently available modalities of 25–100 μm. Its high resolution is offset by higher cost and thus, its utilization rate in the clinic is lower than CT. In MRI, the imaging instrument applies a strong magnetic field on the subject, which aligns the dipoles in hydrogen atoms in biomolecules and generates a radio wave. This radio wave can then be captured by the scanner. An active area of development in MRI technology is to ascertain functional, metabolic information on top at the precise anatomical site (Pinto et al., 2012). For example, cancer cells may exhibit an increased level of choline and citrate metabolism resulting from rapid cell membrane turnover during proliferation. These metabolites can be captured by functional MRI technique based on their characteristic resonance frequency. The conventional approach to develop functional, molecular imaging is largely based on the PET and SPECT (single-photon emission computed tomography) as specific radiolabeled imaging probe can be generated to inform on the activity of a specific metabolic/signaling pathway at the interested site. The combined utilization of anatomic and functional imaging (e.g., PET/CT) is one of the favorite modalities in current clinical practice as it provides most comprehensive information. In the following, we will cover the indirect imaging strategy and the common respective reporter genes used in each modality. As noted above and will be expanded upon later on, the indirect reporter gene-based imaging can be highly selective and easily adapted to interrogate a wide range of molecular process/pathways in living subjects. The discussions below are intended to be concise, introductory remarks, while excellent reviews can be found elsewhere (Kang & Chung, 2008; Ray, 2011; Serganova & Blasberg, 2005). The three common modalities used in cancer research are FI, BLI, and nuclear imaging. Figure 5.1 illustrates the application of these imaging modalities in small animals.

## 2.1. Fluorescent imaging

FI and BLI are two forms of optical imaging techniques that are well suited for preclinical applications. In general, compared to radiolabeled nuclear imaging, optical imaging is less expensive and easier to operate and hence they are widely used in research institutes to establish proof-of-concept, prior to pursuing translational research of clinical-relevant imaging approaches. However, both forms of optical imaging are restrained by their limited capability to image deep tissues as the low energy of visible light

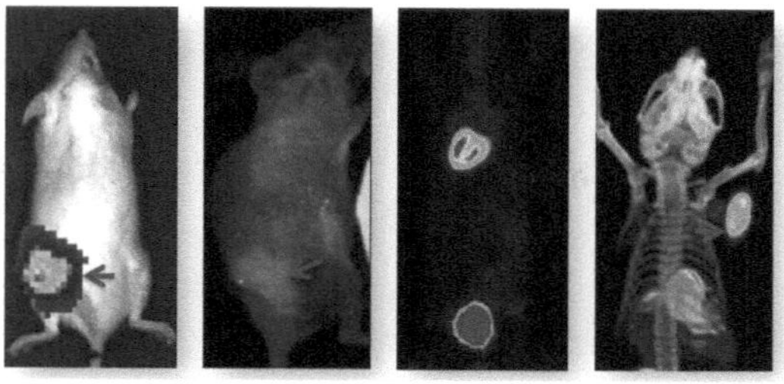


**Figure 5.1** Preclinical imaging modalities. The application of bioluminescent imaging (BLI), fluorescent imaging (FI), and positron emission tomography (PET) in conjunction with computed tomography (CT) in mouse tumor models. Subcutaneous tumor xenograft is indicated by the red arrow. (See Page 6 in Color Section at the back of book.)

emitted is scattered by tissues. The application of optical imaging in the whole-body setting is largely confined to visualization of superficial events, or investigations in small animals such as rats and mice, in which the distance of tissue penetration by the optical source is relatively short.

Fluorescent proteins are a group of self-sufficient proteins that emit light spontaneously owing to their intrinsic chromophores (Stepanenko, Shcherbakova, Kuznetsova, Turoverov, & Verkhusha, 2011). Ever since the discovery of the green fluorescent protein (GFP) from the jellyfish *Aequorea victoria* (Heim, Prasher, & Tsien, 1994; Ward & Cormier, 1979), researchers have took advantage of a tremendous variety of fluorescent proteins and used them as protein trackers, transcription readouts, and cell markers. These proteins produce light when they transit from an excited state into the ground state, and this process does not require cofactors or enzymatic reactions. Therefore, the chromophores in a fluorescent protein molecule can go through this transition, back and forth, for numerous times, generating strong, long-lasting visible/fluorescent light autonomously. Due to their self-sufficiency in generating light signals, fluorescent protein imaging is inexpensive, easy to operate, and has been employed in studies on cellular level, tissue level, and as well as in living complex organisms (Borovjagin et al., 2010; Kishimoto et al., 2006; Tang, Wu, Ugai, Matthews, & Curiel, 2009; Yamamoto, Tsuchiya, & Hoffman, 2011).

The most unique advantage of FI is that numerous fluorescent proteins exist in nature and in reengineered form (Giepmans, Adams, Ellisman, & Tsien, 2006; Shu et al., 2009) that emits light in a rainbow of colors, thus, enabling simultaneous imaging of two or more reporters that reflect the status of multiple molecular and cellular events. The broad spectrum also renders FI convenient to fit different research purposes. For example, GFP has been frequently fused to a protein of interest as a tag to track the subcellular localization and function of the target protein in living cells by sophisticated microscopic techniques (Cormack, 1998; Hutter, 2006; Tsien, 2009). Another favorite technique to follow tumor cell growth or dissemination in animal models is to genetically mark the tumor cells (or other cell of interest) with GFP and enumerate the marked cells in different tissues by histopathological techniques such as immunohistochemistry or flow cytometry (Kiessling, 2008; Mastro & Vogler, 2009; Yates et al., 2007). However, issues such as light scattering by tissues and autofluorescent signal from hair, skin, and food very much preclude sensitive and accurate detection of GFP-marked tumor cells in deeper sites without invasive organ externalization procedures. On the other hand, red and near infrared (NIR) fluorescent proteins have longer wavelength that approaches the lowest light attenuation (1200–1350 nm), and hence exhibit improved live tissue penetration compared to GFP and its blue and cyan variants (Campbell et al., 2002; Shu et al., 2009; Tsai et al., 2006). For instance, whole-body *in vivo* imaging has been applied to monitor cancer progression with RFP/IFP-marked tumor in mice (Contag, Jenkins, Contag, & Negrin, 2000; Hoffman, 2009; Tran Cao et al., 2011). The work of Dadiani et al. (2006) and ongoing research from our group indicated the feasibility of *in vivo* imaging of cancer metastasis occurring in organs that are deep underneath the skin such as lymph nodes with red/NIR fluorescent protein. However, as stated earlier, the genre of optical imaging generally suffers from short range of tissue penetration, and most clinically relevant application of FI is likely confined to intraoperative fluorescent-guided visualization processes (Gioux, Choi, & Frangioni, 2010; Kyo, Takakura, Fujiwara, & Inoue, 2008).

## 2.2. Bioluminescent imaging

BLI is another optical imaging technique that is widely used in preclinical small animal studies. Aside from being inexpensive and easy to operate, BLI gains popularity also because of its high sensitivity and low background signal compared to FI. The cooled charge-coupled device camera attached to light-tight BLI imaging chamber is able to capture light signal generated from as few as several hundreds of cells (Sato, Klaunberg, & Tolwani, 2004). However, it is important to keep in mind that the sensitivity of the imaging depends on various factors such as the robustness of reporter expression, the number of marked cells, tissue depth, light attenuation (both scattering and absorption from local tissue environment), and the time waited between substrate administration and imaging.

The reporter genes for BLI are a series of luciferase proteins. Mostly derived from lower organisms, such as bacteria, insects, and marine crustaceans, luciferase represents a type of enzyme capable of converting chemical energy into light while oxidizing their substrates. The best-studied luciferase is the one derived from firefly *photinus pyralis*, Firefly luciferase (FL), with D-luciferin as substrate. Other widely used luciferases include Renilla luciferase (RL) derived from the sea pansy *Renilla reniformis*, and Gaussia luciferase (GL) derived from the copepod marine animal *Gaussia princeps*, and both of them use coelenterazine as substrate. Despite the shared functionality, there is little sequence homology between these luciferase proteins, and their enzymatic mechanisms are quite diverse too. FL has an excellent quantum yield of >88%, indicating the high efficiency of converting chemical reaction to light production. The quantum yield for RL is only 6%, and the one for GL is still unknown (Tannous, Kim, Fernandez, Weissleder, & Breakefield, 2005). FL requires ATP and $Mg^{2+}$ as cofactors while RL and GL do not. FL exhibits glow kinetics (light signal lasts for up to more than 10 min) whereas coelenterazine luciferase generally present a flash kinetics with the signal rapidly decaying after 1–2 min, rendering a narrow time window for imaging post substrate administration. Compared to RL and GL, whose bioluminescence peak locates at around 480 nm, FL is more suitable for *in vivo* imaging due to its red-shift emission spectrum of over 700 nm. Meanwhile, variants of all these luciferase have been cloned to achieve either enhanced sensitivity or humanization and better cellular localization pattern (Gil, Machado, & Herschman, 2011; Inoue et al., 2011).

## 2.3. Nuclear imaging

PET and SPECT are also typical functional imaging techniques. In PET imaging, a positron-emitting radionuclide is used to label a biological molecule. When the radionuclide decays, it emits a positron that travels within tissue while its energy is dissipated and annihilates with an electron (typically in the submillimeter range) and release energies in the form of two gamma particles that travel in the opposite directions (180°). The two gamma rays can then be detected by the scanner to locate the source of the signal. In SPECT, the target molecule is labeled with a single gamma emitter, and this single gamma ray is detected by the instrument in a single plane and a three-dimensional (3D) image is reconstructed from the sequential planar image sections. These radioactively labeled molecules are imaging probes, and in both PET and SPECT imaging, the biodistribution/tissue-specific retention of the probe is determined by the expression of the corresponding target gene, which could be an endogenously expressed protein or an exogenously delivered reporter gene. There are generally three categories of reporter genes, differ according to their biological function and their mode of interaction with the probes.

### 2.3.1 Cell membrane receptors that selectively bind to the probe

A classic example in this category is the somatostatin receptors (SSTrs) which are a group of specific membrane receptors that mediate the action of the peptide hormone somatostatin. They have been shown to be overexpressed in several types of cancers including brain, kidney, and neuroendocrine tumors. Therefore, one can radiolabel the ligand of SSTrs, namely, somatostatin and analogs, to visualize SSTr-overexpressing cells (Singh, Yang, Ravoori, Han, & Kundra, 2009). A recently engineered fusion receptor in this category is the *DAbR1* (DOTA Antibody reporter 1) gene developed by a collaboration between Claude Meares, Wolfgang Weber, Johannes Czernin, and Anna Wu's group (Corneillie, Lee, et al., 2004; Corneillie, Whetstone, et al., 2004; Meares, 2008; Wei et al., 2008), which is an artificial receptor composed of a transmembrane anchor domain and an extracellular receptor or ligand-binding domain, joined by a linker region. The receptor portion is an antibody that specifically and irreversibly binds DOTA molecules that is chelated to rare-earth metals, which can be radiolabeled with several isotopes for imaging and radiation therapy purposes. For instance, DOTA chelated to gamma/positron emitter 86Yttrium could be a gamma camera, SPECT or PET imaging probe, while beta emitter 90Yttrium would serve for radiotherapy purposes. The infinite affinity and high specificity of this ligand–receptor pair as well as the long half-life of 86Yttrium (more than 14 h) have enabled late post administration PET imaging with low nonspecific background in a mouse glioblastoma xenograft model (Wei et al., 2008).

### 2.3.2 Membrane-anchored transporters with unidirectional transport of the probe into the cell

A well-known reporter in this category is the glucose transporter 1 (GLUT-1), which has enabled the wide application of $^{18}$F-FDG PET imaging in oncology. GLUT-1 actively transports glucose and analogs into highly metabolic cells, thereby enabling imaging of rapidly proliferating cells including cancer and metastatic lesions. The human sodium iodide symporter (*hNIS*) and human norepinephrine transporter (*hNET*) genes are also transporters in this category, but the ability of their probe to be entrapped within the target tissue is less certain (Niu et al., 2006; Serganova, Ponomarev, & Blasberg, 2007). According to their respective biological functions of transporting iodide ions and salvaging norepinephrine at/near the presynaptic terminals, the endogenous expression of hNIS is limited to thyroid and stomach, and the endogenous hNET is exclusively localized to the central and peripheral sympathetic nervous system. Therefore, imaging probes targeting endogenous hNIS could facilitate visualization of thyroid tumors while probes targeting endogenous hNET enable imaging of neuroendocrine tumors. On the other hand, due to the organ-limited natural expression pattern, the transporter genes can also be delivered exogenously as reporter genes to track tumor xenograft, infused immune cells, or induced pluripotent stem cells that are marked with these reporters (Che et al., 2005; Ma, Huang, & Kuang, 2009; Merron et al., 2007; Niu et al., 2006; Rajecki et al., 2012).

### 2.3.3 Enzymes that modify the structure of probe and subsequently trap it inside the cell

The herpes simplex virus thymidine kinase (*HSV-TK*) gene is the prototype of this class of enzymes in the nucleotide salvage pathway that has worked very well as a PET imaging reporter gene. It was first applied for this imaging purpose about 15 years ago (Gambhir et al., 1999, 1998; Tjuvajev et al., 1996, 1995). Over the years, the HSV-TK-based PET reporter imaging has enjoyed widespread popularity and concomitantly, the approach has been improved upon in the form of reengineered HSV-TK enzyme or its probe with higher binding affinity for its cognate partner. To date, the mutant variant of HSV-TK with heightened affinity for guanosine analog, HSV-sr39TK, is still widely used, especially since several F18-labeled probes have been shown to be well suited in HSV-TK-based reporter PET imaging. These probes include FHBG (9-[4-fluoro-3-(hydroxymethyl)butyl] guanine), FIAU (5-iodo-2′-fluoro-2′deoxy-1-β-D-arabino-furanosyl-uracil), and FEAU (5-ethyl-2′-fluoro-2′deoxy-1-β-D-arabino-furanosyl-uracil) (Tjuvajev et al., 2002; Yaghoubi et al., 2001). Since HSV-TK is an exogenous gene, its enzymatic activity is unique and does not overlap with the endogenous thymidine kinase (TK). Hence, HSV-TK-based PET reporter imaging is advantageous for being highly selective with low background activity. However, foreign viral proteins such as HSV-TK are immunogenic and can elicit host immune response that leads to its rapid clearance. Hence, to circumvent this immunogenicity issue recent studies explored several human enzymes in the nucleotide salvage pathway such as the TK and deoxycytidine kinase (dCK) with ability to phosphorylate nucleoside-based probes and then trap them within the cell (Lee, Campbell, Satyamurthy, Czernin, & Radu, 2012; Serganova et al., 2007; Shu et al., 2010). Human mitochondrial TK 2 (Ponomarev et al., 2007) has been demonstrated to function well as a PET reporter gene. However, the human TKs-based reporter imaging could encounter higher background signal as endogenous TK genes could confound the interpretation of imaging results from the introduced reporter gene. Continual development will focus on more specific probes and also active site variants of TK or dCK with distinct substrate specificity than the endogenous gene.

Each of the molecular imaging modalities discussed above has its own advantages and disadvantages (see Table 5.1 for a brief summary). For instance, BLI allows several mice (~3–4) to be imaged in the same chamber as long as the animals' light signals do not cross-interfere. As such, BLI is relatively high throughput compared to the single dedicated study of PET or SPECT. Since the optical imaging methods are also less hazardous and costly than nuclear imaging, they are used in general as the first-line study to establish imaging methodology or proof-of-principle. In turn, PET study is quantitative and provides 3D signal resolution and thus, it is developed in the secondary stage in preparation for clinical translation. Currently, as none of molecular imaging modalities is perfect for all applications, researchers have created fusion reporter gene capable of providing imaging function for multiple modalities. For instance, the triple fusion reporter gene developed by Dr. Gambhir's group can mediate BLI, FI, and PET simultaneously (Ray, Tsien, & Gambhir, 2007). To overcome the instability of large synthetic fusion protein, a fruitful approach is to generate gene-delivery vehicle coexpressing multiple imaging reporter genes simultaneously linked by picornavirus-derived "self-cleaving" 2A-like sequences (Koya et al., 2010). The 2A sequences allow the stoichiometric expression of several genes under a single promoter (Szymczak et al., 2004). Due to their inherent efficiency of gene transfer, viral vectors and especially adenoviral vectors are favorable vehicles to deliver exogenous genes in cancer research. The rest of this review will covers strategies taken to use adenoviral vectors to achieve cancer-specific molecular imaging.

**Table 5.1** Summary of optical and PET reporter imaging modalities

| Modality | Light source | Advantages | Disadvantages | References |
|---|---|---|---|---|
| Fluorescent | Photon emitted during fluorescent protein transitioning from excited to ground state | Simultaneous detection of multiple colors;<br>No need for repeated substrate injection;<br>Translatable to clinics. | Tissue absorption and light scattering;<br>Autofluorescence. | (Kishimoto et al., 2006; Timpson et al., 2011) |
| BLI | Enzymatic conversion of chemical energy to visible light | High sensitivity and high signal-to-noise ratio;<br>Semi-quantitative;<br>Easy-to-use equipment;<br>Short acquisition time and higher throughput (up to 3–4 animals/scan);<br>Inexpensive;<br>Nonradioactive. | Tissue absorption and light scattering;<br>Signal attenuation by melanin and fur;<br>Low spatial resolution;<br>Poor 3D reconstruction. | (Burch et al., 2007; Mouchess et al., 2006; O'Neill et al., 2010; Seethammagari et al., 2006; Wetterwald et al., 2002) |
| PET | A pair of gamma rays emitted from positron annihilation | Excellent 3D signal resolution;<br>Quantitative;<br>Readily translatable. | High cost;<br>Involves radioactive probes;<br>Low sensitivity;<br>Single signal readout at a time;<br>Low throughput. | (Brader et al., 2011; Deroose et al., 2007; Jiang et al., 2011; Johnson et al., 2009) |

# 3. CANCER-TARGETED ADENOVIRAL VECTOR GENE DELIVERY

In the field of indirect gene-based molecular imaging, it is crucial to achieve high tissue/cancer selectivity of reporter gene expression to exclude nonspecific background and increase the rate of on-target diagnostics. Several measures that have been undertaken to augment such specificity include: (1) confined vector delivery, (2) the modification of vector surface to increase affinity for tumor cells (i.e., transductional targeting), (3) restricted gene expression via tumor-specific promoters (i.e., transcriptional targeting), and (4) a combination of the targeting strategies described. To study a locally confined solid tumor, the virus can be delivered intratumorally via guided injection. This attempt would concentrate the Ad vector within the tumor, decreasing off-target reporter gene expression. A valuable imaging application in localized tumor is to query pathways critical to the pathogenesis of the disease. For instance, the AR signaling axis is such a central pathway in prostate cancer. The functional status of this pathway can be assessed during prostate tumor progression or AR blockade therapy in real time by injecting Ad that contains AR-regulated luciferase reporter gene in prostate tumor progression or during AR blockade therapy (Ilagan et al., 2006; Zhang et al., 2003; Fig. 5.2A-B). Further, the direct visualization of AR signaling in the prostate and prostate cancer situated deep in the pelvic cavity required precise 3D localization of imaging signal. In this case, the use of PET imaging directed by the HSV-sr39tk PET reporter gene was able to visualize the AR-regulated PSA gene transcription in both murine and canine models by FHBG PET imaging (Pouliot et al., 2011; Fig. 5.2C).

In the application to detect disseminated metastatic lesions that reside at unknown locations, systemic administration of imaging enabling Ad vector is necessary. In that case, the ability to discriminate target cells from nontarget cells based on cell surface and innate characteristics will be highly conducive to the efficiency of targeted gene delivery. Such specificity can be achieved either by modifying the surface properties of Ad particles to home in on cancer cells (transductional targeting) or by incorporating a cancer-specific promoter in the Ad genome to restrain the expression of the reporter gene within cancer cells (transcriptional targeting). For example, the viral capsid can be modulated to overexpress an RGD moiety to target integrin-expressing cells. Bispecific adaptors can be used to detarget the virus

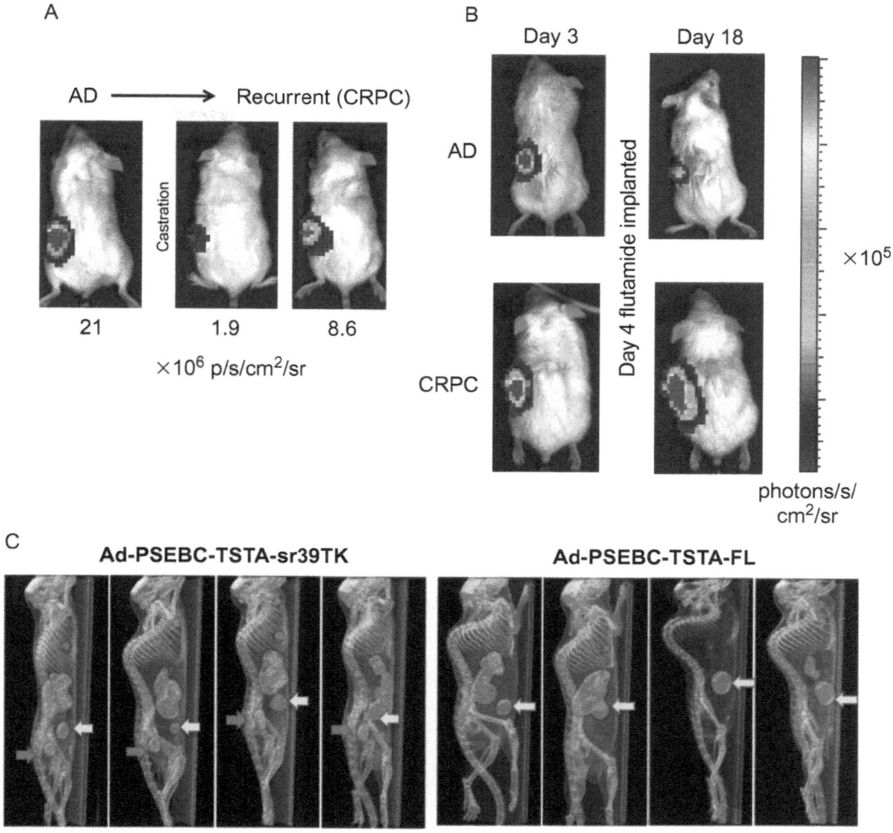


**Figure 5.2** Adenovirus-mediated AR functional imaging of localized tumor. A transcriptional-targeted luciferase reporter vector can monitor efficacy of castration and androgen receptor blockade therapy in prostate tumors. (A) A single mouse-received Ad-PSEBC-TSTA-FL virus injected into the subcutaneous prostate cancer xenograft. BLI imaging demonstrated decrease of AR activity along the course of castration and rebounce when castration-resistant prostate cancer occurred. (B) Mice bearing androgen-dependent (AD) or castration-resistant (CRPC) prostate cancer xenografts received intratumoral injection of Ad-PSEBC-TSTA-FL virus. Upon the treatment of AR antagonist flutamide, AR activity decreased only in the AD tumor, but not in CRPC, demonstrating CRPC tumor's resistance to flutamide treatment. (C) PET imaging of intraprostatic LAPC9 implanted prostate cancer. Fourteen days after implantation, tumors were accessed surgically for intratumoral administration of Ad-PSEBC-TSTA-sr39TK virus or Ad-PSEBC-TSTA-FL virus as control. Four days after viral injection, $^{18}$F-FHBG PET was performed. Distinct prostatic tumor signal was achieved in the sr39TK group indicated by the blue arrow. The probe excreted in the bladder is indicated by the yellow arrow. The 3D signal resolution of PET–CT enables the clear separation between the closely situated prostate- and bladder-originated signal. (See Page 6 in Color Section at the back of book.)

from liver and retarget toward cancer–antigen-expressing tumor cells (Li, Everts, Yamamoto, Curiel, & Herschman, 2009). Such transductional targeting approaches modulate the outside of viral particles toward an altered viral tropism and have generated promising results. This review, however, will focus on transcriptional targeting that manipulates the inside of virus—the viral genome—so as to confine imaging gene expression in tumor cells by cancer or tissue-specific promoters even when Ad viral particles physically distribute outside the target tissue.

Demonstrated by its altered biological functions, a cancer cell usually presents a different profile of transcription, which could result from aberrant activation of transcriptional factors or changes in epigenetic modification of promoters. For instance, the transcriptional factor HIF1$\alpha$ is usually abnormally activated in cancer cells due to local hypoxia and thereby activates promoters that harbor hypoxia responsive elements (Hogg, Garcia, & Gerard, 2010); the human telomerase reverse transcriptase (hTERT) promoter becomes constitutive active leading to constant telomerase activity in cancer cells (reviewed by Kyo et al., 2008); and the survivin promoter contains cell-cycle-regulated elements, resulting in its overactivation in transformed cells (Huang et al., 2011). Thus, one way to endow cancer selectivity is to use such cancer-specific promoters to drive reporter gene expression. For example, Lillehammer et al. (2005) constructed first-generation Ad vectors harboring human tyrosinase promoter and enhancer (2hE-hTyrP) and demonstrated its strict tissue specificity for melanoma cells by *ex vivo* FI. The progression-elevated gene-3 (PEG-3) is found to be overexpressed as a function of cell transformation (Su et al., 2005). The promoter of this gene has been incorporated into Ad vectors to drive cancer-specific GFP expression in breast cancer, glioma, and prostate carcinoma while sparing normal cells (Su et al., 2005). The same promoter was also used in Ad-mediated therapeutic vectors to treat prostate cancer, metastatic breast carcinoma, and distal pancreatic tumors in preclinical mouse models (Sarkar et al., 2007; Sarkar, Su, Vozhilla, Park, Gupta, et al., 2005; Sarkar, Su, Vozhilla, Park, Randolph, et al., 2005). Huyn et al. (2009) took advantage of the Muc1 promoter and generated breast cancer-targeting Ad that can visualize orthotopic and metastatic breast cancer xenograft in a mouse model. The carcinoembryonic antigen promoter was also employed to regulate the expression of NIS gene and thereby guide radioiodide therapy for medullary thyroid cancer (Spitzweg, Baker, Bergert, O'Connor, & Morris, 2007).

Aside from the cancer-specific promoters, tissue-specific promoters can also be exploited in this field when the cancerous organ is dispensable for normal physical functions. A typical illustration of utilizing tissue-specific promoters in gene therapy and molecular imaging is the disease of prostate cancer.

Promoters that have restricted activity toward prostate cells such as prostate-specific antigen (PSA), probasin, prostate-specific membrane antigen (PSMA), and human glandular kallikrein 2 promoters have all been used to drive prostate cancer-specific gene expression (Figueiredo, Kao, & Wu, 2007).

However, one recurring problem for most of these cancer- or tissue-specific promoters is the relatively low transcriptional activity compared to constitutive viral promoters such as cytomegalovirus (CMV) promoter, RSV promoter, and SV40 promoter. For instance, Dwyer, Bergert, O'Connor, Gendler and Morris (2006) showed that compared to the Ad-CMV-NIS, the Muc-1 promoter-driven vector was only able to reveal a less-intensive image of tumor. To achieve adequate imaging signals for *in vivo* detection, transcriptional amplification strategy is generally required. In the following section, we will introduce several strategies to boost the activity from a weak but cell-specific promoter yet maintain its selectivity.

## 4. TRANSCRIPTIONAL AMPLIFICATION STRATEGIES

Several measures have been exploited to overcome the weak transactivation mediated by tissue- or tumor-specific promoters in order to achieve distinct tumor detection.

### 4.1. Optimized chimeric promoter

The very first strategy undertaken by researchers to enhance cancer-selective transcription is constructing modified or chimeric promoters that possess potent transactivation capabilities. The original sequence of a promoter usually has endogenous negative regulatory elements built in as nature's way of incorporating a mixture of cellular stimuli and fine-tuning the expression of that gene. However, for the purpose of gene therapy and molecular imaging, such attenuators/silencer sequences may undercut the amplitude of the promoter. In addition, naturally occurring promoters may harbor long series of nonfunctional sequences. Given the limited spatial capacity of adenoviral genome, such nonfunctional, intervening sequence is usually deleted in order to make a short, concise, and more efficient promoter.

For instance, to visualize prostate cancer progression into advanced stage of castration-resistant metastasis, a prostate-specific enhancing sequence (PSES) was constructed utilizing transcriptional elements derived from the PSA promoter and the PSMA enhancer (Lee et al., 2002). Specifically, one of the AREs (androgen responsive elements) located upstream of the PSA promoter was isolated and linker-scanning mutations were introduced across its sequence. Following assaying each mutant's transcriptional activity,

Lee et al. found a segment that is particularly sensitive to mutation, suggesting this segment's crucial role in activating transcription. This fragment was then chosen as the PSA-derived element to constitute the PSES sequence. A similar tactic was used for the PSMA-derived part of PSES. First, a PSMA enhancer (PSME) that locates in the third intron of the PSMA gene was isolated. Deletion analysis was then conducted on the PSME sequence, which contains nine transcriptional factor-binding sites, a silencer, and a repeat sequence. The deletion analysis identified a truncated section of PSME—PSME(del2)—that demonstrated high activity of transcription resulting from the deletion of the innate silencing sequence (Lee et al., 2002). Therefore, the PSES sequence, composing transcriptional elements from both promoters, is only 500 bp in length, greatly facilitating the incorporation of other gene expression cassettes in a single adenoviral vector.

The same principle of optimizing native promoters is exemplified by several other approaches in the field. Wu et al. (2001) generated a series of PSA-based enhancer constructs by inserting multiple copies of AREs, duplication of the enhancer core, and removal of intervening sequences. The resultant constructs exhibited efficacy that was 7- to 19-fold higher than the basal promoter. An FL-expressing vector driven by one of these optimized promoters—the PSEBC promoter—successfully visualized advanced prostate cancer in living mice via optical imaging (Adams et al., 2002). Zhang, Thomas, Kasper, and Matusik (2000) designed a probasin-based composite promoter that was smaller yet retained high level of prostate-specific gene expression. The basal promoter and enhancer I of α-fetoprotein (AFP) have also been coordinated to form an AFP promoter applicable for liver cancer-specific detection (Klutz et al., 2011; Ma et al., 2009).

## 4.2. Conditional recombinase activation

The versatile and widely utilized Cre–Lox recombination system was derived from bacteriophage P1. The Cre recombinase is able to specifically cut out DNA sequences that are flanked by two *lox P* sites. Interestingly, this system can also be tailored to amplify tissue/cancer-specific promoters. As illustrated in Fig. 5.3A, the idea of the conditional recombinase activation relies on tissue/cancer-specific expression of the Cre recombinase, which then removes the stop cassettes located between a strong promoter and the reporter gene by cleavage mediated by the flanking two *lox P* sites. Therefore, there are two components in this system: tissue-specific Cre expression and the silent-activated reporter gene expression cassette. These two components can be delivered simultaneously via two vectors or be

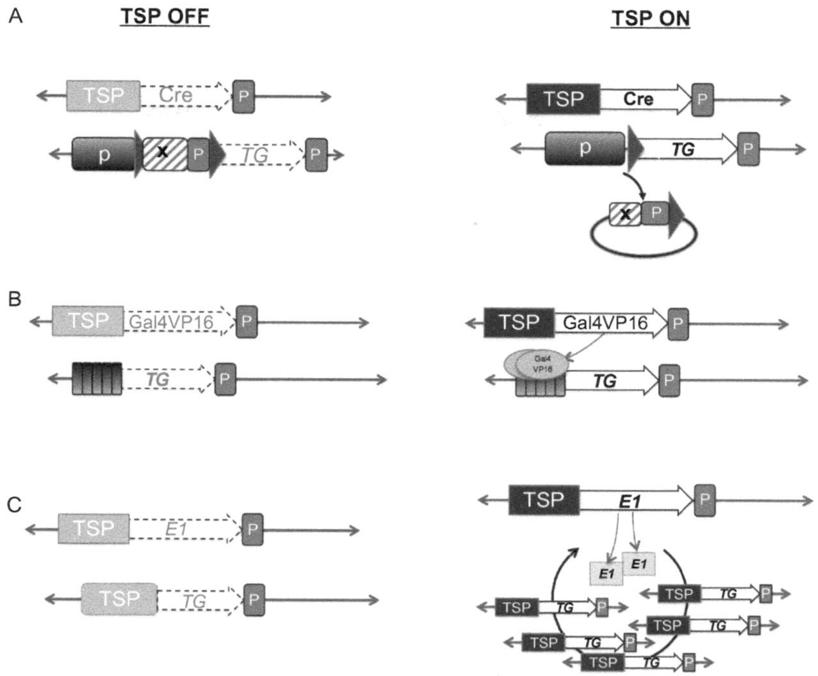


**Figure 5.3** Transcriptional amplification strategies. (A) Conditional recombinase activation. (B) TSTA (two-step transcriptional amplification) system. (C) Genome replication strategy. TSP, tissue- or tumor-specific promoter; P, strong promoter; p, poly-A; TG, transgene; X, stop cassette. (For color version of this figure, the reader is referred to the online version of this chapter.)

incorporated into a single vector. For instance, Sato et al. constructed a regulator Ad that harbored the Cre gene under the control of the AFP promoter. It was then used to coinfect hepatocarcinoma cells with a vector that carried a Cre-activating potent LacZ expression unit. A 50-fold induction was observed compared to single AFP-driven vector (Sato et al., 1998). Alternatively, Kanegae et al. developed a "double-unit" Ad which contained the AFP-Cre recombinase and EFα-fluorescent reporter expression in a single genome. However, the lox P sites were located outside the potent reporter gene cassette, and Cre-mediated site-specific cleavage led to the excision of the whole cassette which then formed a circular DNA properly orienting the poly-A sequence at the end of the expression cassette, supporting reporter expression. Amplification of 40- to 90-fold was achieved in this study (Kanegae et al., 2011). Nevertheless, the imperfect efficiency of the Cre enzyme might pose limitations to the performance of this system.

## 4.3. Two-step transcriptional amplification

As shown in Fig. 5.3B, the TSTA system constitutes two elements. The first part involves expression of a chimeric transcriptional activator under the control of a weak, selective promoter. The activator is a fusion protein of the yeast Gal4 DNA-binding domain with the herpes simplex virus activation domain VP16, which activates transcription by facilitating Pol II initiation and elongation. The second part of this system is composed of a reporter gene placed downstream of five consecutive Gal4-binding sites and a minimal promoter—G5 reporter. In this way, the expression of the Gal4-VP16 protein is restricted to cells permissive to the weak promoter, yet the reporter gene expression level will be greatly elevated attributed to VP16's strong activating effect (Ahn et al., 2011; Huyn et al., 2009; Sato et al., 2005).

There are several variants derived from the original form of TSTA (thoroughly reviewed by Figueiredo, Gambhir, Carey, & Wu, 2009). For example, the G5 sequence is symmetrical so that two reporter genes can be positioned on both sides of G5, flanking it, and can get simultaneous expression. This is called the bidirectional TSTA system (Huyn et al., 2009; Ray et al., 2008). In addition, the two components of TSTA can exist in two separate vectors so that the activator is supplied *in trans*, or they can be cloned into a single adenoviral genome so that they are acting *in cis*. Our previous findings suggested that incorporation of the two elements in a single vector have the highest efficiency in transcriptional activation (Sato et al., 2003).

## 4.4. Genome replication

Another strategy to augment reporter gene expression is to increase the copy number of the gene in cells. Conditional replication-competent Ad can be suitable to fulfill this purpose. There are generally two categories of genome replication strategies that can increase the copy number of a reporter gene: (1) incorporating the viral replication gene (*E1A* and *E1B* genes) and the reporter gene into a single genome, and this amplification effect is *in cis*; (2) administrating the replication-competent virus simultaneously with the reporter vector, and the E1A protein expressed by the replicating virus will act *in trans* to facilitate the replication of the reporter's genome. In both ways, the E1 genes that are essential for viral replication can also be regulated by cancer-selective promoters to further improve the specificity.

The idea of the *in trans* activation is convenient to implement compared to the *in cis* amplification, as viral genome size limitation is often a constraint to accommodate both the replication genes and exogenous reporter genes. For *in trans* amplification, any cancer-selective replication virus can be combined with a reporter vector to increase imaging efficiency. As shown in Fig. 5.3C, the expression of the E1A and E1B genes essential for viral genome replication is controlled by a tissue- or tumor-specific promoter (TSP). Meanwhile, another vector cotransduces the cells, conferring the expression of FL gene under the regulation of the same or different TSP. In this way, the expression of E1 genes in the first replication-competent virus in the permissive cancer cells induce viral genome replication and amplification of both viruses in the same cell, and this positive feedback loop elevates the expression level of the luciferase reporter by genomic amplification. However, such an *in trans* activation system can suffer from heterogeneous expression from cell to cell as the simultaneous transduction of a single cell with two vectors is less likely to occur than if a single virus contains all components of the *in cis* amplification strategy.

A few groups have demonstrated the principle of the *in cis* amplification strategy with oncolytic Ad (Kim et al., 2008; Kishimoto et al., 2006). An example of this approach is the OBP-401 series of conditionally replicating Ad constructed by Fujiwara and colleagues (Kim et al., 2011; Kishimoto et al., 2006, Kishimoto, Urata, Tanaka, Fujiwara, & Hoffman, 2009; Kojima et al., 2009; Maida et al., 2009). The replication of OBP-301 and OBP-401 is controlled by hTERT promoter and thus only takes place in cancer cells that have abnormal constitutive hTERT activity. The OBP-401 virus is further armed with a CMV promoter-driven GFP reporter gene so that the replication and distribution of the virus can be tracked by FI. Compared to replication-deficient GFP-expressing virus, the OBP-401 demonstrated marked amplification. The application of this series of virus will be covered in more detail in the next session.

# 5. SPECIFIC APPLICATIONS

## 5.1. Gene-based imaging seeking out metastatic lesions

Identifying new incidence of cancer metastasis in preclinical models is the aspect where Ad is most advantageous compared to lentivirus or stable transfection. This is attributed to the fact that Ad infection is only transient and weakly symptomatic, so repetitive administration of imaging vectors as surveillance is unlikely to cause undesirable insertional mutations. Several examples of using Ad to detect systemic cancer metastasis are reviewed below.

Steinwaerder et al. fashioned an Ad vector whose replication was selectively activated in hepatic metastases, which resulted in cancer-specific inverted repeats-mediated viral genome rearrangement and subsequent reporter gene (β-gal) expression. This approach successfully marked the metastatic tumor modules upon intravenous viral administration (Steinwaerder et al., 2001). Even though this experiment was not coupled with *in vivo* whole-body imaging, it highlighted the cancer specificity rendered by engineered Ad vectors injected systemically to seek out cancer metastases.

Considering the frequent involvement of lymph nodes in cancer metastasis, it is urgent to develop a noninvasive diagnostic tool to assess the sentinel lymph node without the aggressive lymphoscintigraphy and lymphadenectomy. To this end, Kishimoto et al. (2006) reported the use of a telomerase-specific replication-competent Ad that expressed GFP constitutively (OBP-401) to image lymph node metastasis of rectal cancer. Intratumorally injected into orthotopic tumor, the virus spread into para-aortic lymph nodes and enabled visualization of metastatic cancer cells at laparotomy, manifesting imaging-guided preclinical surgery in removing LN metastases. Notably, the usage of oncolytic other than replication-deficient virus made imaging easier in that a lower amount of virus ($1 \times 10^8$ PFU) and potentially a smaller volume is required. Also, Burton et al. (2008) utilized FL and HSV-sr39tk-expressing Ad vectors and successfully detected lymph node metastases from subcutaneous prostate cancer xenografts by BLI and PET, respectively (Fig. 5.4). The reporter genes in these vectors were driven by the amplified prostate-specific promoter system, PSEBC-TSTA, and the virus was administered either through paws (to the draining lymphatics) or by tumor-directed injections. This study demonstrated the feasibility of exploiting the clinical imaging modality PET in detecting occult lymph node metastasis by Ad-mediated gene transcription. The advantage of using a cancer-specific reporter gene instead of metabolic probes (such as FLT or FDG) lies in that some cancer cells are not metabolically active (Jiang et al., 2011; Rioja et al., 2010).

Liver exhibits the best tropism for serotype 5 Ads in most organisms; it is therefore quite convenient to use Ad to detect liver metastasis from various kinds of cancer. For example, Huyn et al. constructed an optical imaging vector with FL expression controlled by the cancer-specific Muc1 promoter coupled with the TSTA amplification system. Intravenous injection of this vector successfully identified intrahepatic breast cancer metastases that were predicted by FDG PET imaging (Huyn et al., 2009; Fig. 5.5). Also, the OBP-401 virus administered through portal venous or systemic i.v. injection can detect colorectal liver metastases and orthotopic hepatocellular carcinoma by *ex vivo* FI (Kishimoto et al., 2009). However, it remains a challenge for systemically administered Ad to circumvent hepatic transduction and seek out metastases in and outside of the liver.

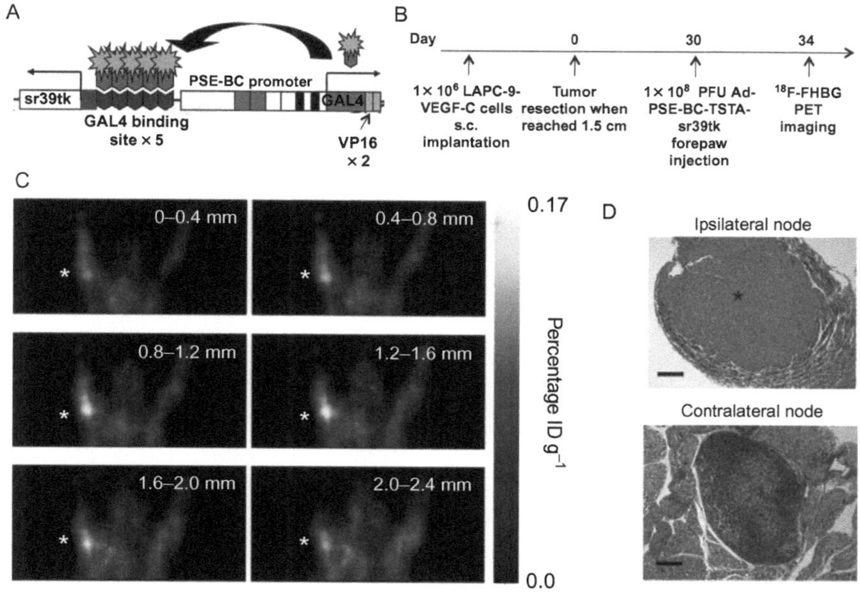


**Figure 5.4** Ad-PSEBC-TSTA-sr39TK/PET imaging detecting prostate cancer lymph node metastasis. (A) The prostate-specific expression cassette of the Ad-PSEBC-TSTA-sr39TK PET reporter virus. (B) The time line of the experiment to detect lymph node metastasis of LAPC-9 prostate tumors. (C) A mouse with prostate xenograft on right upper back that had received bilateral paw injection of Ad-PSEBC-TSTA-sr39TK, 4 days earlier. Sequential coronal sections of this animal showed lymph node metastasis detected by PET in the axillary lymph node (asterisk), ipsilateral to the tumor but not in the contralateral lymph node. ID, injected dose. (D) H&E-staining of the axillary lymph node showed extensive tumor involvement (asterisk) in the ipsilateral but not in the contralateral node. (See Page 7 in Color Section at the back of book.)

In a recent study Jiang et al. took advantage of the TSTA transcriptional amplification strategy and established an androgen-independent promoter system, PSES-TSTA, to specifically detect castration-resistant prostate cancer. This promoter system was incorporated into Ad vectors to drive the expression of BLI and PET imaging reporters (Jiang et al., 2011). Moreover, they established an intratibial prostate cancer metastasis model and demonstrated that intravenous injection of BLI imaging vector could accurately read out the establishment of prostate cancer in the tibial bone marrow. This mimicked the scenario where new incidence of metastasis occurs in the bone after hormonal treatment for prostate cancer, and Ad vectors can be injected systemically to identify the emergence and location of these lesions (Fig. 5.6). More importantly, this study found that Ad-mediated

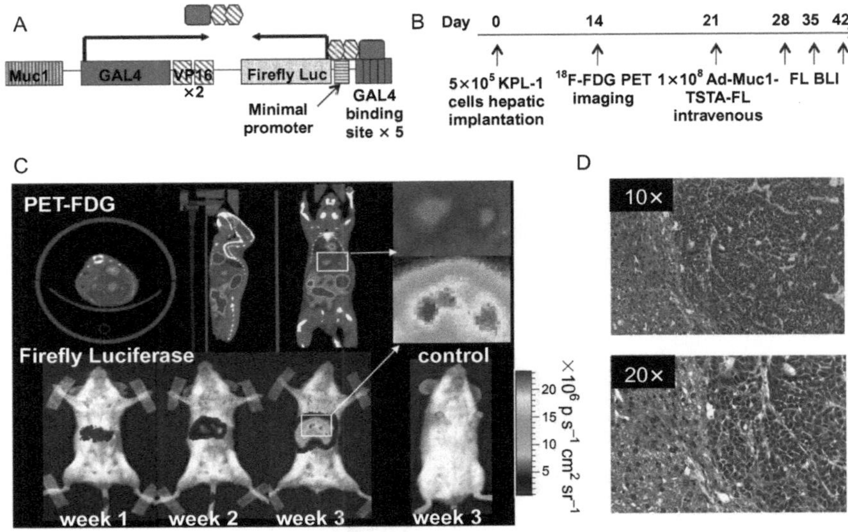


**Figure 5.5** Ad-Muc1-TSTA-FL targeting breast cancer liver metastasis. (A) The breast-specific, mucin-1 promoter directed TSTA expression cassette of Ad-Muc1-TSTA-FL. (B) The time line of the liver metastasis detection experiment. (C) Transverse, sagittal, and coronal image of $^{18}$F-FDG PET imaging are displayed. FL BLI revealed similar metastasis pattern as FDG PET. (D) H&E staining of the left lobes of the liver demonstrating the tumor mass. (For color version of this figure, the reader is referred to the online version of this chapter.)

BLI based on the transcription program intrinsic to cancer cells can detect metastatic lesions at a stage earlier than imaging modalities that target bone remodeling and destruction events (e.g., conventional bone scans and NaF PET). This result was consistent with previous observations that BLI is more sensitive than radiology and thus can detect disease at early phases (Burch, Bisland, Wilson, Whyne, & Yee, 2007; Wetterwald et al., 2002). However, due to the limited translational value of optical imaging, how to apply such imaging vectors into clinically relevant modalities such as PET and SPECT remains an issue worth further investigation.

## 5.2. Monitoring oncolytic viral replication and corresponding therapeutic effects

With intense research on oncolytic virus and their rapid translation into clinical trials (Cerullo et al., 2011; Pesonen et al., 2012), it is important to develop noninvasive imaging tools to monitor the therapeutic effects of such virus so that responsive patients can be identified promptly while

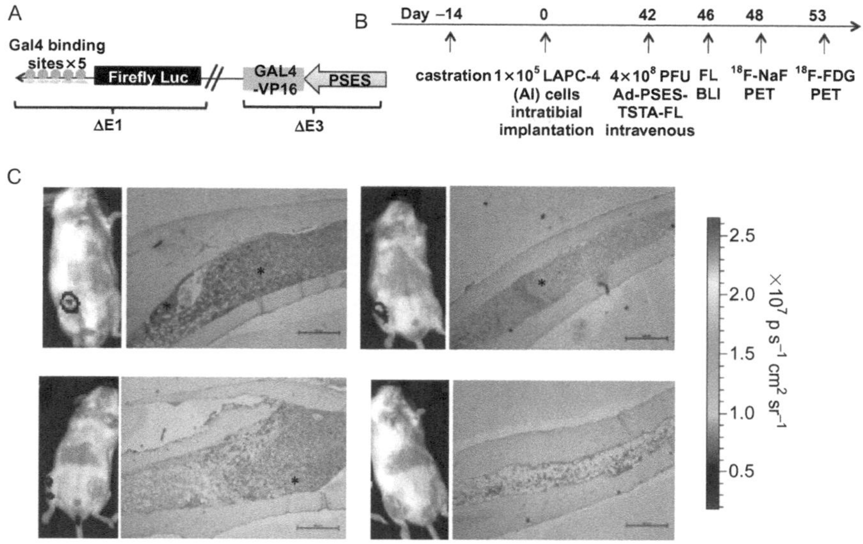


**Figure 5.6** Ad-PSES-TSTA-FL sought out bone metastasis of castration-resistant prostate cancer. (A) The androgen-independent, prostate-specific PSES promoter directed TSTA expression cassette of Ad-PSES-TSTA-FL. (B) The time line of the bone metastasis imaging experiment. (C) BLI indicated the presence or absence of prostate cancer bone metastasis that was confirmed by pan-cytokeratin immunohistochemistry staining of the right tibial bone sections (tumor marked by asterisk). Scale bar = 500 μm. (For color version of this figure, the reader is referred to the online version of this chapter.)

nonresponsive patients and noneffective regimens can be stopped without causing deleterious side effects. To this end, a replication-competent oncolytic virus was constructed with the EGFP gene placed under the control of the major late promoter, which is only active during viral replication (Ono, Le, Davydova, Gavrikova, & Yamamoto, 2005). This vector, even though did not spread sufficiently in the tumor, established a correlation of FI with viral replication. In a similar study, researchers used the major late promoter to drive the expression of the PET-imaging reporter gene TK in a cancer-selective oncolytic Ad, and successfully traced the viral distribution and oncolytic effects in a preclinical model of pancreatic cancer (Abate-Daga et al., 2011). Consistently, Davydova et al. generated an oncolytic Ad vector with COX2 promoter driving E1 expression. A luciferase gene was driven by the major late promoter in this study to reflect later event of cell lysis (Davydova et al., 2010). *In vivo* BLI following systemic viral injection successfully showed dynamic viral replication and, further, predicted therapeutic outcome. Similarly,

Trujillo et al. (2009) fashioned a Muc-1 promoter-driven conditionally oncolytic virus whose replication was traced by SPECT imaging via a constitutively expressed NIS gene in the same vector. Similar imaging strategy utilizing NIS has also been employed to monitor the replication and kinetics of intratumoral viral spread of an hTERT-driven oncolytic virus (Merron et al., 2007).

Aside from these studies of *in cis* monitoring of oncolytic events, Guse et al. coinjected a luciferase-expressing E1-deleted virus with an oncolytic virus intratumorally into murine xenografts. The E1 expression from the coinfected oncolytic virus in the same cell would help the replication of Ad-Luc, which in turn served as readout for oncolytic virus replication *in trans* (Guse et al., 2007). Indeed, they found a significant correlation between infectious viral production and photon emission, suggesting another means of monitoring oncolytic viral gene therapy.

## 6. CONCLUDING REMARKS

The imaging techniques described are widely available and utilized frequently in oncology research nowadays. A clear benefit of noninvasive molecular imaging is to allow repetitive monitoring of tumor growth, dissemination, and treatment response over time in the same subject. Inherently, this form of longitudinal assessment provides relevant and significant information as each subject also serves as its own control. The hope is that noninvasive molecular imaging strategies verified in preclinical setting will be able to translate to clinical setting to guide the detection and treatment of oncology patients.

Based on the extensive research experience on adenoviral gene delivery and imaging gathered from preclinical studies, adenoviral vectors have demonstrated remarkable transduction efficiency, broad infective spectrum, and inherent flexibility for transductional and genetic modifications so as to achieve exquisite tissue/cancer-specific imaging gene expression *in vivo*. It is clear that transcription-targeted molecular imaging strategy can be a highly selective, flexible, and valuable approach to pursue in various oncological applications.

Despite the encouraging progresses made in Ad-mediated transcription-based imaging, we should keep in mind that there are several hurdles that hamper the vector's usage in clinics. For example, tumor-directed delivery of Ad has been shown to have uneven penetration and distribution in the tumor even when injected locally. To this end, factors that help degrading the extracellular matrix, hence facilitate viral spreading, could be delivered with other transgenes to improve viral distribution (Kim et al., 2006; Lavilla-Alonso et al., 2011).

Another challenge for Ad-mediated reporter gene expression attributes to the immunogenicity of the virus (DeMatteo et al., 1999; Dewey et al., 1999; Minagawa, Kawamura, Liu, Govindarajan, & Dennert, 2005; Zirger, Liu, Barcia, Castro, & Lowenstein, 2006). In this regard, various approaches have been taken to reduce the viral immunogenicity and/or to lower the host's response. Protective ionic polymers, for example, can be deposited onto Ad particles and reduce the eliminative effect of anti-Ad antibodies and increase the therapeutic effects (Yoshihara, Hamada, & Koyama, 2010). Alternatively, several research groups including our group are actively investigating the design of short-term immunosuppressive regimens to reduce viral immunity and boost the efficiency and persistence of viral-mediated expression.

Last but not least, the molecular imaging strategies described here are by no means restricted to adenoviral gene-delivery vectors. Similar reporter gene-based imaging is being developed in a wide range of viral and nonviral vehicles. For instance, oncolytic herpes virus was used to image lymph node micrometastases of melanoma (Brader et al., 2009). Watanabe et al. (2011) used adeno-associated virus that harbored amplified hTERT promoter-driven luciferase gene and visualized orthotopic liver cancer. Nonviral vectors such as linear polyethylenimine were chosen to deliver the tumor-specific PEG-3 promoter-driven plasmid systemically to image melanoma and breast cancer metastasis (Bhang, Gabrielson, Laterra, Fisher, & Pomper, 2011). Collectively, the success of all of these strategies in preclinical settings will facilitate their application in clinics to benefit human cancer patients.

## REFERENCES

Abate-Daga, D., Andreu, N., Camacho-Sanchez, J., Alemany, R., Herance, R., Millan, O., et al. (2011). Oncolytic adenoviruses armed with thymidine kinase can be traced by PET imaging and show potent antitumoural effects by ganciclovir dosing. *PLoS One*, *6*, e26142.

Adams, J. Y., Johnson, M., Sato, M., Berger, F., Gambhir, S. S., Carey, M., et al. (2002). Visualization of advanced human prostate cancer lesions in living mice by a targeted gene transfer vector and optical imaging. *Nature Medicine*, *8*, 891–897.

Ahn, B. C., Ronald, J. A., Kim, Y. I., Katzenberg, R., Singh, A., Paulmurugan, R., et al. (2011). Potent, tumor-specific gene expression in an orthotopic hepatoma rat model using a Survivin-targeted, amplifiable adenoviral vector. *Gene Therapy*, *18*, 606–612.

Bhang, H. E., Gabrielson, K. L., Laterra, J., Fisher, P. B., & Pomper, M. G. (2011). Tumor-specific imaging through progression elevated gene-3 promoter-driven gene expression. *Nature Medicine*, *17*, 123–129.

Borovjagin, A. V., McNally, L. R., Wang, M., Curiel, D. T., MacDougall, M. J., & Zinn, K. R. (2010). Noninvasive monitoring of mRFP1- and mCherry-labeled oncolytic adenoviruses in an orthotopic breast cancer model by spectral imaging. *Molecular Imaging, 9*, 59–75.

Brader, P., Kelly, K., Gang, S., Shah, J. P., Wong, R. J., Hricak, H., et al. (2009). Imaging of lymph node micrometastases using an oncolytic herpes virus and [F]FEAU PET. *PLoS One, 4*, e4789.

Brader, P., Wong, R. J., Horowitz, G., & Gil, Z. (2011). Combination of pet imaging with viral vectors for identification of cancer metastases. *Advanced Drug Delivery Reviews, 64*, 749–755.

Burch, S., Bisland, S. K., Wilson, B. C., Whyne, C., & Yee, A. J. (2007). Multimodality imaging for vertebral metastases in a rat osteolytic model. *Clinical Orthopaedics and Related Research, 454*, 230–236.

Burton, J. B., Johnson, M., Sato, M., Koh, S. B., Mulholland, D. J., Stout, D., et al. (2008). Adenovirus-mediated gene expression imaging to directly detect sentinel lymph node metastasis of prostate cancer. *Nature Medicine, 14*, 882–888.

Campbell, R. E., Tour, O., Palmer, A. E., Steinbach, P. A., Baird, G. S., Zacharias, D. A., et al. (2002). A monomeric red fluorescent protein. *Proceedings of the National Academy of Sciences of the United States of America, 99*, 7877–7882.

Cerullo, V., Diaconu, I., Kangasniemi, L., Rajecki, M., Escutenaire, S., Koski, A., et al. (2011). Immunological effects of low-dose cyclophosphamide in cancer patients treated with oncolytic adenovirus. *Molecular Therapy, 19*, 1737–1746.

Che, J., Doubrovin, M., Serganova, I., Ageyeva, L., Zanzonico, P., & Blasberg, R. (2005). hNIS-IRES-eGFP dual reporter gene imaging. *Molecular Imaging, 4*, 128–136.

Contag, C. H., Jenkins, D., Contag, P. R., & Negrin, R. S. (2000). Use of reporter genes for optical measurements of neoplastic disease in vivo. *Neoplasia, 2*, 41–52.

Cormack, B. (1998). Green fluorescent protein as a reporter of transcription and protein localization in fungi. *Current Opinion in Microbiology, 1*, 406–410.

Corneillie, T. M., Lee, K. C., Whetstone, P. A., Wong, J. P., & Meares, C. F. (2004). Irreversible engineering of the multielement-binding antibody 2D12.5 and its complementary ligands. *Bioconjugate Chemistry, 15*, 1392–1402.

Corneillie, T. M., Whetstone, P. A., Lee, K. C., Wong, J. P., & Meares, C. F. (2004). Converting weak binders into infinite binders. *Bioconjugate Chemistry, 15*, 1389–1391.

Dadiani, M., Kalchenko, V., Yosepovich, A., Margalit, R., Hassid, Y., Degani, H., et al. (2006). Real-time imaging of lymphogenic metastasis in orthotopic human breast cancer. *Cancer Research, 66*, 8037–8041.

Davydova, J., Gavrikova, T., Brown, E. J., Luo, X., Curiel, D. T., Vickers, S. M., et al. (2010). In vivo bioimaging tracks conditionally replicative adenoviral replication and provides an early indication of viral antitumor efficacy. *Cancer Science, 101*, 474–481.

DeMatteo, R. P., Yeh, H., Friscia, M., Caparrelli, D., Burke, C., Desai, N., et al. (1999). Cellular immunity delimits adenoviral gene therapy strategies for the treatment of neoplastic diseases. *Annals of Surgical Oncology, 6*, 88–94.

Deroose, C. M., De, A., Loening, A. M., Chow, P. L., Ray, P., Chatziioannou, A. F., et al. (2007). Multimodality imaging of tumor xenografts and metastases in mice with combined small-animal PET, small-animal CT, and bioluminescence imaging. *Journal of Nuclear Medicine, 48*, 295–303.

Dewey, R. A., Morrissey, G., Cowsill, C. M., Stone, D., Bolognani, F., Dodd, N. J., et al. (1999). Chronic brain inflammation and persistent herpes simplex virus 1 thymidine kinase expression in survivors of syngeneic glioma treated by adenovirus-mediated gene therapy: Implications for clinical trials. *Nature Medicine, 5*, 1256–1263.

Dwyer, R. M., Bergert, E. R., O'Connor, M. K., Gendler, S. J., & Morris, J. C. (2006). Sodium iodide symporter-mediated radioiodide imaging and therapy of ovarian tumor xenografts in mice. *Gene Therapy, 13*, 60–66.

Figueiredo, M. L., Gambhir, S. S., Carey, M., & Wu, L. (2010). Cell-specific imaging of reporter gene expression using a two-step transcriptional amplification strategy. New York: Cambridge University Press.

Figueiredo, M. L., Kao, C., & Wu, L. (2007). Advances in preclinical investigation of prostate cancer gene therapy. *Molecular Therapy, 15*, 1053–1064.

Gambhir, S. S., Barrio, J. R., Phelps, M. E., Iyer, M., Namavari, M., Satyamurthy, N., et al. (1999). Imaging adenoviral-directed reporter gene expression in living animals with positron emission tomography. *Proceedings of the National Academy of Sciences of the United States of America, 96*, 2333–2338.

Gambhir, S. S., Barrio, J. R., Wu, L., Iyer, M., Namavari, M., Satyamurthy, N., et al. (1998). Imaging of adenoviral-directed herpes simplex virus type 1 thymidine kinase reporter gene expression in mice with radiolabeled ganciclovir. *Journal of Nuclear Medicine, 39*, 2003–2011.

Giepmans, B. N., Adams, S. R., Ellisman, M. H., & Tsien, R. Y. (2006). The fluorescent toolbox for assessing protein location and function. *Science, 312*, 217–224.

Gil, J. S., Machado, H. B., & Herschman, H. R. (2011). A method to rapidly and accurately compare the relative efficacies of non-invasive imaging reporter genes in a mouse model and its application to luciferase reporters. *Molecular Imaging and Biology, 14(4)*, 462–471.

Gioux, S., Choi, H. S., & Frangioni, J. V. (2010). Image-guided surgery using invisible near-infrared light: Fundamentals of clinical translation. *Molecular Imaging, 9*, 237–255.

Guse, K., Dias, J. D., Bauerschmitz, G. J., Hakkarainen, T., Aavik, E., Ranki, T., et al. (2007). Luciferase imaging for evaluation of oncolytic adenovirus replication in vivo. *Gene Therapy, 14*, 902–911.

Heim, R., Prasher, D. C., & Tsien, R. Y. (1994). Wavelength mutations and posttranslational autoxidation of green fluorescent protein. *Proceedings of the National Academy of Sciences of the United States of America, 91*, 12501–12504.

Hoffman, R. M. (2009). Imaging cancer dynamics in vivo at the tumor and cellular level with fluorescent proteins. *Clinical & Experimental Metastasis, 26*, 345–355.

Hogg, R. T., Garcia, J. A., & Gerard, R. D. (2010). Adenoviral targeting of gene expression to tumors. *Cancer Gene Therapy, 17*, 375–386.

Huang, R., Zhao, Z., Ma, X., Li, S., Gong, R., & Kuang, A. (2011). Targeting of tumor radioiodine therapy by expression of the sodium iodide symporter under control of the survivin promoter. *Cancer Gene Therapy, 18*, 144–152.

Hutter, H. (2006). Fluorescent reporter methods. *Methods in Molecular Biology, 351*, 155–173.

Huyn, S. T., Burton, J. B., Sato, M., Carey, M., Gambhir, S. S., & Wu, L. (2009). A potent, imaging adenoviral vector driven by the cancer-selective mucin-1 promoter that targets breast cancer metastasis. *Clinical Cancer Research, 15*, 3126–3134.

Ilagan, R., Pottratz, J., Le, K., Zhang, L., Wong, S. G., Ayala, R., et al. (2006). Imaging mitogen-activated protein kinase function in xenograft models of prostate cancer. *Cancer Research, 66*, 10778–10785.

Ilagan, R., Zhang, L. J., Pottratz, J., Le, K., Salas, S., Iyer, M., et al. (2005). Imaging androgen receptor function during flutamide treatment in the LAPC9 xenograft model. *Molecular Cancer Therapeutics, 4*, 1662–1669.

Inoue, Y., Sheng, F., Kiryu, S., Watanabe, M., Ratanakanit, H., Izawa, K., et al. (2011). Gaussia luciferase for bioluminescence tumor monitoring in comparison with firefly luciferase. *Molecular Imaging, 10*, 377–385.

Iyer, M., Sato, M., Johnson, M., Gambhir, S. S., & Wu, L. (2005). Applications of molecular imaging in cancer gene therapy. *Current Gene Therapy, 5*, 607–618.

Jiang, Z. K., Sato, M., Wei, L. H., Kao, C., & Wu, L. (2011). Androgen-independent molecular imaging vectors to detect castration-resistant and metastatic prostate cancer. *Cancer Research, 71*, 6250–6260.

Johnson, M., Karanikolas, B. D., Priceman, S. J., Powell, R., Black, M. E., Wu, H. M., et al. (2009). Titration of variant HSV1-tk gene expression to determine the sensitivity of 18F-FHBG PET imaging in a prostate tumor. *Journal of Nuclear Medicine, 50*, 757–764.

Kanegae, Y., Terashima, M., Kondo, S., Fukuda, H., Maekawa, A., Pei, Z., et al. (2011). High-level expression by tissue/cancer-specific promoter with strict specificity using a single-adenoviral vector. *Nucleic Acids Research, 39*, e7.

Kang, J. H., & Chung, J. K. (2008). Molecular-genetic imaging based on reporter gene expression. *Journal of Nuclear Medicine, 49*(Suppl. 2), 164S–179S.

Kiessling, F. (2008). Noninvasive cell tracking. *Handbook of Experimental Pharmacology, 185*, 305–321.

Kim, J. H., Lee, Y. S., Kim, H., Huang, J. H., Yoon, A. R., & Yun, C. O. (2006). Relaxin expression from tumor-targeting adenoviruses and its intratumoral spread, apoptosis induction, and efficacy. *Journal of the National Cancer Institute, 98*, 1482–1493.

Kim, S. J., Masago, A., Tamaki, Y., Akazawa, K., Tsukamoto, F., Sato, J., et al. (2011). A novel approach using telomerase-specific replication-selective adenovirus for detection of circulating tumor cells in breast cancer patients. *Breast Cancer Research and Treatment, 128*, 765–773.

Kim, E. J., Yoo, J. Y., Choi, Y. H., Ahn, K. J., Lee, J. D., Yun, C. O., et al. (2008). Imaging of viral thymidine kinase gene expression by replicating oncolytic adenovirus and prediction of therapeutic efficacy. *Yonsei Medical Journal, 49*, 811–818.

Kishimoto, H., Kojima, T., Watanabe, Y., Kagawa, S., Fujiwara, T., Uno, F., et al. (2006). In vivo imaging of lymph node metastasis with telomerase-specific replication-selective adenovirus. *Nature Medicine, 12*, 1213–1219.

Kishimoto, H., Urata, Y., Tanaka, N., Fujiwara, T., & Hoffman, R. M. (2009). Selective metastatic tumor labeling with green fluorescent protein and killing by systemic administration of telomerase-dependent adenoviruses. *Molecular Cancer Therapeutics, 8*, 3001–3008.

Klutz, K., Willhauck, M. J., Wunderlich, N., Zach, C., Anton, M., Senekowitsch-Schmidtke, R., et al. (2011). Sodium iodide symporter (NIS)-mediated radionuclide ((131)I, (188)Re) therapy of liver cancer after transcriptionally targeted intratumoral in vivo NIS gene delivery. *Human Gene Therapy, 22*, 1403–1412.

Kojima, T., Hashimoto, Y., Watanabe, Y., Kagawa, S., Uno, F., Kuroda, S., et al. (2009). A simple biological imaging system for detecting viable human circulating tumor cells. *The Journal of Clinical Investigation, 119*, 3172–3181.

Korpal, M., Yan, J., Lu, X., Xu, S., Lerit, D. A., & Kang, Y. (2009). Imaging transforming growth factor-beta signaling dynamics and therapeutic response in breast cancer bone metastasis. *Nature Medicine, 15*, 960–966.

Koya, R. C., Mok, S., Comin-Anduix, B., Chodon, T., Radu, C. G., Nishimura, M. I., et al. (2010). Kinetic phases of distribution and tumor targeting by T cell receptor engineered lymphocytes inducing robust antitumor responses. *Proceedings of the National Academy of Sciences of the United States of America, 107*, 14286–14291.

Kyo, S., Takakura, M., Fujiwara, T., & Inoue, M. (2008). Understanding and exploiting hTERT promoter regulation for diagnosis and treatment of human cancers. *Cancer Science, 99*, 1528–1538.

Lavilla-Alonso, S., Bauer, M. M., Abo-Ramadan, U., Ristimaki, A., Halavaara, J., Desmond, R. A., et al. (2011). Macrophage metalloelastase (MME) as adjuvant for intra-tumoral injection of oncolytic adenovirus and its influence on metastases development. *Cancer Gene Therapy, 19*, 126–134.

Lee, J. T., Campbell, D. O., Satyamurthy, N., Czernin, J., & Radu, C. G. (2012). Stratification of nucleoside analog chemotherapy using 1-(2′-deoxy-2′-18F-fluoro-beta-D-arabinofuranosyl)cytosine and 1-(2′-deoxy-2′-18F-fluoro-beta-L-arabinofuranosyl)-5-methylcytosine PET. *Journal of Nuclear Medicine, 53*, 275–280.

Lee, S. J., Kim, H. S., Yu, R., Lee, K., Gardner, T. A., Jung, C., et al. (2002). Novel prostate-specific promoter derived from PSA and PSMA enhancers. *Molecular Therapy, 6,* 415–421.

Li, H. J., Everts, M., Yamamoto, M., Curiel, D. T., & Herschman, H. R. (2009). Combined transductional untargeting/retargeting and transcriptional restriction enhances adenovirus gene targeting and therapy for hepatic colorectal cancer tumors. *Cancer Research, 69,* 554–564.

Lillehammer, T., Tveito, S., Engesaeter, B. O., Fodstad, O., Maelandsmo, G. M., & Engebraaten, O. (2005). Melanoma-specific expression in first-generation adenoviral vectors in vitro and in vivo—Use of the human tyrosinase promoter with human enhancers. *Cancer Gene Therapy, 12,* 864–872.

Ma, X. J., Huang, R., & Kuang, A. R. (2009). AFP promoter enhancer increased specific expression of the human sodium iodide symporter (hNIS) for targeted radioiodine therapy of hepatocellular carcinoma. *Cancer Investigation, 27,* 673–681.

Maida, Y., Kyo, S., Sakaguchi, J., Mizumoto, Y., Hashimoto, M., Mori, N., et al. (2009). Diagnostic potential and limitation of imaging cancer cells in cytological samples using telomerase-specific replicative adenovirus. *International Journal of Oncology, 34,* 1549–1556.

Mastro, A. M., & Vogler, E. A. (2009). A three-dimensional osteogenic tissue model for the study of metastatic tumor cell interactions with bone. *Cancer Research, 69,* 4097–4100.

Meares, C. F. (2008). The chemistry of irreversible capture. *Advanced Drug Delivery Reviews, 60,* 1383–1388.

Merron, A., Peerlinck, I., Martin-Duque, P., Burnet, J., Quintanilla, M., Mather, S., et al. (2007). SPECT/CT imaging of oncolytic adenovirus propagation in tumours in vivo using the Na/I symporter as a reporter gene. *Gene Therapy, 14,* 1731–1738.

Minagawa, M., Kawamura, H., Liu, Z., Govindarajan, S., & Dennert, G. (2005). Suppression of adenoviral gene expression in the liver: Role of innate vs adaptive immunity and their cell lysis mechanisms. *Liver International, 25,* 622–632.

Mouchess, M. L., Sohara, Y., Nelson, M. D., Jr., De, C. Y. A., & Moats, R. A. (2006). Multimodal imaging analysis of tumor progression and bone resorption in a murine cancer model. *Journal of Computer Assisted Tomography, 30,* 525–534.

Niu, G., Anderson, R. D., Madsen, M. T., Graham, M. M., Oberley, L. W., & Domann, F. E. (2006). Dual-expressing adenoviral vectors encoding the sodium iodide symporter for use in noninvasive radiological imaging of therapeutic gene transfer. *Nuclear Medicine and Biology, 33,* 391–398.

Nyati, S., Schinske, K. A., Ray, D., Nyati, M. K., Ross, B. D., & Rehemtulla, A. (2011). Molecular Imaging of TGF{beta}-induced Smad2/3 phosphorylation reveals a role for receptor tyrosine kinases in modulating TGF{beta} signaling. *Clinical Cancer Research, 17,* 7424–7439.

O'Neill, K., Lyons, S. K., Gallagher, W. M., Curran, K. M., & Byrne, A. T. (2010). Bioluminescent imaging: A critical tool in pre-clinical oncology research. *The Journal of Pathology, 220,* 317–327.

Ono, H. A., Le, L. P., Davydova, J. G., Gavrikova, T., & Yamamoto, M. (2005). Noninvasive visualization of adenovirus replication with a fluorescent reporter in the E3 region. *Cancer Research, 65,* 10154–10158.

Penet, M. F., Pathak, A. P., Raman, V., Ballesteros, P., Artemov, D., & Bhujwalla, Z. M. (2009). Noninvasive multiparametric imaging of metastasis-permissive microenvironments in a human prostate cancer xenograft. *Cancer Research, 69,* 8822–8829.

Pesonen, S., Diaconu, I., Kangasniemi, L., Ranki, T., Kanerva, A., Pesonen, S. K., et al. (2012). Oncolytic immunotherapy of advanced solid tumors with a CD40L expressing replicating adenovirus: Assessment of safety and immunological responses in patients. *Cancer Research, 72,* 1621–1631.

Pichler, A., Prior, J. L., Luker, G. D., & Piwnica-Worms, D. (2008). Generation of a highly inducible Gal4→Fluc universal reporter mouse for in vivo bioluminescence imaging. *Proceedings of the National Academy of Sciences of the United States of America*, *105*, 15932–15937.

Pinto, F., Totaro, A., Palermo, G., Calarco, A., Sacco, E., D'Addessi, A., et al. (2012). Imaging in prostate cancer staging: Present role and future perspectives. *Urologia Internationalis*, *88*, 125–136.

Ponomarev, V., Doubrovin, M., Shavrin, A., Serganova, I., Beresten, T., Ageyeva, L., et al. (2007). A human-derived reporter gene for noninvasive imaging in humans: Mitochondrial thymidine kinase type 2. *Journal of Nuclear Medicine*, *48*, 819–826.

Pouliot, F., Karanikolas, B. D., Johnson, M., Sato, M., Priceman, S. J., Stout, D., et al. (2011). In vivo imaging of intraprostatic-specific gene transcription by PET. *Journal of Nuclear Medicine*, *52*, 784–791.

Rajecki, M., Sarparanta, M., Hakkarainen, T., Tenhunen, M., Diaconu, I., Kuhmonen, V., et al. (2012). SPECT/CT imaging of hNIS-expression after intravenous delivery of an oncolytic adenovirus and I. *PLoS One*, 7, e32871.

Ray, P. (2011). Multimodality molecular imaging of disease progression in living subjects. *Journal of Biosciences*, *36*, 499–504.

Ray, P., Lewin, S. A., Mihalko, L. A., Schmidt, B. T., Luker, K. E., & Luker, G. D. (2011). Noninvasive imaging reveals inhibition of ovarian cancer by targeting CXCL12-CXCR4. *Neoplasia*, *13*, 1152–1161.

Ray, S., Paulmurugan, R., Patel, M. R., Ahn, B. C., Wu, L., Carey, M., et al. (2008). Noninvasive imaging of therapeutic gene expression using a bidirectional transcriptional amplification strategy. *Molecular Therapy*, *16*, 1848–1856.

Ray, P., Tsien, R., & Gambhir, S. S. (2007). Construction and validation of improved triple fusion reporter gene vectors for molecular imaging of living subjects. *Cancer Research*, *67*, 3085–3093.

Rioja, J., Rodriguez-Fraile, M., Lima-Favaretto, R., Rincon-Mayans, A., Penuelas-Sanchez, I., Zudaire-Bergera, J. J., et al. (2010). Role of positron emission tomography in urological oncology. *BJU International*, *106*, 1578–1593.

Sarkar, D., Lebedeva, I. V., Su, Z. Z., Park, E. S., Chatman, L., Vozhilla, N., et al. (2007). Eradication of therapy-resistant human prostate tumors using a cancer terminator virus. *Cancer Research*, *67*, 5434–5442.

Sarkar, D., Su, Z. Z., Vozhilla, N., Park, E. S., Gupta, P., & Fisher, P. B. (2005). Dual cancer-specific targeting strategy cures primary and distant breast carcinomas in nude mice. *Proceedings of the National Academy of Sciences of the United States of America*, *102*, 14034–14039.

Sarkar, D., Su, Z. Z., Vozhilla, N., Park, E. S., Randolph, A., Valerie, K., et al. (2005). Targeted virus replication plus immunotherapy eradicates primary and distant pancreatic tumors in nude mice. *Cancer Research*, *65*, 9056–9063.

Sato, M., Johnson, M., Zhang, L., Gambhir, S. S., Carey, M., & Wu, L. (2005). Functionality of androgen receptor-based gene expression imaging in hormone refractory prostate cancer. *Clinical Cancer Research*, *11*, 3743–3749.

Sato, M., Johnson, M., Zhang, L., Zhang, B., Le, K., Gambhir, S. S., et al. (2003). Optimization of adenoviral vectors to direct highly amplified prostate-specific expression for imaging and gene therapy. *Molecular Therapy*, *8*, 726–737.

Sato, A., Klaunberg, B., & Tolwani, R. (2004). In vivo bioluminescence imaging. *Comparative Medicine*, *54*, 631–634.

Sato, Y., Tanaka, K., Lee, G., Kanegae, Y., Sakai, Y., Kaneko, S., et al. (1998). Enhanced and specific gene expression via tissue-specific production of Cre recombinase using adenovirus vector. *Biochemical and Biophysical Research Communications*, *244*, 455–462.

Seethammagari, M. R., Xie, X., Greenberg, N. M., & Spencer, D. M. (2006). EZC-prostate models offer high sensitivity and specificity for noninvasive imaging of prostate cancer progression and androgen receptor action. *Cancer Research, 66*, 6199–6209.

Serganova, I., & Blasberg, R. (2005). Reporter gene imaging: Potential impact on therapy. *Nuclear Medicine and Biology, 32*, 763–780.

Serganova, I., Ponomarev, V., & Blasberg, R. (2007). Human reporter genes: Potential use in clinical studies. *Nuclear Medicine and Biology, 34*, 791–807.

Shu, C. J., Campbell, D. O., Lee, J. T., Tran, A. Q., Wengrod, J. C., Witte, O. N., et al. (2010). Novel PET probes specific for deoxycytidine kinase. *Journal of Nuclear Medicine, 51*, 1092–1098.

Shu, X., Royant, A., Lin, M. Z., Aguilera, T. A., Lev-Ram, V., Steinbach, P. A., et al. (2009). Mammalian expression of infrared fluorescent proteins engineered from a bacterial phytochrome. *Science, 324*, 804–807.

Singh, S. P., Yang, D., Ravoori, M., Han, L., & Kundra, V. (2009). In vivo functional and anatomic imaging for assessment of in vivo gene transfer. *Radiology, 252*, 763–771.

Spitzweg, C., Baker, C. H., Bergert, E. R., O'Connor, M. K., & Morris, J. C. (2007). Image-guided radioiodide therapy of medullary thyroid cancer after carcinoembryonic antigen promoter-targeted sodium iodide symporter gene expression. *Human Gene Therapy, 18*, 916–924.

Steinwaerder, D. S., Carlson, C. A., Otto, D. L., Li, Z. Y., Ni, S., & Lieber, A. (2001). Tumor-specific gene expression in hepatic metastases by a replication-activated adenovirus vector. *Nature Medicine, 7*, 240–243.

Stepanenko, O. V., Shcherbakova, D. M., Kuznetsova, I. M., Turoverov, K. K., & Verkhusha, V. V. (2011). Modern fluorescent proteins: From chromophore formation to novel intracellular applications. *BioTechniques, 51*, 313–314, 316, 318 passim.

Su, Z. Z., Sarkar, D., Emdad, L., Duigou, G. J., Young, C. S., Ware, J., et al. (2005). Targeting gene expression selectively in cancer cells by using the progression-elevated gene-3 promoter. *Proceedings of the National Academy of Sciences of the United States of America, 102*, 1059–1064.

Szymczak, A. L., Workman, C. J., Wang, Y., Vignali, K. M., Dilioglou, S., Vanin, E. F., et al. (2004). Correction of multi-gene deficiency in vivo using a single 'self-cleaving' 2A peptide-based retroviral vector. *Nature Biotechnology, 22*, 589–594.

Tang, Y., Wu, H., Ugai, H., Matthews, Q. L., & Curiel, D. T. (2009). Derivation of a triple mosaic adenovirus for cancer gene therapy. *PLoS One, 4*, e8526.

Tannous, B. A., Kim, D. E., Fernandez, J. L., Weissleder, R., & Breakefield, X. O. (2005). Codon-optimized Gaussia luciferase cDNA for mammalian gene expression in culture and in vivo. *Molecular Therapy, 11*, 435–443.

Timpson, P., McGhee, E. J., & Anderson, K. I. (2011). Imaging molecular dynamics in vivo—From cell biology to animal models. *Journal of Cell Science, 124*, 2877–2890.

Tjuvajev, J. G., Doubrovin, M., Akhurst, T., Cai, S., Balatoni, J., Alauddin, M. M., et al. (2002). Comparison of radiolabeled nucleoside probes (FIAU, FHBG, and FHPG) for PET imaging of HSV1-tk gene expression. *Journal of Nuclear Medicine, 43*, 1072–1083.

Tjuvajev, J. G., Finn, R., Watanabe, K., Joshi, R., Oku, T., Kennedy, J., et al. (1996). Noninvasive imaging of herpes virus thymidine kinase gene transfer and expression: A potential method for monitoring clinical gene therapy. *Cancer Research, 56*, 4087–4095.

Tjuvajev, J. G., Stockhammer, G., Desai, R., Uehara, H., Watanabe, K., Gansbacher, B., et al. (1995). Imaging the expression of transfected genes in vivo. *Cancer Research, 55*, 6126–6132.

Tran Cao, H. S., Kaushal, S., Lee, C., Snyder, C. S., Thompson, K. J., Horgan, S., et al. (2011). Fluorescence laparoscopy imaging of pancreatic tumor progression in an orthotopic mouse model. *Surgical Endoscopy, 25*, 48–54.

Trujillo, M. A., Oneal, M. J., Davydova, J., Bergert, E., Yamamoto, M., & Morris, J. C., 3rd (2009). Construction of an MUC-1 promoter driven, conditionally replicating adenovirus that expresses the sodium iodide symporter for gene therapy of breast cancer. *Breast Cancer Research*, *11*, R53.

Tsai, H. J., Kobayashi, S., Izawa, K., Ishida, T., Watanabe, T., Umezawa, K., et al. (2011). Bioimaging analysis of nuclear factor-kappaB activity in Philadelphia chromosome-positive acute lymphoblastic leukemia cells reveals its synergistic upregulation by tumor necrosis factor-alpha-stimulated changes to the microenvironment. *Cancer Science*, *102*, 2014–2021.

Tsai, T. H., Lin, C. Y., Tsai, H. J., Chen, S. Y., Tai, S. P., Lin, K. H., et al. (2006). Biomolecular imaging based on far-red fluorescent protein with a high two-photon excitation action cross section. *Optics Letters*, *31*, 930–932.

Tsien, R. (2009). Constructing and exploiting the fluorescent protein paintbox (Nobel Lecture). *Angewandte Chemie*, *48*, 215, International Ed. in English.

Ward, W. W., & Cormier, M. J. (1979). An energy transfer protein in coelenterate bioluminescence. Characterization of the Renilla green-fluorescent protein. *The Journal of Biological Chemistry*, *254*, 781–788.

Watanabe, M., Ueki, H., Ochiai, K., Huang, P., Kobayashi, Y., Nasu, Y., et al. (2011). Advanced two-step transcriptional amplification as a novel method for cancer-specific gene expression and imaging. *Oncology Reports*, *26*, 769–775.

Wei, L. H., Olafsen, T., Radu, C., Hildebrandt, I. J., McCoy, M. R., Phelps, M. E., et al. (2008). Engineered antibody fragments with infinite affinity as reporter genes for PET imaging. *Journal of Nuclear Medicine*, *49*, 1828–1835.

Wetterwald, A., van der Pluijm, G., Que, I., Sijmons, B., Buijs, J., Karperien, M., et al. (2002). Optical imaging of cancer metastasis to bone marrow: A mouse model of minimal residual disease. *The American Journal of Pathology*, *160*, 1143–1153.

Wu, L., Matherly, J., Smallwood, A., Adams, J. Y., Billick, E., Belldegrun, A., et al. (2001). Chimeric PSA enhancers exhibit augmented activity in prostate cancer gene therapy vectors. *Gene Therapy*, *8*, 1416–1426.

Yaghoubi, S., Barrio, J. R., Dahlbom, M., Iyer, M., Namavari, M., Satyamurthy, N., et al. (2001). Human pharmacokinetic and dosimetry studies of [(18)F]FHBG: A reporter probe for imaging herpes simplex virus type-1 thymidine kinase reporter gene expression. *Journal of Nuclear Medicine*, *42*, 1225–1234.

Yamamoto, N., Tsuchiya, H., & Hoffman, R. M. (2011). Tumor imaging with multicolor fluorescent protein expression. *International Journal of Clinical Oncology*, *16*, 84–91.

Yates, C., Shepard, C. R., Papworth, G., Dash, A., Beer Stolz, D., Tannenbaum, S., et al. (2007). Novel three-dimensional organotypic liver bioreactor to directly visualize early events in metastatic progression. *Advances in Cancer Research*, *97*, 225–246.

Yoshihara, C., Hamada, K., & Koyama, Y. (2010). Preparation of a novel adenovirus formulation with artificial envelope of multilayer polymer-coatings: Therapeutic effect on metastatic ovarian cancer. *Oncology Reports*, *23*, 733–738.

Zhang, L., Johnson, M., Le, K. H., Sato, M., Ilagan, R., Iyer, M., et al. (2003). Interrogating androgen receptor function in recurrent prostate cancer. *Cancer Research*, *63*, 4552–4560.

Zhang, J., Thomas, T. Z., Kasper, S., & Matusik, R. J. (2000). A small composite probasin promoter confers high levels of prostate-specific gene expression through regulation by androgens and glucocorticoids in vitro and in vivo. *Endocrinology*, *141*, 4698–4710.

Zirger, J. M., Liu, C., Barcia, C., Castro, M. G., & Lowenstein, P. R. (2006). Immune regulation of transgene expression in the brain: B cells regulate an early phase of elimination of transgene expression from adenoviral vectors. *Viral Immunology*, *19*, 508–517.

CHAPTER SIX

# Adenovirus-Based Immunotherapy of Cancer: Promises to Keep

**Tanja D. de Gruijl**[*,1], **Rieneke van de Ven**[*,†]
[*]Department of Medical Oncology, VU University Medical Center/Cancer Center Amsterdam, Amsterdam, The Netherlands
[†]Laboratory of Molecular and Tumor Immunology, Robert W. Franz Cancer Research Center at the Earle A. Chiles Research Institute, Providence Cancer Center, Portland, Oregon, USA
[1]Corresponding author: e-mail address: td.degruijl@vumc.nl

## Contents



*Advances in Cancer Research*, Volume 115
ISSN 0065-230X
http://dx.doi.org/10.1016/B978-0-12-398342-8.00006-9

## Abstract

Progress in vector design and an increased knowledge of mechanisms underlying tumor-induced immune suppression have led to a new and promising generation of Adenovirus (Ad)-based immunotherapies, which are discussed in this review. As vaccine vehicles Ad vectors (AdVs) have been clinically evaluated and proven safe, but a major limitation of the commonly used Ad5 serotype is neutralization by preexistent or rapidly induced immune responses. Genetic modifications in the Ad capsid can reduce intrinsic immunogenicity and facilitate escape from antibody-mediated neutralization. Further modification of the Ad hexon and fiber allows for liver and scavenger detargeting and selective targeting of, for example, dendritic cells. These next-generation Ad vaccines with enhanced efficacy are now becoming available for testing as tumor vaccines. In addition, AdVs encoding immune-modulating products may be used to convert the tumor microenvironment from immune-suppressive and proinvasive to proinflammatory, thus facilitating cell-mediated effector functions that can keep tumor growth and invasion in check. Oncolytic AdVs, that selectively replicate in tumor cells and induce an immunogenic form of cell death, can also be armed with immune-activating transgenes to amplify primed anti-tumor immune responses. These novel immunotherapy strategies, employing highly efficacious AdVs in optimized configurations, show great promise and warrant clinical exploration.

## 1. INTRODUCTION: PROMISES TO KEEP

The immune system can be used as a powerful tool in the treatment of cancer. While surgery, chemotherapy, and radiation therapy remain the first lines of treatment for most tumors, immunotherapy regimens are gaining ground. The recent FDA approvals of Provenge (Brower, 2010; DeFrancesco, 2010) and Yervoy (ipilimumab/anti-CTLA-4) (Hodi et al., 2010; Ledford, 2011) for the treatment of advanced prostate cancer and melanoma, respectively, have set the stage for the approval of other immune-based therapies in the near future as well as for combination therapies with conventional treatment regimens. Nowadays, it is widely appreciated that tumors have developed all sorts of ways to evade immune surveillance, that is, the control of tumor development by the adaptive immune system through recognition of altered autoantigens in tumor cells (Burnet, 1957, 1970; Kaplan et al., 1998; Smyth et al., 2000). It is now recognized that the immune system can act as a double-edged sword in that a variety of immune cells and immune-derived factors can actually promote tumor growth by inducing selective survival of less immunogenic cells and establishment of a tumor-permissive environment. This dual role of the immune system is referred to as immuno-editing (Dunn, Bruce, Ikeda, Old, & Schreiber, 2002; Vesely, Kershaw, Schreiber, & Smyth, 2011). Finding ways to promote antitumor immunity, while counteracting tumor-induced immune suppression, has become an important focus in the search for new effective anticancer immunotherapies. Ideally, vaccine strategies should evoke specific cell-mediated and humoral immune responses against tumor-associated antigens (TAAs) by targeting and activating potent antigen-presenting cells (APCs), while avoiding off-target effects such as tolerance induction or promotion of severe autoimmunity. Adenovirus (Ad)-based approaches may be of use in this regard.

Ads are powerful *in vivo* and *ex vivo* transduction tools for a wide variety of human cell types and as such are very useful for various strategies aimed at the immune modulation of tumors, their microenvironment, and their host. Their high cloning capacity and transduction efficiency have made Ad vectors (AdVs) preferred preclinical research tools as well as clinical vehicles for therapeutic gene transfer. In this review, we give an overview of the variety of approaches pursued over time to utilize and optimize AdVs for cancer immunotherapy. AdVs have long been regarded as promising vaccine vehicles and have been tried and tested in a large number of clinical trials (approximating 400; Seregin & Amalfitano, 2010). Clinical exploration of the use of AdVs as vaccine vehicles (most notably for HIV) has uncovered complications, ascribed either to (pre)existent serological immunity to the employed Ad2 and Ad5 serotypes, both belonging to the C-type Ad species or to induced Ad-specific cellular immunity (Buchbinder et al., 2008; Rosenberg et al., 1998; Seregin & Amalfitano, 2010). This has led to a general perception of Ad vaccines as clinically failed. However, as outlined in this review, newly developed AdV configurations bypass these problems and offer more powerful vaccination opportunities. Within the foreseeable future, it should become clear whether these next-generation Ad vaccines will finally deliver on their long-held promise of clinical efficacy against cancer.

## 2. AdVs AS VACCINE VEHICLES

Effective cancer vaccines consist of at least two components: (1) a tumor-specific antigen or antigenic sequence and (2) an immune-activating adjuvant. APCs, of which dendritic cells (DCs) are the most powerful, play a pivotal role in generating and orchestrating the antitumor immune response. They should ingest and process the antigen of interest for subsequent presentation to tumor-specific effector T cells. In order to achieve this,

APCs must first be sufficiently activated, for example, by cytokines or pathogen-associated or danger-associated molecular patterns (so-called PAMPs and DAMPs, respectively). Upon activation, APCs migrate to the draining lymph nodes (LNs) and simultaneously upregulate all the necessary co-stimulatory molecules and cytokines to prime both helper (Th) and cytotoxic T lymphocytes (CTLs) in the paracortical areas of the LN. In this way, DCs provide the link between immediate, innate immunity, and long-term, adaptive immunity. Classically, tumor vaccines have consisted of proteins or peptides derived from TAAs. In contrast to the use of proteins or peptides, genetic TAA vaccines provide a long-lived and continuous source of antigen, facilitating durable presentation of TAA-derived epitopes to both CTL and Th cells. Their natural ability to efficiently infect target cells and hijack their transcriptional and translational machinery for production of viral antigens makes viral vectors very attractive vehicles for genetic vaccines (Bråve, Ljungberg, Wahren, & Liu, 2007). Also, viral vectors trigger innate immune responses (e.g., by activating infected APCs) that may prove useful for generating cell-mediated anticancer immunity (Harrop, John, & Carroll, 2006; Sorensen, Holst, Pircher, Christensen, & Thomsen, 2009). In this way, cleverly selected (or designed) viral vectors may act both as antigen vehicles and as vaccine adjuvant. Some viruses (e.g., lentiviruses) have even developed a natural tropism for DCs, making them ideal vehicles for DC-targeted vaccination strategies (Pincha, Sundarasetty, & Stripecke, 2010). Viruses have developed mechanisms to efficiently transfer their genetic cargo to the host cytoplasm and nucleus in order to take over the host replication and transcription machinery and thus ensure high-level expression of the transgenes they carry. Finally, viral transduction results in durable endogenous antigen expression which facilitates antigen processing for MHC class I presentation and continuous presentation of antigen to specific CTLs, the main effectors of anticancer immunity, over prolonged periods of time (Oosterhoff, Sluijter, Hangalapura, & de Gruijl, 2012; Sloan et al., 2002). In contrast, nonviral systems, for example, liposome- or nanoparticle-based vaccines, often need to be specifically tailored to achieve the above listed natural traits of viral vectors, which are advantageous to genetic antitumor vaccination (Oosterhoff et al., 2012).

All of the above listed advantages of viral vaccines also apply to AdVs. AdVs are icosahedral nonenveloped DNA viruses with a 90- to 100-nm diameter (Khare, Chen, Weaver, & Barry, 2011). Ad infection is very common among humans, but usually only manifests with mild, mostly respiratory symptoms. As outlined in the next section, AdVs can easily be made replication defective, further increasing their safety for *in vivo* use, although their safety track record is such that even replication competent AdVs have been used clinically for vaccination (Robert-Guroff, 2007). Ad genomes can carry large transgenes and are very stable. Upon infection, Ad genomes are not integrated into the host genome but persist episomally, thus minimizing the risk of insertional mutagenesis (Khare, Chen, et al., 2011). AdVs can efficiently transduce both dividing and nondividing cells, and in the latter, their genomes can persist for years (Khare, Chen, et al., 2011). In direct comparisons to DNA- or vaccinia virus-based vaccines, AdVs have shown an exceptional ability to generate cellular immunity against SIV, providing the most powerful protection against infection in nonhuman primates (Shiver et al., 2002). This superior ability to induce robust immune responses was echoed by data obtained in humans (Bett et al., 2010). In addition, AdVs have been used successfully to generate antitumor T cell responses, both *in vitro* and *in vivo*, and vaccination with AdV has been shown to lead to tumor rejection in various mouse models (Lotem et al., 2006; Steitz et al., 2000; Tuettenberg et al., 2003; Zhai et al., 1996). Finally, AdVs can be grown to high titers *in vitro* and can now be dried and thermo-stabilized to such an extent that they can be kept stable for months, even at tropical temperatures (Alcock et al., 2010). All these features have made AdVs preferred viral vaccine vehicles and have made them the most widely explored viral vectors for cancer immunotherapy, despite misgivings about preexistent or induced neutralizing immunity that might limit their clinical efficacy in the long run. Increased knowledge of the structure of AdVs and how this relates to preexistent or induced immune reactivity has led to new developments in vectorology whereby AdVs are engineered to avoid immune-mediated neutralization while optimizing their transduction efficiency and cloning capacity.

## 2.1. Engineering therapeutic AdVs: Structure–function relationships

The Ad structure has been exquisitely shaped by its functional infectivity, as comprehensively discussed in a recent review by Smith, Wiethoff, Stewart, and Nemerow (2010). Ads consist of an icosahedral capsid with protruding trimeric fibers at vertices. The fiber trimer comprises an N-terminal tail, a central shaft, and a globular knob. The Ad fiber knob domain mediates binding to the primary docking receptor on the target cell surface of Ad-susceptible cells. Subsequent interaction of the Ad penton base (i.e., where the protruding fibers connect to the viral capsid) with integrins on the cell surface facilitates internalization of the virion and also involves the induction of signaling cascades that results in reorganization of the intracellular matrix. After receptor-mediated endocytosis, the virion and its genomic contents are translocated to the cytoplasm and transported to the nuclear pores. There, the viral genome is released into the nucleus where viral transcription ensues. Transport of Ad from endosomes to the nucleus involves both the Ad penton and hexon proteins (the latter being the most abundant Ad capsid protein) and is accompanied by a gradual "uncoating" of the virus. All viral components—both structural and transcriptional—may be altered or ablated to optimize the *in vivo* transduction efficiency and selective tropism of AdV for (immuno)therapeutic applications. A more extended overview of Ad target proteins to consider in this regard is given in the next sections. How these considerations have led to concrete new strategies to optimize immunotherapeutic AdVs will be discussed in Section 3.

### 2.1.1 First-generation replication-defective AdVs

The majority of preclinical and clinical studies have been undertaken with first-generation AdVs, which have deletions in the E1 and E3 gene regions (Bett, Haddara, Prevec, & Graham, 1994; Danthinne & Imperiale, 2000; Mizuguchi, Kay, & Hayakawa, 2001). These viruses are rendered replication deficient by deletion of the E1 viral genomic region, as E1 gene products are required for viral replication. Removal of the E1A and E1B gene and replacement by a foreign transgene of choice make it necessary for these viruses to be propagated in cell lines that can *trans*-complement the E1 region like 293 cells, 911, A549 or Per.C6 cells (Graham, Smiley, Russell, & Nairn, 1977; Fallaux et al., 1998, 1996; Imler et al., 1996). Additional removal of the E3 region allows for the incorporation of more foreign gene products, maximizing the AdV's cloning capacity. The maximum total length of foreign DNA that can be incorporated in E1- and E3-deleted adenoviral genomes without interfering with virus production is 8.2 kb (Bett, Prevec, & Graham, 1993). The proteins encoded by the viral E3 region counteract the host immune response, for example, by interfering with MHC-I processing and trafficking (Schowalter, Tubb, Liu, Wilson, & Kay, 1997). For cancer immunotherapy purposes, it thus makes sense to remove this region.

### 2.1.2 Second-generation and helper-dependent AdVs

First-generation AdVs have a disadvantage of often inducing a potent anti-Ad immune response against the remaining viral proteins (even in the absence of preexisting neutralizing antibodies) and inducing toxicity in the liver upon systemic injection. In second-generation AdVs, the transcription and subsequent presentation of immunogenic viral proteins have been reduced through the deletion of the E2A, E2B, and E4 regions in addition to E1 and E3 (Amalfitano et al., 1998; Gorziglia et al., 1999). These deletions make room for incorporation of even larger, or multiple, foreign genes up to a maximum of 14 kb. Importantly, reduced transcription, processing, and presentation of viral products by transduced APCs will likely result in the enhanced presentation of the transgene products; in the case of TAA, this will promote an antitumor immune response rather than an antiviral response (Ji, Bouvet, Price, Roth, & Fang, 1999). Schirmbeck, Reimann, Kochanek, and Kreppel (2008) showed that preexisting cellular responses against viral proteins could negatively influence the induction of cell-mediated immunity against transgene products, lending support to the use of second-generation AdVs over first-generation AdVs for vaccination purposes. The development of helper-dependent (HD) or so-called "gutted" or high-capacity (HC) adenoviral vectors has taken this even further. These vectors have been stripped of the majority of viral protein-encoding genes and solely contain nonencoding sequences, that is, the inverted terminal repeats and the packaging sequence at the 5′-end of the virus genome as well as potential stuffer DNA to help stabilize transgene expression or viral packaging. A total of 37 kb of foreign genetic material can be incorporated in these viruses. HD-adenoviral vectors are dependent on helper viruses for their growth and propagation (Dormond & Kamen, 2011; Dormond, Perrier, & Kamen, 2009; Jager et al., 2009). HD-AdVs exert reduced liver toxicity and cause minimized anti-AdV immunity. Interestingly, HD-Ad do still provoke innate immune responses, maintaining their endogenous danger traits: not necessarily a bad quality for a vaccine vehicle (Cerullo et al., 2007; Muruve et al., 2004).

### 2.1.3 Ad serotypes

The family of the *Adenoviridae* counts almost 60 serotypes that can infect human cells, spanning seven species (A–G) (Arnberg, 2009; Chen et al., 2011; de Jong, Osterhaus, Jones, & Harrach, 2008; Jones et al., 2007; Kaneko et al., 2011; Seto, Chodosh, Brister, & Jones, 2011; Walsh et al., 2011). Ad classification is based on specific hypervariable regions (HVR) within the hexon protein and type-specific variations in the fiber knob (Eiz & Pring-Akerblom, 1997; Madisch, Harste, Pommer, & Heim, 2005). The Ad structure has been described in detail in previous reviews (Nemerow, Pache, Reddy, & Stewart, 2009; Russell, 2009; Zhang & Bergelson, 2005) and will therefore not be extensively described here. Important for this review are the interactions that viral components like the viral fiber knob, the shaft, penton base, and hexon of various Ad serotypes have with cells of the immune system. Table 6.1 provides an overview of the natural receptors (primary and co-receptors) that have been recognized as binding receptors for the various Ad species, the viral component that was shown to be important for binding of these receptors, their described expression on cells of the human immune system, and reported binding or (*in vitro*) transduction efficiency for specific immune cells. In the next section, these viral components will be discussed further with regard to ways in which to modify viral tropism and avoid immune neutralization for cancer immunotherapy purposes.

### 2.1.4 Ad-binding receptors on immune effector cells

Prior to infection, AdVs bind to a primary docking receptor. For the majority of serotypes, including the most widely used Ad2 and Ad5, this is the cocksackie- and adenovirus receptor (CAR; see Table 6.1). The recognition of the primary docking receptor is mediated through the Ad fiber knob domain (Bergelson et al., 1997; Roelvink, Mi Lee, Einfeld, Kovesdi, & Wickham, 1999; Smith et al., 2010; Tomko, Xu, & Philipson, 1997; Xia, Henry, Gerard, & Deisenhofer, 1994). CAR is expressed on many cells and tissues in the body at varying levels. For a therapeutic vaccine, which requires the selective transduction of professional APCs, this natural tropism to CAR might not be beneficial, as the vaccine will bind and infect many bystander cells, giving rise to off-target effects and reducing the APC transduction efficiency. Expression of CAR on most immune cells is low; hence transduction by serotypes that naturally target CAR, like Ad5, is generally poor. Furthermore, there is some controversy in the literature over CAR expression on certain immune cells like dermal DCs (DDCs) and epidermal Langerhans cells (LCs) in the skin. Whereas DDCs and LCs were found to lack CAR expression in our hands (de Gruijl et al., 2002), others reported high CAR expression, particularly on LCs (Adams et al., 2009). However, in both cases, transduction efficiency of LCs (and DDCs) by the CAR-binding Ad5 did not exceed 10–15% of total cells.

Unlike species A and C–F, several species B and selected species D (Ad8, Ad19, Ad37) Ads do not bind to CAR but use CD46 as their primary receptor on target cells (Gaggar et al., 2003; Segerman et al., 2003; Wu et al., 2004). CD46 is ubiquitously expressed on virtually all cell types in humans, including immune cells (Seya, Hirano, Matsumoto, Nomura, & Ueda, 1999) and is also used for binding by other viruses like the measles

**Table 6.1** Natural tropism of adenoviral serotypes

| Ad binding receptor | Ad serotype | Ad species | Viral binding structure | Expression on immune cells | Reported transduction efficiency (*in vitro*) | References |
|---|---|---|---|---|---|---|
| CAR | Ad1, 2, 4–10, 12, 13, 15, 17–19, 20, 22–27 29–33, 36–49, 51 | A, C–F | Fiber knob | Erythrocytes<br>skin LC, skin DDC[a]<br>platelets[a] | Ad5 ± 10%<br>Ad5 ± 10% | Bergelson et al. (1997), de Gruijl, Ophorst, et al. (2006), de Gruijl, Sombroek, et al. (2006), Adams et al. (2009), Carlisle et al. (2009), Seiradake et al. (2009), Othman, Labelle, Mazzetti, Elbatarny, and Lillicrap (2007), Shimony et al. (2009), Gupalo, Buriachkovskaia, and Othman (2011),<br>Roelvink et al. (1998) |
| $\alpha_v$-Integrins | Ad1–39, 42–51 | A–E | Penton base (RGD motif) | Monocytes<br>Lymphocytes<br>Platelets | Ad5 ± 20% | Wickham, Mathias, Cheresh, and Nemerow (1993), Roelvink, Kovesdi, and Wickham (1996) |
| $\beta_2$-Integrins | Ad2 | C | Penton base | Monocytes/macrophages | Ad2 binding | Huang, Kamata, Takada, Ruggeri, and Nemerow (1996) |

*Continued*

**Table 6.1** Natural tropism of adenoviral serotypes—cont'd

| Ad binding receptor | Ad serotype | Ad species | Viral binding structure | Expression on immune cells | Reported transduction efficiency (*in vitro*) | References |
|---|---|---|---|---|---|---|
| CD46 | Ad8, 11, 14, 16, 19, 21, 34, 35, 37, 50 | B1, B2 (not Ad3, 7), D | Fiber knob | Nucleated cells | Ad35 ± 30% | Gaggar, Shayakhmetov, and Lieber (2003), Sirena et al. (2004), de Gruijl, Ophorst, et al. (2006), de Gruijl, Sombroek, et al. (2006), Segerman et al. (2003), Segerman, Lindman, Mei, Allard, and Wadell (2006) Loré et al. (2007), van de Ven et al. (2009), Adams et al. (2011) |
| | | | | CD34 + cells | Ad35 ± 70% | |
| | | | | monocytes blood DC | Ad35 ± 70% | |
| | | | | CD83 + skin DC | Ad35 ± 30% | |
| | | | | pDC, (CD4 +) T cells | Ad35 ± 0% | |
| | | | | B cells | Ad35 ± 0% | |
| Sialic acid | Ad8, 19, 37 | D | Fiber knob | Erythrocytes | | Arnberg, Edlund, Kidd, and Wadell (2000), Arnberg, Kidd, Edlund, Olfat, and Wadell (2006), Arnberg (2009), Seiradake et al. (2009) |
| Heparin sulfate proteoglycans (HSPG) | Ad1, 2, 5, 6 (Ad3, 35) | C (B) | Fiber shaft (KKTK motif) | | | Dechecchi, Tamanini, Bonizzato, and Cabrini (2000), Dechecchi et al. (2001), Cheng et al. (2007), Tuve et al. (2008) |

| | | | | | | |
|---|---|---|---|---|---|---|
| CD80/CD86 | Ad3, 7, 11, 14, 16, 21, 34, 35, 50 | B | Fiber knob | CD1a+ skin DC | Ad5/3 ± 50% | Short (2004), Short, Vasu, Holterman, Curiel, and Pereboev (2006), van de Ven et al. (2009) |
| | | | | CD1a− skin DC | Ad5/3 ± 30% | |
| | | | | B cells | Ad5/3 < 5% | |
| Desmoglein 2 | Ad3, 7, 11, 14 | B (2/3) | Fiber | Platelets Granulocytes | Ad3 ± 30% | Wang et al. (2011) |
| VCAM-1 | Ad1, 2, 5, 6 | C | Fiber knob | | | Chu, Heistad, Cybulsky, and Davidson (2001) |
| MHC-$I_{\alpha 2}$ | Ad1, 2, 5, 6 | C | Fiber knob | B lymphoblasts | | Hong, Karayan, Tournier, Curiel, and Boulanger (1997) |
| Scavenger receptor | Ad1, 2, 5, 6 | C | ? | | | Arnberg (2009), Khare, Chen, et al. (2011), Khare, Reddy, et al. (2011) |

| Indirect binding | Ad serotype | Ad species | Viral binding structure | Indirect receptor | References |
|---|---|---|---|---|---|
| Co-agulation factor IX | Ad1, 2, 5, 6, 12, 18, 31 | A, C | Hexon | HSPG | Johnson, Langdown, and Huntington (2010) |
| Co-agulation factor X | Ad1–3, 5–39 | A–D | Hexon, fiber | HSPG | Waddington et al. (2008), Johnson et al. (2010), Bradshaw et al. (2010), Corjon et al. (2011) |
| Lactoferrin | Ad1, 2, 5, 6, Ad35 | B, C | | DC-SIGN | Johansson et al. (2007), Adams et al. (2009) |
| Dipalmitoyl phosphatidylcholine | Ad1, 2, 5, 6 | C | Hexon | | Balakireva et al. (2003) |

[a]controversy in the literature: de Gruijl, Ophorst, et al. (2006) and de Gruijl, Sombroek, et al. (2006) find negligible expression CAR on skin-migrated cells from full skin explants. Adams et al. (2009) find high CAR expression on skin LC migrated from epidermal sheet. Othman et al. (2007) and Gupalo et al. (2011) described CAR expression on human platelets, whereas Shimony et al. (2009) reported no CAR expression on human platelets.

virus (Dörig, Marcil, & Richardson, 1994). Several type B Ads have been shown able to efficiently transduce immune cells like monocytes, conventional DCs (cDCs), plasmacytoid DCs (pDCs), and T- and B-lymphocytes (Adams et al., 2011; de Gruijl, Ophorst, et al., 2006; Loré et al., 2007; Segerman et al., 2006; van de Ven et al., 2009). The fiber knob of some species B viruses has also been shown to bind the co-stimulatory molecules CD80 (B7.1) and CD86 (B7.2) and to transduce human skin and LN DCs (Short, 2004; Short et al., 2006; van de Ven et al., 2009). Other natural targets that have been described for Ad fiber knobs are sialic acid for species D (Arnberg, 2009; Arnberg, Edlund, et al., 2000; Arnberg, Kidd, et al., 2000; Seiradake et al., 2009) and VCAM-1 and MHC-$1_{\alpha 2}$ for the species C viruses Ad1, Ad2, Ad5, and Ad6 (Chu et al., 2001; Hong et al., 1997). These species C viruses were also shown to bind scavenger receptors through specific hexon regions (Arnberg, 2009; Khare, Reddy, Nemerow, & Barry, 2011).

After recognition of the primary docking receptor (e.g., CAR, CD46, CD80) by the fiber knob, viral binding and entry are facilitated through the binding of the RGD motif in the penton base to $\alpha_v$-integrins (species A–E) or $\beta_2$-integrins (species C) (Huang et al., 1996; Li et al., 2001; Roelvink et al., 1996; Wickham et al., 1993). Integrin binding triggers viral internalization via clathrin-coated pits (Bai, Harfe, & Freimuth, 1993; Wang, Huang, Kapoor-Munshi, & Nemerow, 1998; Wickham et al., 1993). Integrins are expressed on various immune cells, including LCs and DDCs, monocytes, macrophages, platelets, and lymphocytes, and their presence can often explain viral binding and entry in the absence of CAR surface expression.

In addition to the direct binding of Ad components to receptors on target cells, several indirect binding strategies have been identified. For instance, interactions of the viral hexon with coagulation factor IX (Jonsson et al., 2009) and coagulation factor X (Bradshaw et al., 2010; Corjon et al., 2011; Jonsson et al., 2009; Waddington et al., 2008) facilitate binding of species A–D viruses to heparan sulfate proteoglycans (HSPG). Serotypes of both species B and C were shown to bind to cells in a lactoferrin-dependent manner (Adams et al., 2009; Johansson et al., 2007). Adams et al. (2009) showed that Ad5 can infect human DCs through interaction between lactoferrin and the C-type lectin receptor DC-SIGN and in addition showed that the presence of bovine or human lactoferrin enhanced the Ad5 transduction of primary human DDCs and LCs. Transduction of the latter occurred in a DC-SIGN-independent way.

# 3. OPTIMIZING AdVs FOR CLINICAL TRANSLATION

For optimal Ad vaccine efficacy, several aspects are of importance (Liu, 2010), including (a) the expression level of the transgene (heterologous antigen); (b) the cell type that presents the antigen (professional APCs vs. bystander non-APCs); this will determine the strength and type of immune response; (c) the innate response directed against the AdV and the impact of this response on the Ag-specific immune response; and (d) the adaptive immune response directed against the AdV, which will determine the ability to prime and boost with the same vector. Several ways in which to optimize these aspects are discussed in the next sections.

## 3.1. Innate immunogenicity of Ad: A blessing or a curse?

Systemic AdV administration elicits an innate immune response against the vector, and this is seen as a major hurdle in the clinical applicability of these vectors, but is this also a negative feature for Ad-based vaccines? For years, it has been unclear which processes lead to Ad-induced innate responses. A recent review by Thaci, Ulasov, Wainwright, and Lesniak (2011) provides an elaborate overview of the Ad-induced signaling events that lead up to cytokine and chemokine secretion and subsequent recruitment and activation of cells of the innate and adaptive immune system. A point well taken from the review by Thaci et al. is that our understanding of the innate immune responses induced by AdVs is confounded by the multiplicity of different experimental setups and clinical trial designs. Two pathways that have been a research focus are the type-I interferon (IFN)/interferon receptor and IL-β/IL-1 receptor (IL-1R) pathways (Zhu, Huang, & Yang, 2008). Several recent studies have shed light on some of the mechanisms underlying the observed IFN and IL-1β responses elicited by AdVs. Innate responses to AdVs were shown to be mediated both in a TLR-dependent and in a TLR-independent manner (Zhu, Huang, & Yang, 2007). Uptake of AdVs by APCs, that is, pDCs, cDCs, and macrophages, led to the induction of type-I IFNs. In pDCs, this IFN response was shown to depend on TLR9 activation and on MyD88, whereas in cDCs and macrophages, Ad-induced IFN responses were independent of TLR signaling and were believed to involve the sensing of Ad DNA in the cytoplasm (Zhu et al., 2007). The same group went on to show that the induction of type-I IFN was responsible for the recruitment and activation of natural killer (NK) cells and that these NK cells were involved in the elimination of AdVs by the innate immune system

(Zhu et al., 2008). NK cell activation in response to AdVs appeared to depend on the ligation of the NKG2D receptor on NK cells and AdV-related induction of NKG2D-ligands on APCs (Zhu, Huang, & Yang, 2010). Importantly, this Ad-induced NK cell activation may well be beneficial in terms of antitumor immunity (Terme, Ullrich, Delahaye, Chaput, & Zitvogel, 2008; Zamai et al., 2007). In addition to the induction of type-I IFNs, Ad5 infection of macrophages was reported to induce the release of reactive oxygen species (ROS) (McGuire, Barlan, Griffin, & Wiethoff, 2011). Ad-induced ROS production supported cytokine secretion, followed by further activation of innate immune cells. Barlan, Griffin, McGuire, and Wiethoff (2011) recently showed that human, but not mouse, macrophages released IL-1β upon Ad5-transduction and that this release was dependent on the activation of the NLRP3 inflammasome. Ad5-induced activation of the NLRP3 inflammasome was also responsible for the induction of necrotic cell death and the release of high-mobility group box 1 (HMGB1). HMBG1 release has been reported to contribute to antigen presentation by DC via binding to TLR4 and to promote antigen processing and the activation of CTLs (Kepp et al., 2009). Several of the above listed Ad-induced innate responses are likely to promote an antitumor immune response. They could lead to improved TAA processing and (cross-)presentation by DCs as well as recruitment and activation of immune effector cells. Understandably, when AdVs are exploited for gene therapy, such responses are generally not desired. However, unlike when Ad-based gene therapy is used as gene replacement therapy for genetic diseases (Seiler, Cerullo, & Lee, 2007), the aim of Ad-based vaccination is to promote immunity to the transgene product. Hence, an innate immune response to AdV, promoting transgene recognition by the adaptive immune system, can actually be a blessing for Ad-based vaccination, whereas it can be a curse for other applications.

## 3.2. Circumventing neutralizing immunity and toxicity

Whereas the innate response to AdV can be a positive feature of Ad vaccines, preexisting neutralizing antibodies against the most common Ad serotypes as well as rapid induction of neutralizing antibodies to AdV and liver toxicity upon systemic administration do represent a hurdle for successful Ad-based immunotherapy. A strategy to reduce Ad neutralization and toxicity is by covalent incorporation of synthetic polypeptide sequences in the viral capsid. One such polypeptide-linking agent is polyethylene glycol (PEG), which in various murine experimental studies has been shown to reduce neutralizing antibody responses (Mok, Palmer, Ng, & Barry, 2005; O'Riordan et al., 1999). Mok et al. (2005) showed that PEGylation could reduce *in vitro* transduction of target cells but surprisingly had no detrimental effect on *in vivo* transgene expression or distribution. PEGylation reduced the uptake of AdVs by tissue macrophages and Kupffer cells *in vivo*, indicating that innate immunity to AdVs was also affected. Similar to PEGylation, coating with the biodegradable polymer polylactic–glycolic acid (PLGA) was shown to reduce AdV neutralization by preexisting antibodies (Beer et al., 1998; Chillón, Lee, Fasbender, & Welsh, 1998). Another polymer coat, based on poly-[*N*-(2-hydroxypropyl)methacrylamide] (pHPMA), was shown to detarget AdVs and reduce host–vector interactions (Fisher et al., 2001). In addition to the detargeting from CAR-expressing cells and reducing innate immune responses to AdVs, polymer coats like PEG and pHPMA can be linked to targeting moieties like antibodies which allows for specific retargeting of these vectors (Bachtarzi, Stevenson, Šubr, Seymour, & Fisher, 2011; Bachtarzi, Stevenson, Šubr, Ulbrich, et al., 2011; Fisher et al., 2001; Ogawara et al., 2004). In terms of vaccine efficacy, it will be interesting to see whether reduction of recognition by neutralizing antibodies will outweigh the collateral ablation of innate immunogenicity of the modified AdVs.

Not only humoral responses are to blame for neutralization of AdVs or AdV-transduced cells. A conserved 15-mer (H910-924), recognized by CD4 (T helper) cells in the context of the common HLA-DP4 allele (expressed by 75% of the Caucasian population), was found to be present within the Ad hexon protein and to be conserved among various serotypes (Olive, Eisenlohr, Flomenberg, Hsu, & Flomenberg, 2002; Tang et al., 2004). The majority of individuals tested had generated both memory B and T cell responses against this epitope. Since this epitope is still present in HD-AdV, preexistent cellular and humoral immunity against it will hamper most Ad-based vaccines. MHC-I-restricted $CD8^+$ T cell responses against AdVs in healthy volunteers were found to be directed against the hexon protein, whereas the fiber and penton base seemed less immunogenic. The MHC-I-restricted epitopes recognized by the $CD8^+$ T cells were in close proximity to, and even overlapped with, the previously identified MHC class II epitope for HLA-DP (Tang et al., 2006). Hexon modifications, as discussed in Section 3.4, might be a method to circumvent the induction of these unwanted T cell responses.

## 3.3. Serotype knob-swap

As most individuals display strong neutralizing antibody responses against the most common serotypes of species C (Ad5, Ad2), many studies have been conducted to replace the Ad5 fiber domain (or just the fiber knob) with a fiber of less common human serotypes with lower seroprevalence in the general population (Barouch et al., 2011; Kahl et al., 2010; Mast et al., 2010; Seshidhar Reddy et al., 2003) or fibers of nonhuman serotypes (Bru, Salinas, & Kremer, 2010; McCoy et al., 2007; Stoff-Khalili et al., 2005; Tatsis et al., 2007, 2006). The most widely studied serotypes for altering Ad5 tropism are the species B Ad3, Ad11, and Ad35 serotypes, all with relatively low seroprevalence (Abbink et al., 2007; Brouwer et al., 2007; Holterman et al., 2004; Sakurai, Kawabata, & Mizuguchi, 2007). Through binding to alternative docking receptors (CD46, CD80/CD86; see Table 6.1), chimeric vectors containing fibers from these serotypes display enhanced transduction of certain immune cells (Adams et al., 2011; de Gruijl, Ophorst, et al., 2006; Loré et al., 2007; Segerman et al., 2006; van de Ven et al., 2009). Importantly, these knob-swaps still do not preclude neutralization of the AdVs by Ad5-specific antibodies, which mostly bind the Ad5 hexon protein (Ophorst et al., 2004; Sumida et al., 2005). For clinical purposes, it might therefore be preferable to use the actual Ad35/Ad11/Ad3 vectors rather than their Ad5 knob-swap variants (Sakurai et al., 2007).

## 3.4. Capsid modifications

Not only the fiber proteins but also hexon and penton base proteins have been exploited for reducing anti-AdV immune responses as well as for targeting purposes. Gall, Crystal, and Falck-Pedersen (1998) showed that it is possible to substitute the hexon proteins of one serotype for those of a different serotype. A hexon-chimera of Ad5 and Ad12 was shown to be resistant against cross-neutralization in Ad5-immune mice (Roy, Shirley, McClelland, & Kaleko, 1998) but was in a different study found to be relatively inefficient in promoting transgene expression upon intramuscular (i.m.) administration (Youil et al., 2002). An Ad5/Ad48 hexon-chimera maintained infectivity in mice and nonhuman primates, despite preexisting immune responses against Ad5 (Roberts et al., 2006).

For targeting purposes (see also Section 5.1), incorporation of the integrin-binding ligand RGD in an HVR of the Ad5 hexon facilitated infection of $\alpha_v$-integrin-expressing cells in a fiber knob-independent manner

(Vigne et al., 1999). A second protein that has emerged as a site for modification is the minor capsid protein pIX (Parks, 2005). Insertion of heterologous peptides at the C-terminus of pIX was shown to improve infection of CAR-negative cells (Dmitriev, Kashentseva, & Curiel, 2002). Unlike in fiber, rather large sequences can be inserted at the pIX site. de Vrij et al. (2008) showed that fusion of a scFv T cell receptor to pIX, which was able to recognize a MAGE-A1-derived peptide in the context of HLA-A1, resulted in targeting to HLA-A1$^{+}$MAGE-A1$^{+}$ tumor cells and did not affect viral production. Incorporation of a biotin-acceptor peptide sequence at the C-terminal end of pIX enabled the generation of metabolically biotinylated AdVs, providing a versatile system to bind various biotin-labeled peptides, scFv, or antibodies (Campos, Parrott, & Barry, 2004).

# 4. Ad-BASED TUMOR IMMUNOTHERAPY APPROACHES: VACCINATION AND BEYOND

AdVs have proven very flexible and multipurpose transduction tools, not just in aid of vaccination but also of other immunological approaches to the therapy of cancer. A few examples are listed and discussed in the next sections. A schematic overview of Ad-based tumor immunotherapy approaches is provided in Fig. 6.1.

## 4.1. Vaccination

Compared to other genetic vaccines, Ad-based vaccines have proven superior in inducing cellular immunity (Bett et al., 2010; Shiver et al., 2002) and also in translating in the generation of robust antitumor immunity in preclinical models (Lotem et al., 2006; Steitz et al., 2000; Tuettenberg et al., 2003; Zhai et al., 1996). *In vivo* mouse studies have shown that AdVs carrying human TAA genes with homology to murine TAA genes (e.g., the melanoma antigen Trp-2) can be directly injected and elicit protective tumor immunity (Hangalapura, Oosterhoff, Gupta, et al., 2011; Steitz et al., 2000). Comparing different routes of administration, intravenous (i.v.) and intradermal (i.d.) delivery of AdVs proved to be the most efficacious in terms of inducing antitumor immunity (Hangalapura, Oosterhoff, Gupta, et al., 2011). In the case of i.v. vaccination, this may involve antigen transfer from marginal metallophilic macrophages to DCs for cross-priming of CD8$^{+}$ T cells in the spleen (Backer et al., 2010). In contrast, i.d. vaccination efficacy appears to be independent of

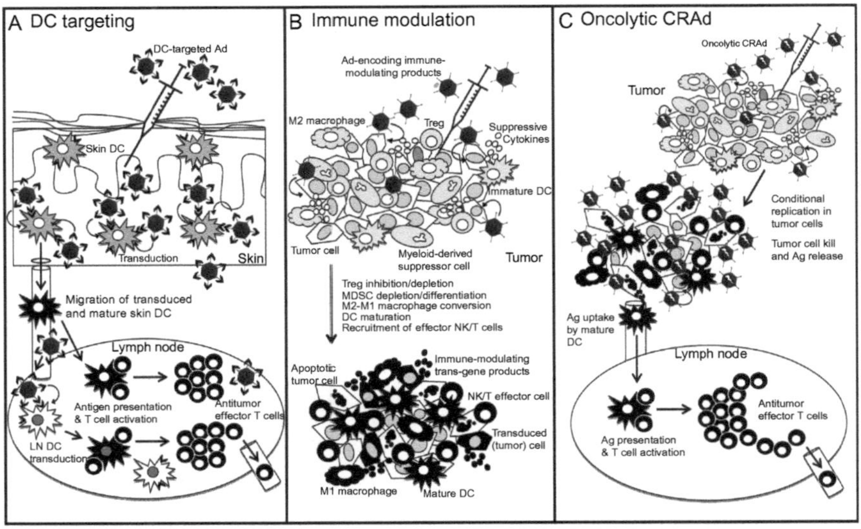


**Figure 6.1** Three approaches to Adenovirus (Ad)-based immunotherapy of cancer: (A) Intradermally injected dendritic cell (DC)-targeted Ad vaccines can transduce both skin- and lymph node-resident DCs, which upon transduction and appropriate maturation (achieved directly either through binding to an activating targeting motif or through admixed adjuvant) can prime T cells in the paracortical areas of the lymph node against the tumor antigen, carried as transgene by the DC-targeted Ad vaccine; (B) modulation of the tumor microenvironment by intratumoral (i.t.) delivery of an Ad vector (AdV) encoding (an) immune-stimulatory transgene(s) results in conversion from an immuno-suppressed state to a proinflammatory state with infiltrates of mature DCs, M1-type macrophages and effector NK/T cells. Of note, AdVs may also be tumor targeted and systemically delivered; (C) oncolytic conditionally replicating Ad (CRAd) vectors, i.t. delivered or targeted to the tumor and systemically administered, infect and selectively replicate in tumor cells, leading to their immunogenic lysis, the uptake of tumor fragments by DCs, and subsequent priming of antitumor effector T cells in the tumor-draining lymph nodes. Immune response induction can be enhanced by arming the CRAds by immune-stimulating transgenes.

macrophages but to require MyD88 signaling and cross-priming by DCs in skin-draining LNs (Hangalapura, Oosterhoff, Gupta, et al., 2011). Early studies on DNA vaccination showed that direct transgene delivery to DCs results in more potent immunization than indirect uptake of transgene-expressed antigens by DCs (Akbari et al., 1999; Porgador et al., 1998). This finding, together with the observation that AdVs encoding weakly immunogenic antigens, to which considerable tolerance is maintained (like TAA), were relatively ineffective in inducing protective antitumor immunity (Hangalapura, Oosterhoff, de Groot, et al., 2011;

Steitz et al., 2000), sparked interest in DC-based Ad vaccines. As shown by a plethora of preclinical studies conducted from the late 1990s onward and as recently reviewed by Butterfield and Vujanovic (2010), AdVs can (1) transduce DCs at high efficiency *ex vivo* when used at high multiplicities of infection and achieve unparalleled levels of transgene expression, (2) induce a certain level of DC activation, (3) maintain antigen presentation by infected DCs for at least 10 days, and (4) prime both MHC-I-restricted $CD8^+$ T cells and MHC-II-restricted Th cells upon DC transduction. Importantly, *ex vivo* Ad-transduced DC vaccines showed superior efficacy in terms of *in vivo* $CD8^+$ and $CD4^+$ T cell induction and tumor protection over directly injected Ad vaccines and in addition were superior to peptide-loaded DCs (Tuting et al., 1999; Steitz, Tormo, Schweichel, & Tuting, 2006). In addition to TAA transduction, AdVs may also be employed to increase the maturation state and T cell stimulatory potential of DCs by delivery of immune-potentiating genes, for example, coding for hCD40L or IL-12p70 (Knippertz et al., 2009; Tatsumi et al., 2007). Transduction of DCs by Ad-hCD40L led to maturation with an enhanced LN-migratory potential and prolonged IL-12 release. By co-transduction with an AdV encoding the melanoma antigen MART-1, effective CTL priming was observed *in vitro* (Knippertz et al., 2009). The latter observation points the way to employing this strategy in combination with any TAA-encoding Ad vaccine to enhance its clinical efficacy. Beside activation of the adaptive arm of the immune system, Ad-transduced DCs also have the ability to activate NK cells, the main effectors of innate cell-mediated immunity (Wargo et al., 2005). This may not only result in direct tumor cell lysis by the NK cells but also lead to further DC activation and potentiation of Th1/CTL responses through cross talk with the activated NK cells (Terme et al., 2008; Zamai et al., 2007). Finally, *ex vivo* Ad-transduced DC vaccines have the added advantage of avoiding Ad neutralization by preexistent or vaccination-induced antibodies that can otherwise compromise efficacy (Lindsay et al., 2010; Ribas et al., 1999, 1997). All these advantages have led to clinical testing of Ad-based DC vaccines in cancer patients as further discussed in Section 4.4.

In all the above-discussed studies, DCs were *ex vivo* transduced using relatively high titers of Ad5. Novel approaches circumvent inefficient Ad5 transduction of DC by redirecting Ad5 entry via alternative cell surface molecules expressed on DC. These novel DC-targeted approaches are further discussed in Section 5.1.

## 4.2. Immune modulation

As previously mentioned, tumors have developed multiple mechanisms to escape the immune system in order to invade surrounding tissues or metastasize to distant sites. Tumors can profoundly suppress immunity through the release of soluble factors that interfere with myeloid differentiation and/or activation, or that sabotage effector T cell activation (reviewed by Lindenberg, Fehres, van Cruijsen, Oosterhoff, & de Gruijl, 2011). Tumor-derived suppressive factors like interleukin (IL)-10, IL-6, vascular endothelial growth factor (VEGF), or prostaglandins can systemically block the differentiation of myeloid cells like macrophages and DCs, leading to the accumulation of immature myeloid precursors, that under the influence of the same suppressive factors can instead develop into so-called myeloid-derived suppressor cells (MDSCs). These MDSCs are potent inhibitors of T cell activation and proliferation and are found in high frequencies in tumor infiltrates (Gabrilovich & Nagaraj, 2009). As a result of this aberrant myeloid differentiation and trafficking, decreased numbers of DCs are present in the blood, tumors, and tumor-draining LNs of cancer patients as compared to healthy donors. Both pDCs and cDCs can be affected (van Cruijsen et al., 2008). Moreover, these tumor-associated DCs are minimally activated and have low levels of co-stimulatory molecules like CD80 and CD86 at their cell surface. They consequently induce T cell anergy or tolerance rather than T cell activation and promote the activation and outgrowth of regulatory T cells (Tregs) that can actively suppress conventional effector T cells (Gabrilovich, 2004). In keeping with this notion, low numbers of activated DCs in cancer patients have been associated with a poor prognosis (Gabrilovich, 2004). Clinical experience with tumor vaccines has made it clear that cancer immunotherapy can only be successful if the immune system is operating at optimal efficiency. More and more the realization is taking hold that, for cancer immunotherapy to be effective, it will have to be combined with therapies aimed at the elimination of tumor-related immune suppression (Gabrilovich & Nagaraj, 2009). AdVs can be employed to achieve local immune modulation of the tumor microenvironment in such a way that the balance is shifted from immune suppression to immune activation (Fig. 6.1B). To achieve this, AdVs may be injected intratumorally (i.t.) or retargeted to the tumor and i.v. injected. This tactic has been used, so far mostly preclinically, to enforce i.t. expression of immune-stimulatory transgenes, like cytokines (e.g., IFNα, TNFα, GM-CSF) (Hara et al., 2009, 2007; Park et al., 2003; Ye et al., 2007), chemokines that attract an immune infiltrate (e.g., CCL19, CCL21, stromal-derived factor-1α) (Okada et al., 2006; Fushimi, O'Connor, & Crystal, 2006; Yang et al., 2004), T cell stimulatory antibodies or ligands (e.g., to 4-1BB, OX40, or CTLA-4) (Andarini et al., 2004; Liu et al., 2011, 2009, 2008), or to deliver shRNAs that modulate signaling pathways involved in tumor-induced immune suppression (e.g., JAK2/STAT3) (Iwata-Kajihara et al., 2011). For most of these immune-modulating transgene products, it does not really matter what cell types in the tumor microenvironment are transduced, as long as effective i.t. concentrations are reached. However, to avoid Ad-specific immunity interfering with i.t. transduction efficiency, in some instances macrophages and DCs have been *ex vivo* transduced and i.t. injected (Satoh et al., 2003; Yang et al., 2004). In this regard, properly matured DCs carry the added advantage of favorably influencing the i.t. cytokine balance with the ability to reverse T cell tolerance and to further support the T cell effector phase (Higham, Shen, Wittrup, & Chen, 2010). Alternatively, more generally applicable approaches are to use either second-generation or HD-AdVs with diminished immunogenicity or oncolytic AdVs (Dias et al., 2011; Dormond & Kamen, 2011). In various mouse tumor models, these immune-modulating approaches have proven effective. Modulation of the tumor milieu may strengthen the effector phase of naturally occurring antitumor immune responses, of immune responses resulting from antigen release upon conventional cancer therapies like chemo- or radiotherapy, or it may make the tumors more amenable to vaccination. Indeed, depending on the immunogenicity of the employed tumor model, immune-mediated tumor rejection was achieved by immune modulation alone or in addition required tumor vaccination. Below a few examples of effective Ad-mediated immune modulation are discussed in more detail.

1. *Ad-IFNα*: Emerging evidence on the immunological effects of IFNα argues in favor of exploring its use as a local immune modulator (Glaspy, Ribas, & Chmielowski, 2009). It stimulates T and NK cells (Rizza, Capone, Moretti, Proietti, & Belardelli, 2011) and has strong DC-maturing effects, both *in vitro* and *in vivo* (Grabbe, Bruvers, & Granstein, 1992). IFNα-exposed monocytes quickly differentiate into mature DCs with the ability to boost memory T cells (Gerlini et al., 2008), while IFNα-exposed DCs are superior inducers of type-1 cell-mediated immunity (Gigante et al., 2008). IFNα also enhances the cross-priming abilities of DCs, facilitating the activation of $CD8^+$ T cells against tumor-derived antigens (Lapenta et al., 2006). IFNα treatment can also interfere with melanoma-induced immune suppression (Wang et al.,

2007), and transcriptional analyses of tumor samples have shown that type-I IFN response signatures are associated with a favorable clinical outcome (Gajewski, Fuertes, Spaapen, Zheng, & Kline, 2011). In keeping with all these observations, i.t. injection of AdVs encoding IFNα resulted in T and NK cell-mediated tumor rejection in a pancreatic hamster model (Hara et al., 2007) and effective T cell-mediated antitumor immunity in renal and colon tumor models in mice (Hara et al., 2009).

2. *Ad-IL-12*: IL-12p70, secreted by professional APCs, is a pivotal cytokine for programming T cells for an effective antitumor response characterized by a type-1 cytokine profile, a full-blown CTL effector response and long-term memory (Mescher et al., 2006). While systemic doses of IL-12 in a phase II trial led to unacceptable toxicities in patients with renal cell cancer, localized i.t. IL-12 gene therapy approaches have yielded promising results, both as monotherapy and in combination with other immunotherapies (Mazzolini, Prieto, & Melero, 2003). i.t. administration of AdVs encoding IL-12 was reported to result in protective antitumor immunity with long-term T cell memory (Gambotto et al., 1999; Gao et al., 2008), which, at least in part, might have been directed against antigens in the tumor stroma (Zhao et al., 2011). Cyclophosphamide together with an IL-12 encoding AdV reversed immune suppression in a colorectal cancer model, with reduced IL-10 and TGFβ levels as well as Treg and MDSC frequencies (Malvicini et al., 2011). Finally, macrophages, which were Ad-transduced with IL-12 and i.t. administered, induced increased T cell infiltration and inhibited tumor growth and metastasis in a prostate cancer model (Satoh et al., 2003).
3. *Ad-GM-CSF*: Granulocyte/macrophage-colony-stimulating factor (GM-CSF) is known to promote the proliferation, survival, maturation, and migration of DCs and is a widely used cytokine to boost antitumor immunity, both as a stand-alone immune modulator (Vuylsteke et al., 2004, 2006) and as a vaccine adjuvant (Clive et al., 2010; Waller, 2007). AdVs carrying the GM-CSF gene have been used to transduce tumor cells and DCs prior to delivery to tumor hosts but have also been directly injected *in vivo* (Dranoff, 2003; Lee et al., 1997). These approaches resulted in effective T cell-mediated antitumor immunity in murine models (Dranoff, 2003; Kass, Panicali, Mazzara, Schlom, & Greiner, 2001; Wang, Qi, Sun, Liang, & Ju, 2002). Combinations with other cytokines (e.g., IL-4 and FLT3L) or maturational agents

(e.g., CpG, CD40L) can further enhance its efficacy as an immune adjuvant (Oosterhoff et al., 2012; Park et al., 2003).

4. *Ad-CD40L*: Numerous preclinical studies have shown agonistic CD40 stimulation to result in favorable immune activation for tumor rejection. CD40-mediated activation of DCs results in their phenotypic and functional maturation and renders them capable of activating antitumor effector CTLs with long-term memory (Bennett et al., 1998; Melief, van Der Burg, Toes, Ossendorp, & Offringa, 2002; Ridge, Di Rosa, & Matzinger, 1998; Rieger & Kipps, 2003; Schoenberger, Toes, van der Voort, Offringa, & Melief, 1998; Toka, Gieryńska, Suvas, Schoenberger, & Rouse, 2005). CD40 stimulation results in the reversal of T cell tolerance and renders DCs resistant to the suppressive effects of (tumor-derived) IL-10 (Haenssle et al., 2008; Ichikawa, Williams, & Segal, 2002). A number of mouse studies showed that i.t. delivery of AdVs encoding CD40L or of DCs transduced by CD40L-encoding AdVs, induced potent tumor-specific CTL responses, and tumor rejection, both of primary tumors and of distant metastases (Iida et al., 2010; Kikuchi & Crystal, 1999; Kikuchi, Moore, & Crystal, 2000; Zhang, Kedl, & Xiang, 2009). Some CD40-expressing tumors are also directly sensitive to CD40L so that i.t. administration of AdVs encoding CD40L can induce tumor cell apoptosis by direct interaction with the expressed CD40L (Vardouli, Lindqvist, Vlahou, Loskog, & Eliopoulos, 2009). Similarly, Ad-transduced DCs expressing CD40L can directly bind and kill tumor cells through IFN$\gamma$ release (Tomihara et al., 2007). Of note, although systemic anti-CD40 therapy has shown clinical activity (Beatty et al., 2011; Vonderheide et al., 2007), a preclinical study suggested that local low-dose administration may be equally effective without eliciting unwanted side effects such as liver toxicity (Fransen, Sluijter, Morreau, Arens, & Melief, 2011). This too would argue in favor of local administration or targeted delivery of AdVs encoding CD40L or anti-CD40.

5. *Ad-anti-CTLA-4*: Important progress has been made in cancer immunotherapy in recent years by blocking T cell checkpoints that prevent unbridled T cell proliferation and maintain tolerance to self-antigens, such as cytotoxic T lymphocyte antigen-4 (CTLA-4) and programmed death-1 (PD-1) (Weber, 2010). Blockade of CTLA-4 and PD-1 signaling through the systemic administration of monoclonal antibodies has shown great promise in various phases I–III clinical trials, with clinical

activity observed in a variety of tumor types (Brahmer et al., 2010; Hodi et al., 2010; Small et al., 2007). However, CTLA-4 checkpoint blockade has also resulted in the breaking of tolerance to self-antigens, thereby inducing immune-related adverse events like colitis, dermatitis, hepatitis, alveolitis, and hypophysitis (Bouwhuis, Hagen, Ten, Suciu, & Eggermont, 2011; Weber, 2009). These severe autoimmune side effects may be prevented by localized administration of anti-CTLA-4 rather than the now applied systemic high-dose administration. Simmons et al. (2008) reported equivalent antitumor activity with GM-CSF- and anti-CTLA-4-secreting tumor cells with reduced systemic autoimmunity, as compared to systemic administration of high-dose anti-CTLA-4. Similarly, Tuve et al. (2007) showed that tumor-expressed anti-CTLA-4 (achieved by retroviral transduction) resulted in significantly delayed tumor outgrowth, which was mediated by $CD8^+$ effector T cells. Interestingly, i.t. delivery of anti-CTLA-4 encoding and replication-deficient Ad in breast and cervical cancer tumor models was shown to be effective only in combination with Treg depletion through anti-CD25 and cyclophosphamide treatment (Liu et al., 2011). Antitumor efficacy involved both NK and $CD8^+$ effector T cells and, importantly, was not accompanied by any autoimmune effects. Novel oncolytic AdVs that encode anti-CTLA-4 have now also been developed in an effort to further enhance their therapeutic efficacy (Dias et al., 2011).

6. *Ad-CCL21*: Effector chemokines can be used to attract an immune infiltrate to the tumor microenvironment and thus effect T cell-mediated tumor rejection (Hensbergen et al., 2005). Alternatively, sustained i.t. expression of LN-homing chemokines may be employed to attract $CCR7^+$ mature DCs and naive T cells that together form LN-like structures. Indeed, i.t. injection of DCs transduced by an AdV-encoding CCL21 resulted in the recruitment of both T cells and DCs, and in subsequent CTL-mediated tumor rejection (Kirk et al., 2001). CCL21 was shown to provide the DCs with a suitable microenvironment to generate an antitumor immune response extranodally, with ready access to TAA and differentiation and activation of effector CTL in the exact place where they need to be (Kirk, Hartigan-O'Connor, & Mule, 2001; Kirk et al., 2001). Systemic tumor protection elicited by this approach was accompanied by increased tumor infiltration by effector T cells and DCs but decreased infiltration by Tregs. In keeping with this

observation, i.t. expression of IFNγ and IL-12 was increased, whereas expression of TGFβ and prostaglandin-E2 was decreased (Yang et al., 2004). This approach is now being tested in a phase I clinical trial in patients with non-small cell lung cancer (NSCLC) and, promisingly, is showing immune enhancement (Baratelli et al., 2008; Kar et al., 2011).

**7.** *Ad-shSTAT3*: STAT3 signaling is frequently hyperactivated in cancer and simultaneously promotes tumor growth, angiogenesis, and immune suppression (Bromberg, 2002; Yu, Kortylewski, & Pardoll, 2007; Yu, Pardoll, & Jove, 2009). Most of the tumor-derived factors that induce disturbed myeloid development (e.g., IL-10, VEGF, IL-6) trigger signaling pathways in DCs and MDSCs that converge at the level of STAT3 (Yu et al., 2007, 2009). Like tumor-conditioned DCs, MDSCs from tumor-bearing mice have increased levels of phosphorylated STAT3 (Nefedova et al., 2007), as do human $CD14^{+}HLA\text{-}DR^{lo}$ MDSCs (Poschke, Mougiakakos, Hansson, Masucci, & Kiessling, 2010). Interference with STAT3 signaling led to a reduction in MDSC expansion and an increase in effector T cell responses in tumor-bearing mice (Kortylewski et al., 2005; Nefedova et al., 2005). Accordingly, inhibition of STAT3 activity abrogated the suppressive effects of tumor-conditioned medium on DC differentiation (Bharadwaj, Li, Zhang, Chen, & Yao, 2007; Nefedova et al., 2004; Wang et al., 2004). These findings clearly make STAT3 an attractive therapeutic target to overcome tumor-mediated immune suppression and improve the chances of success for immunotherapy. However, JAK2/STAT3 signaling also mediates vital physiological processes, such as early DC differentiation (Smithgall et al., 2000). To avoid unwanted side effects, tumor-localized interference in STAT3 signaling may be preferable to break immune suppression. Indeed, i.t. delivery of STAT3 knockout DCs loaded with TAA-derived peptides resulted in effective antitumor immunity, a systemic Th1 response, and delayed tumor growth (Iwata-Kajihara et al., 2011). Similarly, Ad-mediated STAT3 knockdown in human monocyte-derived DCs (MoDCs), through the expression of short hairpin (sh)RNA, resulted in increased cytokine release in response to TLR stimulation, enhanced activation of tumor-specific T cells, and increased resistance to the suppressive effects of tumor-derived factors (Iwata-Kajihara et al., 2011). Another appealing approach is to utilize oncolytic viruses carrying shSTAT3 sequences (Han et al., 2009).

8. *Ad-MyD88*: TLRs expressed on APCs or accessory cells are much-studied targets for immune potentiation. Eleven TLRs have so far been identified, each with their own class of pathogen-derived ligands: for example, TLR1, -2, and -4 bind bacterial glycoproteins and endotoxins, while TLR3, -7, -8, and -9 bind viral or bacterial RNA or DNA (Palm & Medzhitov, 2009). Most TLRs are present on the cell surface, with the exception of TLR7–9, which are located in the endosomal compartment, and TLR3, which is an intracellular receptor. TLR-Ls are instrumental in inducing proinflammatory cytokine release from APCs in support of long-term effector T cell-mediated immunity (Palm & Medzhitov, 2009). This characteristic has generated interest for their implementation in antitumor immunotherapy, alone, as vaccine adjuvants, or in combination with other immune modulators. With the exception of TLR3, all TLRs elicit a response via MyD88 signaling, eventually leading to nuclear translocation of NF-κB and transcriptional activation of a plethora of proinflammatory mediators (Palm & Medzhitov, 2009). As a viral pathogen, it stands to reason that AdVs in themselves harbor TLR-Ls and thus activate innate immunity. This may account for the intriguing observation that i.t. injection of empty replication-defective Ad5 (i.e., not encoding any transgene) in a *neu*-induced murine tumor model *per se* induced both an Ad5- and *neu*-specific immune response and inhibited tumor outgrowth (Tuve et al., 2009). In line with this intrinsic immunogenicity, AdVs can activate DCs and induce their phenotypic maturation: this effect is generally stronger in murine DCs and requires high viral particle-to-DC ratios for human DCs (de Gruijl et al., 2002; Miller, Lahrs, Pillarisetty, Shah, & DeMatteo, 2002; Tillman et al., 1999). In view of these apparent interspecies differences, differential modes of DC activation may be in place for humans and mice (Butterfield & Vujanovic, 2010). Both MyD88 and TLR-dependent and -independent mechanisms in Ad5-induced DC activation, proinflammatory cytokine secretion, and (antitumor) T cell activation have previously been reported (Appledorn et al., 2008; Hangalapura, Oosterhoff, Gupta, et al., 2011; Hartman et al., 2007; Yamaguchi et al., 2007; Zhu et al., 2007). Whereas pDCs are activated in a TLR9-dependent manner (Basner-Tschakarjan et al., 2006), cDCs can also be activated in a TLR-independent fashion (Hensley & Amalfitano, 2007; Nociari, Ocheretina, Schoggins & Falck-Pedersen, 2007). It is conceivable that in humans, Ad-mediated cDC activation may similarly depend on an as yet unknown TLR-independent cytoplasmic dsDNA-sensing mechanism

(Nociari et al., 2007). This would be in keeping with the finding that Ad5 infection of human DCs does not elicit a transcriptional program typifying MyD88 signaling (Hartman et al., 2010). To nevertheless ensure strong phenotypic and functional DC activation, an AdV-encoding MyD88 may be employed. Indeed, Ad-MyD88 infection was reported to elicit a Th1-associated transcriptional and secreted cytokine profile in both murine and human DCs. Importantly, direct i.t. injection of Ad-MyD88 *in vivo* enhanced adaptive antitumor immunity and inhibited local tumor-associated immune suppression, resulting in the inhibition of both local and distant tumor growth (Hartman et al., 2010).

## 4.3. Oncolytic AdVs

In preclinical models oncolytic, conditionally replicating AdVs (CRAds) have proven a powerful tool in the elimination of tumors, not only by their direct lytic effects but also by their triggering of a subsequent immune response, providing further systemic and potentially long-term protection against tumor outgrowth (see Fig. 6.1C; Cheong et al., 2008). A key feature of oncolytic CRAds is that they selectively replicate in, and induce lysis of, tumor tissues, without affecting normal tissues. Two main strategies are applied to achieve this (1) inactivation or deletion of Ad genes vital to replication (e.g., E1B, E3) that are compensated for by selective aberrant expression of oncogenic/mutated tumor suppressor genes (e.g., p53, RB) in cancer cells and (2) transcriptional control of Ad genes involved in replication by tumor-specific promoters (e.g., the PSA promoter for prostate cancer or progression-elevated gene-3 [PEG-3]) (Fukuhara, Homma, & Todo, 2010; Sarkar, Su, Vozhilla, Park, Gupta, et al., 2005; Toth & Wold, 2010; Yamamoto & Curiel, 2010). Alternatively, (bystander) tumor elimination may be achieved through tumor-selective expression of transcriptionally controlled tumoricidal products or through gene-directed enzyme/prodrug therapy, for example, horseradish peroxidase under the control of the hTERT (telomerase reverse transcriptase in humans) promoter, combined with indole-3-acetic acid delivery (Xiong et al., 2012; Xu et al., 2011). Such strategies may also be combined with a CRAd platform, for example, by PEG-3-mediated transcriptional control of the E1A gene combined with replication-conditional expression of the cytokine melanoma-differentiation-associated gene-7/interleukin-24 (*mda*-7/IL24), which has known tumor-suppressive and apoptosis-inducing properties; the latter CRAd configuration has been evocatively dubbed "cancer terminator virus" (Sarkar et al., 2007, 2008;

Sauane et al., 2008). Clinical testing in various phases I and II trials showed oncolytic viruses to be very safe but not very efficacious with clinical response rates rarely exceeding 15% (Toth & Wold, 2010; Yamamoto & Curiel, 2010).

At least part of the observed preclinical efficacy of oncolytic Ad-based therapies appears to lie in the amplification of their therapeutic effect by secondarily induced immune responses, which can be enhanced by local attraction and activation of DCs (Bernt, Ni, Tieu, & Lieber, 2005; Edukulla et al., 2009; Halldén et al., 2003; Lapteva, Aldrich, Rollins, et al., 2009; Lapteva, Aldrich, Weksberg, et al., 2009; Robinson et al., 2009). Indeed, human MoDCs were shown to remain viable even in the presence of high doses of oncolytic CRAds and to be able to ingest CRAd-induced oncolysates (Schierer et al., 2012, 2007). Moreover, DCs that had phagocytosed CRAd-oncolysates were readily matured by transduction with Ad5-hCD40L and able to prime tumor-specific $CD8^+$ T cells upon *in vitro* loading with MART-1 peptides (Schierer et al., 2012). Melanoma cells lysed by CRAds underwent necrosis and released HSP70 and HMBG1, signs of so-called immunogenic cell death, and indicative of their utility as an *in vivo* generated source for TAAs, readily available for cross-priming by DCs (Schierer et al., 2012; Tesniere et al., 2008). Such cross-priming of CTLs against numerous epitopes released from the tumor ensures polyclonal T cell activation, minimizing the risk of immune escape by TAA downregulation. Interestingly, oncolytic CRAds were also shown to induce autophagy (Jiang et al., 2011; Rodriguez-Rocha et al., 2011), which further facilitates cross-presentation by DCs (Li et al., 2008). These findings provide a rationale for combining oncolytic viruses with immunotherapy. Zhang et al. (2011) showed that delivery to tumors of oncolytic AdVs encoding the respective DC- and CTL-activating cytokines GM-CSF and IL-12 inhibited immune suppression in the tumor microenvironment and enhanced the efficacy of a subsequently administered DC vaccine.

The inherently immunogenic properties of tumor kill by CRAds can be further exploited by arming them with immune-modulating transgenes that can (1) recruit and/or activate DCs and promote antigen presentation or processing, for example, by CD40L (Diaconu et al., 2012; Fernandes et al., 2009; Pesonen, Diaconu, Kangasniemi, et al., 2012), GM-CSF (Choi et al., 2006; Choi, Zhang, Choi, Kim, & Yun, 2011; Koski et al., 2010; Zhang et al., 2011), FLT3-L (Edukulla et al., 2009), IFNγ

(Sarkar, Su, Vozhilla, Park, Randolph, et al., 2005), or RANTES (Lapteva, Aldrich, Weksberg, et al., 2009); (2) release the brakes on effector CTLs, for example, by anti-CTLA4 (Dias et al., 2011); (3) activate CTLs, for example, by B7.1 (Choi et al., 2006; Lee et al., 2006), IL-18 (Choi et al., 2011), or IL-12 (Choi et al., 2011; Huang et al., 2010; Lee et al., 2006; Zhang et al., 2011); or (4) sustain activation and survival of tumor-infiltrating CTLs, for example, by 4-1BBL (Huang et al., 2010). Indeed, this pairing of what is essentially an *in vivo* vaccination method (i.e., enforcing antigen release from tumors in an immunogenic manner) with immune potentiation of the tumor microenvironment (including the immune infiltrate) makes for a potentially very powerful approach to the treatment of cancer.

## 4.4. Clinical studies

Ad-based immunotherapy of cancer has been clinically tested in numerous trials over the past decade. Initial phase I clinical studies, in which patients with metastatic melanoma received Ad2 vaccines encoding MART-1 ($n=36$) or gp100 ($n=18$), showed AdV vaccination to be safe but failed to show immunological or clinical efficacy (Rosenberg et al., 1998). Specific immune responses to MART-1 or gp100 were not induced, and in only one patient, receiving an Ad-MART-1 vaccine, a complete response (CR) was observed that could be attributed to the vaccination (Rosenberg et al., 1998). The induction of neutralizing antibody responses to the Ad2 vaccine was offered as an explanation for this poor efficacy. Similarly, in a phase I trial, testing a p53 encoding Ad vaccine in six advanced-stage cancer patients, safety was demonstrated but no specific immune responses were induced (Kuball et al., 2002). Both prevailing T cell tolerance to wt p53 and neutralizing immunity to Ad5 might have been responsible for this lack of vaccine efficacy. In another phase I trial, 13 patients with NSCLC received sequential DNA and Ad vaccines coding for the lung tumor antigen L523S (Nemunaitis et al., 2006). In only one patient, increased seroreactivity to L523S was observed. It was suggested that the route of administration should be reconsidered: rather than the employed i.m. route, i.d. delivery might have been more effective. More success was obtained with Ad vaccines encoding the prostate cancer antigens PSMA and PSA. Prostate cancer patients receiving i.d. Ad-PSMA vaccines (with prior or subsequent DNA vaccination in a prime-boost setup) showed postvaccination immune reactivity by delayed-type hypersensitivity (DTH) skin reactions, that was stronger than in patients receiving DNA vaccines only (Mincheff et al., 2000). Remarkably, 86% of the vaccinated patients showed postvaccination seroreactivity to PSMA that was not detectable prior to vaccination (Todorova et al., 2005). Unfortunately, clinical efficacy could not be assessed in these patients as they simultaneously received hormone therapy. A phase I study of vaccination with AdVs encoding PSA in a group of patients with hormone-resistant prostate cancer (HRPC) demonstrated not only safety and feasibility but also vaccine-induced PSA-specific antibody responses in 34% and T cell responses in 68% of the patients (Lubaroff et al., 2009). Moreover, approximately half of the vaccinated men showed increases in PSA doubling time and experienced longer survival than predicted by their Halabi nomogram scores.

To enhance immune efficacy of Ad vaccines and to avoid antibody-mediated neutralization, autologous DCs have been generated and infected *in vitro* prior to readministration to patients. In a first clinical trial, advanced melanoma patients received DCs transduced by both Ad-MART-1 and Ad-gp100; 1 of 17 patients experienced a CR and 3 patients were reported to develop postvaccination vitiligo, a clear sign of tolerance breakthrough (Butterfield & Vujanovic, 2010; Tsao et al., 2002). In another phase I/II trial, metastatic melanoma patients received three injections of $10^6$ or $10^7$ i.d. injected DCs (Butterfield et al., 2008). A total of 14 of the 23 enrolled patients received all 3 vaccines according to schedule. Vaccination-induced $CD8^+$ and $CD4^+$ T cell responses to MART-1 were found in 6/11 and 2/4 evaluable patients, respectively. Evidence of epitope spreading was obtained in two patients. Out of the patients receiving all three vaccines, one had a suspected CR, four had durable stable disease (SD) exceeding 2 years, and one remained disease-free after becoming eligible for a surgical resection. This trial thus showed the Ad-MART-1 DC vaccine to be safe and immunogenic, with a promise of clinical efficacy (Butterfield et al., 2008; Butterfield & Vujanovic, 2010). A recent report described results from a phase I trial in small cell lung cancer patients, who received multiple vaccines of autologous DCs transduced by Ad-p53 (INGN-225) (Chiappori, Soliman, Janssen, Antonia, & Gabrilovich, 2010). The vaccine was well tolerated, and in 18 of 43 patients, increased T cell reactivity to full-length p53 was observed in an IFNγ Elispot readout. Twenty-eight percent of the treated patients showed PR or SD. Interestingly, 78.6% of patients with a positive immune response experienced a clinical response to second-line chemotherapy, compared to 33.3% with a negative immune response ($p=0.014$) (Chiappori et al., 2010). A randomized phase II trial is now planned comparing observation (i.e., standard of care) with INGN-225 or with INGN-225 combined with all-*trans*-retinoic acid (ATRA). ATRA is expected to lower MDSC levels, thereby improving vaccine efficacy. Of note, none of the DC vaccines tested in the above listed trials were exposed to additional maturational signals prior to vaccination, which might have further increased their efficacy.

AdV-based immune modulation has been tested in a number of phase I and II clinical trials in various advanced-stage tumors. Replication-defective AdVs encoding IL-12 (Mazzolini et al., 2005; Sangro et al., 2004), IL-2 (Trudel et al., 2003), IFNβ (Sterman et al., 2010), IFNγ (Dummer et al., 2010), or CD40L (Malmström et al., 2010) were delivered through i.t. injections, intrapleural dosing or bladder instillations. All AdVs were well tolerated, and all showed signs of clinical efficacy with usually heavy immune (T cell) infiltrates in the treated tumors with associated Th1 cytokine profiles. For some, induced tumor-specific antibody reactivity was observed, while for others, regression of distant metastases was reported, both evidence of the induction of systemic immunity. Since oncolytic AdVs as monotherapy generally lacked clinical efficacy, recent phase I/II trials have explored the use of oncolytic CRAds armed with immune-modulating transgenes. An oncolytic Ad5 with a genetically modified Ad3 knob, encoding GM-CSF, was delivered i.t. and i.v. in 21 patients with advanced solid tumors, combined with low-dose cyclophosphamide to reduce Treg levels (Koski et al., 2010). Biological activity of the vector was confirmed in 13 patients, and radiological clinical responses were observed in 8 of 12 evaluated patients. Both anti-Ad and antitumor immune responses were induced. Similar results were obtained for an Ad5-based oncolytic vector encoding GM-CSF and carrying the integrin-binding RGD sequence in its knob, with disease stabilization in three of six evaluable patients (which was not observed for the vector without the GM-CSF transgene) and induced anti-Ad and antitumor immune responses (Pesonen, Diaconu, Cerullo, et al., 2012). Li et al. (2009) showed that i.t. administration of an HSP70-coding oncolytic AdV resulted in clinical benefit for 48% of the treated patients ($n=27$). They also reported increased NK and T cell rates as well as transient partial remissions in uninjected tumors in three patients. Altogether these clinical studies have shown the safety of Ad-based immune modulation, both for replication-defective AdVs and for oncolytic CRAds. Promising signs of immunological and clinical efficacy warrant the clinical progression of these therapies.

In the field of Ad-based tumor vaccines, much is to be learned from clinical trials testing vaccines against infectious diseases, most notably malaria and HIV. Like cancer, these intracellular pathogens require the induction of balanced humoral and cell-mediated immunity for their effective elimination. The field experienced a setback a few years ago when a large randomized phase IIb trial (the so-called STEP trial) failed to meet preset efficacy goals and was halted (Gray, Buchbinder, & Duerr, 2010; Kresge, 2007). Unexpectedly, a higher infection rate was observed in uncircumcised men with high preexistent antibody titers to Ad5 receiving the vaccine than in men who received the placebo. One hypothesis offered in explanation for this observation was more vigorous Ad5-specific Th cell activation in preimmune men, providing an ideal target cell pool for HIV infection. Another hypothesis held that resulting immune complexes might have bound DCs, leading to their activation and thus altering their permissiveness to HIV infection. There is also the possibility of a statistical fluke, particularly as the observed difference was not significant ($p=0.08$) (Gray et al., 2010; Liu, 2010). Important clinical developments in the infectious vaccine field since then have included (1) the use of alternative Ad serotypes to circumvent preexistent immunity, for example, Ad35, Ad26, and chimpanzee serotypes such as Ad63 (Draper et al., 2010; Sheehy et al., 2011; von Bubnoff, 2010); (2) DNA-prime and Ad-boost regimens to optimize vaccine efficacy, resulting in the induction of humoral responses as well as specific $CD8^+$ and $CD4^+$ T cells, which display a central and effector memory phenotype and persist in circulation for over 6 months (Churchyard et al., 2011; De Rosa et al., 2011; Kibuuka et al., 2010; Koup et al., 2010). Despite earlier setbacks, these recent clinical successes have set viral vectors back on track as vaccine vehicles (Rollier et al., 2011).

## 5. PRECLINICAL MODELS TO TEST Ad CANCER VACCINES

### 5.1. Murine *in vivo* models

Most preclinical *in vivo* testing of Ad vaccines has been done with inbred mouse strains. As with other immunotherapies, the antitumor efficacy observed in these studies could not be replicated in subsequent clinical trials. Indeed, one of the hurdles to clinical translation of immunotherapies, recently identified in a consensus paper by experts in the field, is the limited relevance of available murine tumor models for the effects of immunotherapy in man (Fox et al., 2011). This may have several causes: (1) the use of inbred mouse strains; (2) the use of transplantable tumors which lack the complexity of naturally arising tumors; (3) the relative immunogenicity of the employed tumor models; and (4) differences between species in relevant tumor antigens, antigen processing, and MHC binding of epitopes, which may translate not only into differences in efficacy but also into differences in autoimmune side effects (Fox et al., 2011). Meaningful analysis of efficacy and toxic side effects of Ad vaccines may be further hampered by the fact that, apart from liver cells, most mouse cells do not express high levels of CAR (Fechner et al., 1999; Tomko et al., 1997), nor do they express CD46 (Seya et al., 1998; Tsujimura et al., 1998). In order to study the effects of gene transfer by CAR- or CD46-binding Ad serotypes to various cell types, human CAR transgenic mice (Tallone et al., 2001) and hCD46-Tg mice (Kemper et al., 2001) have been developed. However, anatomical barriers, which result in differences in biodistribution characteristics, may still render these preclinical *in vivo* models for type B and type C AdVs unrepresentative for the human situation (Fechner et al., 1999; Ni et al., 2005; Verhaagh et al., 2006). Indeed, Ad serotypes that *in vitro* transduce DCs at high efficacy do not necessarily prove immunogenic *in vivo* (Hsu et al., 2005). The exploration of alternative preclinical models to test the dynamics of *in vivo* transduction by Ad vaccines is therefore warranted. Beside dogs or nonhuman primates (Bett et al., 2010; Casal & Haskins, 2006; Thacker et al., 2009), which provide *in vivo* outbred models with Ad biodistribution patterns that may be more comparable to humans, there is an increasingly important role for *ex vivo* human model systems in this regard.

## 5.2. *In vitro* human model systems

Results obtained from *in vivo* studies on Ad vaccines in mice can be misleading due to interspecies differences in immune components. Most notably, the DC compartment, which is a focus of Ad vaccine targeting studies, is very different between mice and humans. Various DC subsets with specialized functions have been identified in mice (reviewed by Merad & Manz, 2009), but often their human counterparts remain unidentified or express different sets of surface markers, for example, TLRs (Bachem et al., 2010; Crozat et al., 2010; Jongbloed et al., 2010; Palucka, Ueno, Fay, & Banchereau, 2009;

Poulin et al., 2010; Villadangos & Shortman, 2010). As a result, Ad-based DC-targeting therapies that are effective in mouse models may fail in human clinical trials as the targeting moieties identified for murine DCs are not expressed on the same subset of human DCs, and the human DCs that might express these targeting molecules could have completely different biological functions. Human MoDCs and DCs derived from $CD34^+$ progenitors are good *in vitro* targets to study the DC transduction efficiency of different Ad vaccine candidates as well as their ability to induce DC activation (Arthur et al., 1997; Dietz & Vuk-Pavlović, 1998; Hangalapura et al., 2010; Rea et al., 1999; Short et al., 2006; Tillman et al., 1999; van de Ven et al., 2009). However, as these *in vitro* generated DCs are not necessarily representative of subsets found *in vivo*, these experiments will not give insight into the *in vivo* targeting efficacy of AdVs for human DC subsets. Nor can they predict how the studied Ad vaccines will behave in the complex context of human tissues. Particularly, the skin is a tissue of interest in this regard as it is populated by a dense network of DCs and vaccination through the skin has proven very effective in generating cell-mediated antitumor immunity (Oosterhoff et al., 2012). The human skin explant model provides clinically relevant information about the performance of human DC-targeted vaccines in an intact tissue context. Indeed, an ever-growing appreciation of differences in cutaneous DC subset activities between mice and man, for example, poor cross-priming by murine LCs versus efficient cross-priming and induction of $CD8^+$ T cells by human LCs (Romani, Clausen, & Stoitzner, 2010; Ueno, Palucka, & Banchereau, 2010), supports the use of the human skin explant model. Selective and/or efficient *in situ* transduction of LCs and DDCs by DC-targeted Ad vaccines was demonstrated in this organotypic model in the presence of all epidermal and dermal cellular and matrix components that would also be accessible to Ad vaccines upon their *in vivo* delivery (de Gruijl et al., 2002; de Gruijl, Ophorst, et al., 2006; Short, 2004; van de Ven et al., 2009). Moreover, this model can provide vital proof of increased DC maturation induction, functional DC migration and transgene-specific $CD8^+$ T cell activation by *in situ* transduced and subsequently migrated DCs (de Gruijl et al., 2002; de Gruijl, Ophorst, et al., 2006; de Gruijl, Sombroek, et al., 2006; van de Ven et al., 2008). All these events are vital steps toward efficient antitumor T cell priming and, ultimately, clinical efficacy. Similarly, single-cell suspensions derived from melanoma-draining LNs were utilized to demonstrate selective DC transduction in LNs, as well as DC maturation and enhanced activation and expansion of TAA-specific T cells derived from the LN suspensions (Hangalapura et al., 2010; van de Ven et al., 2009). These human *ex vivo* models represent two vital *in vivo* compartments where induction of a T cell response takes place upon i.d. delivery of a vaccine: the skin with its DCs and dense lymphatic network and the draining LNs where the migrated DCs come into contact with naive T cells and where specific T cell priming and expansion takes place. Together, these human models provide valuable information on how the targeted virus will behave in tissue microenvironments in a clinical setting. This is important since oft-used mouse models differ considerably from human skin in anatomy, cutaneous lymphatic drainage patterns, and DC subset definitions (Steinman, 2008). Anatomically, canine skin and its lymphatic drainage system more closely resemble human skin, and dogs may therefore provide a very relevant transitional animal model for preclinical validation of i.d. administered (DC-targeted) Ad vaccines. We recently showed the utility of a canine model for immunological efficacy assessment of an i.d. injected CD40-targeted Ad vaccine encoding the TAA CEA (Thacker et al., 2009). Canine tumors closely resemble human tumors in histology, molecular makeup, and biological behavior (Paoloni & Khanna, 2008), and since dogs, in contrast to immune-competent mouse strains employed in tumor models, are outbred, canine tumor models may provide a highly relevant setting in which to test the antitumor efficacy of DC-targeted vaccines such as CD40-targeted Ad5. Thus, in obtaining preclinical proof of the efficacy of Ad-based vaccines, the limitations of murine models should be recognized, specifically in regard to differences in anatomy and DC subset definition and functionality. Complementary human *ex vivo* models and transitional animal models should therefore additionally be considered for preclinical validation.

## 6. NEXT-GENERATION Ad VACCINES AGAINST CANCER: THE WAY FORWARD

### 6.1. DC targeting

As mentioned previously, *ex vivo* transduction and subsequent administration of autologous DCs have greatly enhanced the efficacy of Ad-based vaccines against cancer (Butterfield & Vujanovic, 2010). This approach holds great promise, but wide-scale implementation is hampered by its labor intensiveness and the complicated logistics and high costs involved. Importantly, the observed suboptimal migration of the administered DCs to the draining LNs may hinder vaccine efficacy. Direct *in vivo* targeting of vectors carrying the genetic code of tumor antigens to DCs in patients may thus represent a more attractive and standardized alternative (Tacken & Figdor, 2011). By careful selection of the targeting motif and the route of administration, DCs may be selectively targeted and simultaneously activated by vaccine vectors. This is particularly important for cancer vaccines, the efficacy of which is challenged by immunosuppressive conditions in tumor hosts. Examples of DC-activating receptors that may be targeted *in vivo* are CD40 (Hangalapura, Oosterhoff, de Groot, et al., 2011) and TLR4 (Aranda et al., 2011; Lasarte et al., 2007).

While some viruses display a natural DC-targeting ability, this is unfortunately not the case for AdVs. DCs are relatively resistant to the commonly used C-type AdVs Ad5 and Ad2 due to their low levels of surface CAR expression, whereas CAR is abundantly expressed on many normal cell types. A logical approach to circumvent inefficient CAR-mediated Ad5 transduction *in vivo* is by redirecting Ad5 entry (i.e., targeting) via alternative cell surface molecules abundantly expressed on DCs (Fig. 6.1A). In order to transform an AdV into a selective gene therapy vector, the interaction with its natural receptor will have to be ablated and it will have to be retargeted to a new receptor of choice (Wickham, 2000). Roughly, two different strategies have been employed to promote retargeting of Ad to DCs: (1) adaptor-mediated targeting, also known as the two-component strategy; (2) genetic targeting, also referred to as the single-component strategy (Wickham, 2000; Wu & Curiel, 2008).

### *6.1.1 Adaptor-mediated targeting*

In adaptor-mediated targeting, a bispecific adaptor molecule is complexed with an AdV of choice to simultaneously block native receptor binding and retarget the AdV to a receptor expressed on the target cell of interest, for example, a DC. The first manuscript reporting on this technique used a CAR-neutralizing Fab fragment coupled to folate, allowing retargeting of Ad5 to folate-receptor-expressing cells (Douglas et al., 1996). The CAR-blocking moiety can also be a soluble form of the ectodomain of CAR (sCAR) or a high-affinity peptide that can bind to the viral knob. The retargeting portion of the bispecific adaptor molecule can be an antibody or a fragment thereof (e.g., a Fab fragment or a single chain (sc)Fv), a receptor ligand, or a specific peptide sequence (Dmitriev, Kashentseva,

Rogers, Krasnykh, & Curiel, 2000; Dreier et al., 2011; Harvey et al., 2010; Kashentseva et al., 2002; Kim et al., 2002; Korn et al., 2004; Nettelbeck et al., 2001, 2004; Pereboev et al., 2002; van Beusechem et al., 2002).

The advantage of the two-component strategy is that it is flexible, that is, one viral vector can be targeted to various different receptors on different target cells by changing the bispecific adaptor used, or the other way around, multiple viral vectors encoding various transgenes of interest can be targeted to the same cell using the same bispecific adaptor molecule. This allows for efficient preclinical screening of multiple receptors of interest, or multiple transgenes of interest, without difficult cloning steps or problems in propagating and rescuing a genetically modified, and potentially destabilized, AdV. From a clinical perspective, the adaptor approach carries the disadvantage that two separate clinical grade components need to be developed and end-product reproducibility can be an issue due to possible variations in AdV:adaptor ratios in the resulting mixtures. Using bispecific adaptor molecules consisting of CD40-targeting moieties (scFv or the TNF-like domain of CD40L) and Ad5 fiber knob-neutralizing moieties (scFv or sCAR), a string of studies have shown the utility of this two-component approach to selectively target and transduce DCs in skin and skin-draining LNs, to induce superior T cell-mediated antitumor immunity, and to induce effective tumor protection (Brandão et al., 2003; de Gruijl et al., 2002; Hangalapura et al., 2010; Hangalapura, Oosterhoff, de Groot, et al., 2011; Kim et al., 2010; Pereboev et al., 2002). Beside targeted transduction, this approach also achieved simultaneous activation of the DCs, resulting in the upregulated expression of co-stimulatory molecules and IL-12p70 release. A recent review provides a comprehensive overview of these studies (Hangalapura et al., 2012).

### 6.1.2 Genetic targeting

The single-component approach entails genetic alteration of the Ad fiber knob and fiber proteins in order to ablate binding capacity to its natural receptor and to gain the ability to target a new receptor (Krasnykh et al., 2000). CAR-binding ablation is required for true retargeting of Ad5. Genetic mutations that affect binding of the Ad5 viral knob to CAR as well as viral liver toxicity have been reviewed by Nicklin et al. (2005) and will not be further discussed here. Research on fiber modifications for retargeting purposes has mostly focused on two domains within the Ad fiber knob protein, that is, the HI-loop and the C-terminal region (Belousova, Krendelchtchikova, Curiel, & Krasnykh, 2002; Dmitriev et al., 1998;

Krasnykh et al., 1998; Magnusson et al., 2012; Wu et al., 2002). The choice of ligand incorporated in the viral fiber is restricted to its size and its effect on (1) viral folding, (2) proper ligand folding and glycosylation, and (3) viral propagation. It was shown that for peptides incorporated in the fiber HI-loop, the maximum length could reach up to 83 amino acids (Belousova et al., 2002). Incorporation of the integrin-binding RGD motif, as well as of the pK7 peptide that targets heparin-sulfate containing surface proteins, at the site of either the HI-loop or the C-terminal fiber region, has been extensively explored for improving tumor cell targeting, since expression of these proteins is often enhanced during malignant transformation (Dehari et al., 2003; Dmitriev et al., 1998; Gonzalez et al., 1999; Koizumi et al., 2007; Wickham et al., 1997; Wu et al., 2002). RGD-mediated targeting does not only enhance transduction of various tumor cells but also enhances transduction of DCs (Harui, Roth, Sanghvi, et al., 2006; Harui, Roth, Vira, et al., 2006; Sas, Chan, Sami, El-Gayed, & Xiang, 2008; van de Ven et al., 2009; Worgall et al., 2004). However, whereas an RGD-modified AdV improved transduction of *in vitro* cultured immature LCs, no enhanced transduction of primary skin DCs was observed upon i.d. injection in the human skin explant model, suggesting that the *in vivo* DC-targeting ability of RGD-modified viruses might be limited due to high integrin expression on surrounding tissue cells (van de Ven et al., 2009). Belousova et al. (2003) constructed a chimeric AdV containing the CAR-binding ablated wild-type Ad5 fiber and a bacteriophage T4 fibritin fiber in which CD40L was incorporated. Intradermal injection of this vector in human skin explants resulted in targeted, enhanced gene transfer to migrating DCs, as well as their phenotypic maturation (Korokhov, Noureddini, et al., 2005). Finally, another way by which retargeting of AdVs to DCs may be achieved is through knob-swap between Ad serotypes. Replacement of the Ad5 fiber knob with the CD46-binding Ad35 fiber knob resulted in enhanced gene transfer to human DCs, both *in vitro* and *in situ* in a human skin explant model (de Gruijl, Ophorst, et al., 2006; Rea et al., 2001). Similarly, CD46-binding Ad11 vectors were shown to transduce DCs at high efficiency (Stone et al., 2005). We have also shown that replacement of the Ad5 fiber knob with that of Ad3 resulted in increased transduction efficiency of human skin DCs *in situ* (through targeting of CD80/CD86) and that this Ad5/3 vector was more specific for mature, T cell-stimulatory DCs than Ad5/35 (van de Ven et al., 2009).

### 6.1.3 Alternative DC retargeting strategies

While most studies use either the two-component or the single-component targeting strategies, there are some examples where the two are combined. Korokhov et al. (2003) constructed genetically engineered AdVs with the Fc-binding domain of *Staphylococcus aureus* protein-A incorporated within the fiber protein. These AdVs were subsequently combined with a bispecific fusion protein, which consisted of the Fc-domain portion of human IgG1 fused to a scFv derived from a human anti-CD40 antibody (scFv G28-5), resulting in efficient DC transduction. The same strategy was successfully employed to achieve enhanced transduction of DCs through targeting of DC-SIGN (Korokhov, de Gruijl, et al., 2005). Glasgow, Mikheeva, Krasnykh, and Curiel (2009) developed soluble leucine zipper-scFv-G28-5 molecules, which during viral propagation dimerized with leucine zipper-moieties in the HI-loop of genetically altered AdVs, yielding CD40-targeting vectors. Others reported targeting of EGFR-expressing cells by combining metabolically biotinylated AdVs with a bispecific protein consisting of epidermal growth factor (EGF) linked to streptavidin or neutravidin or by combining a genetically modified avidin-AdV with a biotin–PEG–EGF adaptor (Chen, Mok, Pflugfelder, Li, & Barry, 2006; Park, Mok, & Park, 2008; Pereboeva, Komarova, Roth, Ponnazhagan, & Curiel, 2007). These approaches could also be utilized for DC targeting. Finally, Ding et al. (2007) described the use of a reductive amination methodology to link mannan residues to the Ad capsid. Mannan-modified AdVs encoding the TAA hTERT were shown to induce enhanced T cell-mediated tumor protection in a mouse model. The authors hypothesized that this was most likely due to targeted *in vivo* transduction of DCs through binding of the mannan-AdVs to mannose receptors on their surface (Ding et al., 2007). Finally, in the presence of (bovine) lactoferrin, Ad5 and Ad2 vectors can bind DC-SIGN on the surface of cDCs (Adams et al., 2009; Günther et al., 2011), leading to enhanced transduction and presenting yet another opportunity for DC targeting.

### 6.1.4 Transcriptional targeting

Another way to ensure DC-restricted transgene expression is through transcriptional targeting. Several DC-specific promoters have been identified that can be incorporated into genetic vaccines to ensure selective *in vivo* expression of antigens and/or cytokines in DCs: CD11c (Tian, Liu, Donahue, Falo, & You, 2011; Tian et al., 2009), DC-STAMP (Dresch,

Edelmann, Marconi, & Brocker, 2008), Dectin-2 (Bonkobara et al., 2001; Lopes et al., 2008; Morita et al., 2001), and Fascin (Bros, Ross, Pautz, Reske-Kunz, & Ross, 2003; Sudowe, Ludwig-Portugall, Montermann, Ross, & Reske-Kunz, 2003). Interestingly, Dectin-2-directed transcription can effect specific transgene expression in LCs, which is of potential interest for vaccines delivered through the skin (Morita et al., 2001). By placing genes under the control of the Dectin-2 promoter, selective expression in mature DCs is achieved, a particularly attractive trait for vaccination purposes (Sudowe et al., 2003). Transcriptional targeting may be combined with physical retargeting of AdVs as a fail–safe mechanism to guarantee selective expression of antigens in DCs.

## 6.2. Incorporation of immunogenic sequences in the hexon

An interesting development for the field has been the incorporation of immunogenic peptide sequences in the adenoviral capsid, the so-called antigen capsid-incorporation strategy, reviewed by Matthews (2011). The rationale behind this strategy is that viral capsid proteins are key inducers of strong humoral responses as well as T cell responses (Molinier-Frenkel et al., 2002; Olive et al., 2002; Tang et al., 2006). As humoral immune responses against Ad capsid proteins are increased upon readministration of the AdV, humoral responses directed against the incorporated foreign capsid antigen may also be expected to become stronger upon boosting. If such a foreign capsid antigen is a TAA, this may result in enhanced antitumor antibody responses. Antigen incorporation in the hexon protein has been studied in most detail (Crompton, Toogood, Wallis, & Hay, 1994; Matthews et al., 2008; McConnell, Danthinne, & Imperiale, 2006; Wu et al., 2005). Beside the induction of potent humoral responses, these immunogenic peptides present on the viral capsid could also be available for cross-presentation by DCs (Heath & Carbone, 2001; Pooley, Heath, & Shortman, 2001). Cross-presentation is important for the priming of antigen-specific cytotoxic $CD8^+$ T cells, a crucial step in the generation of antitumor immunity (Flinsenberg, Compeer, Boelens, & Boes, 2011). Incorporation of Th epitopes will further aid optimal priming of Ag-specific $CD8^+$ T cells (Rosa, Ribeiro, & Cunha-Neto, 2010; Worgall et al., 2005). This may be of particular relevance as AdV-mediated intracellular transgene expression will favor MHC-I processing and presentation, rather than MHC-II-mediated presentation, leading to CTL rather than Th stimulation (Matthews, 2011). Using the hypervariable region 5 (HVR5) of the Ad hexon to incorporate a B cell epitope from *Bacillus anthracis* protective antigen (PA), McConnell et al. (2006) showed that this strategy induced strong Th1- and Th2-directed antibody responses, whereas co-injection of a wt AdV together with the synthetic PA did not elicit any antigen-specific antibody response, supporting the advantage of incorporating such immunogenic peptides within the viral capsid. Comparing hexon HVR2 and HVR5, it was shown that viruses could be propagated when epitopes with a length of 53 amino acids plus a 12-amino acid linker sequence were incorporated into HVR5, whereas the maximum length for HVR2 comprised 33 amino acids plus linker (Matthews et al., 2008). Incorporation of a 14-amino acid epitope of the outer membrane protein F (OprF) of the respiratory pathogen *Pseudomonas aeruginosa* in HVR5 was shown to induce humoral as well as OprF-specific $CD4^+$ and $CD8^+$ T cell responses in mice and was able to protect against a lethal challenge with *P. aeruginosa* (Worgall et al., 2005). Responses were shown to differ between mouse strains, suggesting involvement of specific MHC haplotypes, and could be strengthened through boosting with OprF-expressing AdVs. Krause et al. compared the immunogenicity of influenza HA-peptides located at the hexon, fiber, penton, or pIX and found that the most profound anti-HA response (both humoral and Th) was obtained when the HA-peptide was located in the fiber protein (Krause et al., 2006). On a final note, beside incorporation of additional epitopes in the Ad hexon, linkage of TLR agonists might also be a promising approach to increase the adjuvanticity of Ad vaccines, particularly in view of a recent report showing that co-targeting of TLR-Ls to DCs through encapsulation in DC-targeted nanoparticles increased adjuvanticity of a tumor vaccine to a larger extent than separate admixing of the same TLR-L (Tacken et al., 2011).

## 6.3. A choice of vaccine adjuvants

Adjuvants increase the immunogenicity of vaccines by direct immunostimulatory effects, by attraction of immune effector cells to the vaccination site, or by slow and gradual release of the vaccine, thus ensuring prolonged immune activation. By these immune-enhancing activities, adjuvants can exert a dose-sparing effect, lowering the effective vaccine dose. Adjuvants coadministered with Ad vaccines may boost *in vivo* immunogenicity of the delivered transgene product but preferably should not boost neutralizing immunity against the virus, as was reported for an Ad-based Venezuelan equine encephalitis virus vaccine that was co-injected with a CpG adjuvant (Perkins, Williams, O'Brien, Laws, & Phillpotts, 2008). Any such co-injected vaccine adjuvants should therefore be carefully selected and tested prior to clinical use.

One way to focus the immune response on the (tumor) antigen of interest rather than the Ad vaccine vehicle may be to incorporate immune-modulatory transgenes in the AdV so that upon target cell infection, they are chronologically and spatially coexpressed with the transgene encoding the antigen. Examples of this are Ad vaccines encoding GM-CSF, which upon expression will attract DCs to the vaccination site and activate them, thus boosting T cell-mediated immunity against the antigen (Ojima et al., 2007; Tenbusch et al., 2008). Another approach is to design AdVs that encode a fusion protein consisting of CD40L and the tumor antigen of interest. Upon *in vivo* transduction, the expressed fusion protein will locally target and activate DCs. This strategy has been shown to effectively induce protective antitumor immunity *in vivo* (Zhang et al., 2003).

Coadministration of classic adjuvants like alum-based compounds can enhance Ad vaccine efficacy (Ophorst et al., 2007), but the past decade has seen the rise of various novel immune-stimulatory agents that may also serve as vaccine adjuvants. Co-injected CD40L and TLR-L will activate DCs at the vaccination site, thus enhancing antigen-processing and specific T cell activation (Grossmann et al., 2009; Karan, Krieg, & Lubaroff, 2007). Combinations of TLR9 and TLR3 agonists or TLR4 agonists have been shown to enhance the efficacy of Ad vaccines (Grossmann et al., 2009; Rhee et al., 2010). Blockade of T cell checkpoints like CTLA-4 and PD-1 can be combined with Ad vaccines to enhance and prolong specific T cell responses (Finnefrock et al., 2009; Song, Park, Nam, Choi, & Sung, 2011; Sorensen, Holst, Steffensen, Christensen, & Thomsen, 2010), while 4-1BB and OX40 agonists can have similar effects, promoting the survival and poly-functionality of vaccine-primed memory T cells (Arribillaga et al., 2005; Bassett et al., 2012). α-Galactosyl-ceramide, the CD1d-restricted agonistic ligand for invariant NKT cells, can be co-delivered with Ad vaccines to boost both NK cells and antigen-specific Th1 cells and CTLs and thus provide enhanced tumor protection; this is likely achieved through cytokine release by the activated NKT cells (Kim et al., 2008; Ko et al., 2005). Another interesting adjuvant approach is the i.d. delivery of nanorods that contain the chemokine MIP-3β and can attract mature DCs/LCs, thus priming the dermal site for subsequent delivery of a (DC-targeted) vaccine (Kumamoto et al., 2002). That caution is warranted in the selection of an adjuvant was demonstrated by Lemke et al., who combined an Ad vaccine with the polysaccharide adjuvant chitosan, generally known as a powerful stimulator of cell-mediated immunity, and found a reduced tumor-specific CTL response (Lemke et al., 2011). *In vitro* studies indicated that decreased transgene expression and APC activation might be the underlying causes of this unexpected observation. Thus, careful selection of (a combination of) adjuvant(s) from the ever-increasing number of available agents and compounds is of the essence.

## 6.4. To prime or to boost, that is the question

### *6.4.1 Homologous versus heterologous adenoviral prime-boost regimens*

Vaccines based on Ad5 or Ad2 serotypes can often only be given once, since a booster vaccine will be rapidly neutralized by an induced or enhanced Ad-specific immune response. Compared to first-generation AdVs, second-generation AdVs were shown to improve efficacy of homologous prime-boost regimens. In Ad5-immune BALB/c mice, as well as in Ad5-immune cynomolgus macaques, an E1/E2b-deleted Ad5V encoding HIV-gag improved cell-mediated immune responses in the presence of Ad5-neutralizing antibodies significantly when tested against E1-deleted first-generation Ad5V-gag (Gabitzsch et al., 2009). In Ad5-immune mice, a similar E1/E2b-deleted Ad5 containing the CEA gene (Ad5 (E1-,E2b-)-CEA) was shown to be superior over Ad5 (E1)-CEA in inducing anti-CEA cellular immune responses. Multiple injections with Ad5 (E1-E2b-)-CEA were able to reduce the outgrowth of subcutaneously injected CEA-expressing MC38-CEA tumor cells, whereas multiple injections with Ad5 (E1-)-CEA were not (Gabitzsch et al., 2010). Similarly, an Ad5 (E1-,E2b-)-HER2/neu vector in a homologous prime-boost setting could reduce the size of established HER2/neu-expressing tumors (CT-26-HER2/neu) (Gabitzsch, Xu, Balcaitis, Balint, & Jones, 2011).

Heterologous prime-boost strategies whereby AdVs of different serotypes are used for the priming or the boosting step are another alternative and have been shown to be more effective compared to homologous prime-boost (even with second-generation AdVs) and to induce more potent Th1 cellular responses in mice and nonhuman primates (Radosevic et al., 2010; Rodríguez et al., 2009). Using AdVs encoding the influenza nuclear protein (NP), Kahl et al. showed that, whereas a homologous prime-boost with Ad5 (second-generation: E1/E3/E4-deleted) gave significantly higher percentages of IFNγ or TNF-α producing $CD8^+$ NP-specific T cells compared to a single dose of NP-encoding Ad5, even higher percentages of cytokine-secreting $CD8^+$ CTL could be induced upon Ad14 (E1-deleted), Ad28 (E1-deleted), or Ad35 (E1/E3-deleted) priming in combination with Ad5 as a booster vaccine (Kahl et al., 2010). Priming with Ad5 was most efficient when followed by a boost vaccine using a rare Ad serotype with low seroprevalence like the species D Ad28 (Kahl et al., 2010). With respect to antigen-specific humoral responses, several studies showed that not every heterologous AdV combination is as effective, indicating that heterologous prime-boost pairs cannot be randomly selected (Kahl et al., 2010; Lemckert et al., 2005; Thorner et al., 2006). In nonhuman primates, a heterologous prime-boost regimen with Ad26 encoding Ebolavirus glycoprotein as primary vaccine, followed by a booster vaccine with Ad35 encoding the same transgene, resulted in complete protection against Ebolavirus and significantly increased both B and T cell responses against the Ebola glycoprotein (Geisbert et al., 2011). In cancer patients, using a prime-boost regimen with different knob-chimeric oncolytic CRAds (Ad5 vs. Ad5-RGD vs. Ad5/3), Raki et al. showed that even minor changes in the fiber knob could circumvent neutralization of the boosting CRAd (Raki et al., 2011). Not only the rare human serotypes might be more efficacious than Ad5/Ad5 homologous prime-boost regimens, but also the use of AdVs of nonhuman origins like chimpanzee, canine, ovine, or porcine could be employed to improve therapeutic efficacy while circumventing the issue of preexisting or induced Ad neutralizing immune responses (Tatsis et al., 2009).

With more rare serotypes becoming available as recombinant AdVs, future clinical trials using these vectors will have to prove what prime-boost combinations will generate the best and long-anticipated clinical results. In this age of personalized treatment where patients are screened for specific mutations in their tumors (Jiang & Wang, 2010; Kasaian & Jones, 2011) or certain immune signatures (Gajewski et al., 2011; Zitvogel, Kepp, & Kroemer, 2011), it seems only natural to screen patients for preexisting seroreactivity against Ad serotypes before selecting an optimal AdV prime-boost combination.

### 6.4.2 Combining Ad vaccines with non-Ad vaccines

A different way to circumvent the negative effects of Ad vaccine-neutralizing antibodies is to combine AdVs with protein, DNA or RNA vaccines, bacterial vaccines, or other viral vectors based on, for example, herpes virus, pox virus, vesicular stomatitis virus, or vaccinia viruses (Liu, 2010). Particularly in the HIV/AIDS field, extensive experience has been gained with heterologous gene-based prime-boost vaccine combinations. Clinically most widely studied is the DNA-prime–Ad-boost regimen, showing superior immune activation against transgene products over regimens solely based on DNA or Ad vaccines (Churchyard et al., 2011; De Rosa et al., 2011; Kibuuka et al., 2010; Koup et al., 2010). Collectively, these clinical studies showed that DNA-prime–Ad-boost vaccination resulted in higher titers of specific antibodies, higher frequencies of specific $CD4^+$ and $CD8^+$ T cells, more differentiated and polyfunctional effector-memory T cells, and long-term memory, without being affected by titers of preexisting Ad5-neutralizing antibodies (the latter in contrast to vaccination with Ad5 alone). Other combinations explored in infectious disease settings have included synthetic long peptide-prime Ad63-boost vaccines (Rosario et al., 2012) and Ad-prime with modified vaccinia virus Ankara (MVA)- and/or adjuvanted protein-boost vaccination, the latter regimen resulting in superior humoral and T cell immunity against the transgene product, again without being affected by preexistent immunity (Douglas et al., 2010; Draper et al., 2010). In a tumor model, addition of Ad-targeted chemotherapy to a prime-boost regimen of an AdV-prime vaccine encoding for a secreted fusion protein of human MUC1 and CD40L (Ad-sig-hMUC-1/ecdCD40L) and a protein-boost vaccine consisting of the hMUC-1/ecdCD40L protein, significantly prolonged the survival of mice with established Lewis Lung tumors (Akbulut, Tang, Akbulut, Maynard, & Deisseroth, 2010). In scheduling such Ad-prime protein-boost vaccinations, it is important to take into account a recently reported transient shutdown of cross-priming by DCs upon Ad infection (Sutherland et al., 2011). In conclusion, learning from clinical experience in the infectious vaccine field, next-generation Ad-based tumor vaccines should consider adopting novel heterologous prime-boost regimens with proven efficacy in terms of provoking both humoral and cellular immune responses.

## 7. CONCLUDING REMARKS: MILES TO GO

After some bad press (Marshall, 1999; Kresge, 2007; Sekaly, 2008), therapeutic AdVs and Ad-based vaccines are making a comeback (Patterson, 2011). New developments in vectorology, including hexon and fiber modifications to enable DC targeting and scavenger and liver detargeting, to circumvent neutralizing immunity, and to enhance TAA immunogenicity, have resulted in Ad configurations with enhanced vaccine efficacy. In addition, increased understanding of how to manipulate the immune system to overcome mechanisms of tumor-induced tolerance and suppression, both systemically and in the tumor microenvironment, and new therapeutic tools to target immune checkpoints, Tregs, MDSCs, and suppression-related cytokines and signaling pathways facilitate the design of clever immunotherapeutic approaches that fight cancer on multiple fronts. In a similar vein, powerful combinations of adjuvants with different mechanisms of action can now be exploited to boost the immune system (Aranda et al., 2011), both at the vaccination site and in the tumor microenvironment. In this regard, it is important to keep in mind that the ideal adjuvant combination may differ for protein vaccines and Ad-based vaccines and should therefore be carefully assessed in relevant preclinical models.

What are the most promising approaches to Ad-based immunotherapy of cancer in this new age? As hypothetical immunotherapeutic platforms, we would like to propose two innovative "lean and mean" AdV prototypes: (1) a DC-targeted Ad vaccine and (2) an oncolytic CRAd, armed with immune-modulatory transgenes (see Fig. 6.2A and B for a description).

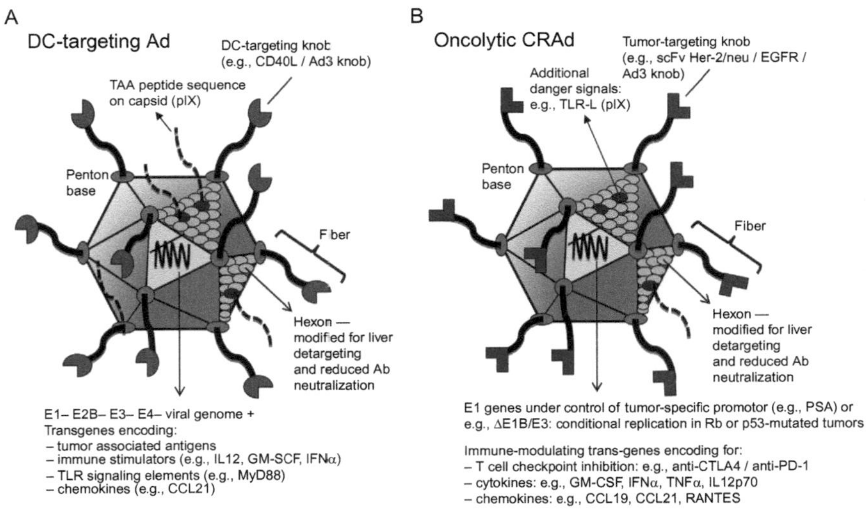


**Figure 6.2** The next generation of Adenoviral (Ad) vectors for immuno-gene therapy of cancer: blueprints of (A) a dendritic cell (DC)-targeted Ad and (B) an oncolytic conditionally replicating Ad (CRAd). Modifications allow for liver/scavenger detargeting, escape from immune neutralization, DC targeting, tumor targeting, and adjuvant immune stimulation.

"Lean" because they are "gutted" to carry as little virus-derived immunogenic proteins as possible and "mean" because they have been modified to detarget liver and scavengers and have been outfitted with additional TAA-derived (Th) epitopes in their capsid and with immune-stimulatory transgenes.

Beside vectorological considerations and a choice of transgenic cargo, we now also have a better appreciation of the effects of the scheduling and routing of Ad vaccine administration. The i.d. delivery route has proven very effective, possibly due to a combination of a dense network of DCs and draining lymph vessels and lower levels of persistent antibodies, minimizing the risk of vector neutralization (Hangalapura, Oosterhoff, Gupta, et al., 2011; Khare, Chen, et al., 2011). As for vaccination schedules, it is clear that heterologous prime-boost regimens should be further explored for Ad-based tumor vaccination.

All these new insights and developments lead to a high number of possible combinations of treatment modalities and delivery schemes. This calls for careful preclinical evaluation in relevant *ex vivo* human models and transitional *in vivo* tumor models, before clinical translation. To bring optimized combinations of (re)targeted Ad configurations and adjuvants in optimized heterologous prime-boost schedules to the clinic will be a major effort, both in terms of time and costs. The best approach may be to start with small-scale phase 0 studies to obtain proof of biological efficacy, followed by randomized phase II trials to ascertain clinical efficacy. Only then should the most promising combinations be progressed to phase III testing.

Although it would now seem that Ad-based cancer vaccines may yet deliver on their promise of clinical efficacy, there clearly are many miles left to go.

## ACKNOWLEDGMENTS

T. D. d. G.'s work was supported by grants from the Netherlands Organization for Scientific Research (NWO VIDI Grant 917-56-321) and Stichting Cancer Center Amsterdam (CCA) and R. v. d. V. is a fellow of the Dutch Cancer Society (KWF BUIT-4643).

## REFERENCES

Abbink, P., Lemckert, A. A. C., Ewald, B. A., Lynch, D. M., Denholtz, M., Smits, S., et al. (2007). Comparative seroprevalence and immunogenicity of six rare serotype recombinant adenovirus vaccine vectors from subgroups B and D. *Journal of Virology*, *81*, 4654–4663.

Adams, W. C., Bond, E., Havenga, M. J. E., Holterman, L., Goudsmit, J., Karlsson Hedestam, G. B., et al. (2009). Adenovirus serotype 5 infects human dendritic cells via a coxsackievirus-adenovirus receptor-independent receptor pathway mediated by lactoferrin and DC-SIGN. *The Journal of General Virology*, *90*, 1600–1610.

Adams, W. C., Gujer, C., McInerney, G., Gall, J. G. D., Petrovas, C., Karlsson Hedestam, G. B., et al. (2011). Adenovirus type-35 vectors block human CD4+ T-cell activation via CD46 ligation. *Proceedings of the National Academy of Sciences of the United States of America, 108*, 7499–7504.

Akbari, O., Panjwani, N., Garcia, S., Tascon, R., Lowrie, D., & Stockinger, B. (1999). DNA vaccination: Transfection and activation of dendritic cells as key events for immunity. *The Journal of Experimental Medicine, 189*, 169–178.

Akbulut, H., Tang, Y., Akbulut, K. G., Maynard, J., & Deisseroth, A. (2010). Addition of adenoviral vector targeting of chemotherapy to the MUC-1/ecdCD40L VPPP vector prime protein boost vaccine prolongs survival of mice carrying growing subcutaneous deposits of Lewis lung cancer cells. *Gene Therapy, 17*, 1333–1340.

Alcock, R., Cottingham, M. G., Rollier, C. S., Furze, J., De Costa, S. D., Hanlon, M., et al. (2010). Long-term thermostabilization of live poxviral and adenoviral vaccine vectors at supraphysiological temperatures in carbohydrate glass. *Science Translational Medicine, 2*, 19ra12.

Amalfitano, A., Hauser, M. A., Hu, H., Serra, D., Begy, C. R., & Chamberlain, J. S. (1998). Production and characterization of improved adenovirus vectors with the E1, E2b, and E3 genes deleted. *Journal of Virology, 72*, 926–933.

Andarini, S., Kikuchi, T., Nukiwa, M., Pradono, P., Suzuki, T., Ohkouchi, S., et al. (2004). Adenovirus vector-mediated in vivo gene transfer of OX40 ligand to tumor cells enhances antitumor immunity of tumor-bearing hosts. *Cancer Research, 64*, 3281–3287.

Appledorn, D. M., Patial, S., McBride, A., Godbehere, S., van Rooijen, N., Parameswaran, N., et al. (2008). Adenovirus vector-induced innate inflammatory mediators, MAPK signaling, as well as adaptive immune responses are dependent upon both TLR2 and TLR9 in vivo. *Journal of Immunology, 181*, 2134–2144.

Aranda, F., Llopiz, D., Díaz-Valdés, N., Riezu-Boj, J. I., Bezunartea, J., Ruiz, M., et al. (2011). Adjuvant combination and antigen targeting as a strategy to induce polyfunctional and high-avidity T-cell responses against poorly immunogenic tumors. *Cancer Research, 71*, 3214–3224.

Arnberg, N. (2009). Adenovirus receptors: Implications for tropism, treatment and targeting. *Reviews in Medical Virology, 19*, 165–178.

Arnberg, N., Edlund, K., Kidd, A. H., & Wadell, G. (2000). Adenovirus type 37 uses sialic acid as a cellular receptor. *Journal of Virology, 74*, 42–48.

Arnberg, N., Kidd, A. H., Edlund, K., Olfat, F., & Wadell, G. (2000). Initial interactions of subgenus D adenoviruses with A549 cellular receptors: Sialic acid versus alpha(v) integrins. *Journal of Virology, 74*, 7691–7693.

Arribillaga, L., Sarobe, P., Arina, A., Gorraiz, M., Borrás-Cuesta, F., Ruiz, J., et al. (2005). Enhancement of CD4 and CD8 immunity by anti-CD137 (4-1BB) monoclonal antibodies during hepatitis C vaccination with recombinant adenovirus. *Vaccine, 23*, 3493–3499.

Arthur, J. F., Butterfield, L. H., Roth, M. D., Bui, L. A., Kiertscher, S. M., Lau, R., et al. (1997). A comparison of gene transfer methods in human dendritic cells. *Cancer Gene Therapy, 4*, 17–25.

Bachem, A., Güttler, S., Hartung, E., Ebstein, F., Schaefer, M., Tannert, A., et al. (2010). Superior antigen cross-presentation and XCR1 expression define human CD11c+CD141+ cells as homologues of mouse CD8+ dendritic cells. *The Journal of Experimental Medicine, 207*, 1273–1281.

Bachtarzi, H., Stevenson, M., Šubr, V., Seymour, L. W., & Fisher, K. D. (2011). E-selectin is a viable route of infection for polymer-coated adenovirus retargeting in TNF-α-activated human umbilical vein endothelial cells. *Journal of Drug Targeting, 19*, 690–700.

Bachtarzi, H., Stevenson, M., Šubr, V., Ulbrich, K., Seymour, L. W., & Fisher, K. D. (2011). Targeting adenovirus gene delivery to activated tumour-associated vasculature via endothelial selectins. *Journal of Controlled Release, 150*, 196–203.

Backer, R., Schwandt, T., Greuter, M., Oosting, M., Jüngerkes, F., Tüting, T., et al. (2010). Effective collaboration between marginal metallophilic macrophages and CD8+ dendritic cells in the generation of cytotoxic T cells. *Proceedings of the National Academy of Sciences of the United States of America*, *107*, 216–221.

Bai, M., Harfe, B., & Freimuth, P. (1993). Mutations that alter an Arg-Gly-Asp (RGD) sequence in the adenovirus type 2 penton base protein abolish its cell-rounding activity and delay virus reproduction in flat cells. *Journal of Virology*, *67*, 5198–5205.

Balakireva, L., Schoehn, G., Thouvenin, E., & Chroboczek, J. (2003). Binding of adenovirus capsid to dipalmitoyl phosphatidylcholine provides a novel pathway for virus entry. *Journal of Virology*, *77*, 4858–4866.

Baratelli, F., Takedatsu, H., Hazra, S., Peebles, K., Luo, J., Kurimoto, P. S., et al. (2008). Pre-clinical characterization of GMP grade CCL21-gene modified dendritic cells for application in a phase I trial in non-small cell lung cancer. *Journal of Translational Medicine*, *6*, 38.

Barlan, A. U., Griffin, T. M., McGuire, K. A., & Wiethoff, C. M. (2011). Adenovirus membrane penetration activates the NLRP3 inflammasome. *Journal of Virology*, *85*, 146–155.

Barouch, D. H., Kik, S. V., Weverling, G. J., Dilan, R., King, S. L., Maxfield, L. F., et al. (2011). International seroepidemiology of adenovirus serotypes 5, 26, 35, and 48 in pediatric and adult populations. *Vaccine*, *29*, 5203–5209.

Basner-Tschakarjan, E., Gaffal, E., O'Keeffe, M., Tormo, D., Limmer, A., Wagner, H., et al. (2006). Adenovirus efficiently transduces plasmacytoid dendritic cells resulting in TLR9-dependent maturation and IFN-alpha production. *The Journal of Gene Medicine*, *8*, 1300–1306.

Bassett, J. D., Swift, S. L., Vanseggelen, H., Hammill, J. A., McGray, A. R., Evelegh, C., et al. (2012). Combined mTOR inhibition and OX40 agonism enhances CD8(+) T cell memory and protective immunity produced by recombinant adenovirus vaccines. *Molecular Therapy*, *20*, 860–869.

Beatty, G. L., Chiorean, E. G., Fishman, M. P., Saboury, B., Teitelbaum, U. R., Sun, W., et al. (2011). CD40 agonists alter tumor stroma and show efficacy against pancreatic carcinoma in mice and humans. *Science*, *331*, 1612–1616.

Beer, S. J., Matthews, C. B., Stein, C. S., Ross, B. D., Hilfinger, J. M., & Davidson, B. L. (1998). Poly (lactic-glycolic) acid copolymer encapsulation of recombinant adenovirus reduces immunogenicity in vivo. *Gene Therapy*, *5*, 740–746.

Belousova, N., Korokhov, N., Krendelshchikova, V., Simonenko, V., Mikheeva, G., Triozzi, P. L., et al. (2003). Genetically targeted adenovirus vector directed to CD40-expressing cells. *Journal of Virology*, *77*, 11367–11377.

Belousova, N., Krendelchtchikova, V., Curiel, D. T., & Krasnykh, V. (2002). Modulation of adenovirus vector tropism via incorporation of polypeptide ligands into the fiber protein. *Journal of Virology*, *76*, 8621–8631.

Bennett, S. R., Carbone, F. R., Karamalis, F., Flavell, R. A., Miller, J. F., & Heath, W. R. (1998). Help for cytotoxic-T-cell responses is mediated by CD40 signalling. *Nature*, *393*, 478–480.

Bergelson, J. M., Cunningham, J. A., Droguett, G., Kurt-Jones, E. A., Krithivas, A., Hong, J. S., et al. (1997). Isolation of a common receptor for Coxsackie B viruses and adenoviruses 2 and 5. *Science*, *275*, 1320–1323.

Bernt, K. M., Ni, S., Tieu, A.-T., & Lieber, A. (2005). Assessment of a combined, adenovirus-mediated oncolytic and immunostimulatory tumor therapy. *Cancer Research*, *65*, 4343–4352.

Bett, A. J., Dubey, S. A., Mehrotra, D. V., Guan, L., Long, R., Anderson, K., et al. (2010). Comparison of T cell immune responses induced by vectored HIV vaccines in non-human primates and humans. *Vaccine*, *28*, 7881–7889.

Bett, A. J., Haddara, W., Prevec, L., & Graham, F. L. (1994). An efficient and flexible system for construction of adenovirus vectors with insertions or deletions in early regions 1 and 3. *Proceedings of the National Academy of Sciences of the United States of America, 91*, 8802–8806.

Bett, A. J., Prevec, L., & Graham, F. L. (1993). Packaging capacity and stability of human adenovirus type 5 vectors. *Journal of Virology, 67*, 5911–5921.

Bharadwaj, U., Li, M., Zhang, R., Chen, C., & Yao, Q. (2007). Elevated interleukin-6 and G-CSF in human pancreatic cancer cell conditioned medium suppress dendritic cell differentiation and activation. *Cancer Research, 67*, 5479–5488.

Bonkobara, M., Zukas, P. K., Shikano, S., Nakamura, S., Cruz, P. D., & Ariizumi, K. (2001). Epidermal Langerhans cell-targeted gene expression by a dectin-2 promoter. *Journal of Immunology, 167*, 6893–6900.

Bouwhuis, M. G., Ten Hagen, T. L., Suciu, S., & Eggermont, A. M. (2011). Autoimmunity and treatment outcome in melanoma. *Current Opinion in Oncology, 23*, 170–176.

Bradshaw, A. C., Parker, A. L., Duffy, M. R., Coughlan, L., van Rooijen, N., Kähäri, V.-M., et al. (2010). Requirements for receptor engagement during infection by adenovirus complexed with blood coagulation factor X. *PLoS Pathogens, 6*, e1001142.

Brahmer, J. R., Drake, C. G., Wollner, I., Powderly, J. D., Picus, J., Sharfman, W. H., et al. (2010). Phase I study of single-agent anti-programmed death-1 (MDX-1106) in refractory solid tumors: Safety, clinical activity, pharmacodynamics, and immunologic correlates. *Journal of Clinical Oncology, 28*, 3167–3175.

Brandão, J. G., Scheper, R. J., Lougheed, S. M., Curiel, D. T., Tillman, B. W., Gerritsen, W. R., et al. (2003). CD40-targeted adenoviral gene transfer to dendritic cells through the use of a novel bispecific single-chain Fv antibody enhances cytotoxic T cell activation. *Vaccine, 21*, 2268–2272.

Bråve, A., Ljungberg, K., Wahren, B., & Liu, M. A. (2007). Vaccine delivery methods using viral vectors. *Molecular Pharmaceutics, 4*, 18–32.

Bromberg, J. (2002). Stat proteins and oncogenesis. *The Journal of Clinical Investigation, 109*, 1139–1142.

Bros, M., Ross, X.-L., Pautz, A., Reske-Kunz, A. B., & Ross, R. (2003). The human fascin gene promoter is highly active in mature dendritic cells due to a stage-specific enhancer. *Journal of Immunology, 171*, 1825–1834.

Brouwer, E., Havenga, M. J., Ophorst, O., de Leeuw, B., Gijsbers, L., Gillissen, G., et al. (2007). Human adenovirus type 35 vector for gene therapy of brain cancer: Improved transduction and bypass of pre-existing anti-vector immunity in cancer patients. *Cancer Gene Therapy, 14*, 211–219.

Brower, V. (2010). Approval of provenge seen as first step for cancer treatment vaccines. *Journal of the National Cancer Institute, 102*, 1108–1110.

Bru, T., Salinas, S., & Kremer, E. J. (2010). An update on canine adenovirus type 2 and its vectors. *Viruses, 2*, 2134–2153.

Buchbinder, S. P., Mehrotra, D. V., Duerr, A., Fitzgerald, D. W., Mogg, R., Li, D., et al. (2008). Efficacy assessment of a cell-mediated immunity HIV-1 vaccine (the Step Study): A double-blind, randomised, placebo-controlled, test-of-concept trial. *The Lancet, 372*, 1881–1893.

Burnet, M. (1957). Cancer; a biological approach. I. The processes of control. *British Medical Journal, 1*, 779–786.

Burnet, F. M. (1970). The concept of immunological surveillance. *Progress in Experimental Tumor Research, 13*, 1–27.

Butterfield, L. H., Comin-Anduix, B., Vujanovic, L., Lee, Y., Dissette, V. B., Yang, J.-Q., et al. (2008). Adenovirus MART-1-engineered autologous dendritic cell vaccine for metastatic melanoma. *Journal of Immunotherapy, 31*, 294–309.

Butterfield, L. H., & Vujanovic, L. (2010). New approaches to the development of adenoviral dendritic cell vaccines in melanoma. *Current Opinion in Investigational Drugs, 11*, 1399–1408.

Campos, S. K., Parrott, M. B., & Barry, M. A. (2004). Avidin-based targeting and purification of a protein IX-modified, metabolically biotinylated adenoviral vector. *Molecular Therapy, 9*, 942–954.

Carlisle, R. C., Di, Y., Cerny, A. M., Sonnen, A. F. P., Sim, R. B., Green, N. K., et al. (2009). Human erythrocytes bind and inactivate type 5 adenovirus by presenting Coxsackie virus-adenovirus receptor and complement receptor 1. *Blood, 113*, 1909–1918.

Casal, M., & Haskins, M. (2006). Large animal models and gene therapy. *European Journal of Human Genetics, 14*, 266–272.

Cerullo, V., Seiler, M. P., Mane, V., Brunetti-Pierri, N., Clarke, C., Bertin, T. K., et al. (2007). Toll-like receptor 9 triggers an innate immune response to helper-dependent adenoviral vectors. *Molecular Therapy, 15*, 378–385.

Chen, Z., Mok, H., Pflugfelder, S. C., Li, D.-Q., & Barry, M. A. (2006). Improved transduction of human corneal epithelial progenitor cells with cell-targeting adenoviral vectors. *Experimental Eye Research, 83*, 798–806.

Chen, C. Y., Senac, J. S., Weaver, E. A., May, S. M., Jelinek, D. F., Greipp, P., et al. (2011). Species D adenoviruses as oncolytics against B-cell cancers. *Clinical Cancer Research, 17*, 6712–6722.

Cheng, C., Gall, J. G. D., Kong, W.-P., Sheets, R. L., Gomez, P. L., King, C. R., et al. (2007). Mechanism of Ad5 vaccine immunity and toxicity: Fiber shaft targeting of dendritic cells. *PLoS Pathogens, 3*, e25.

Cheong, S. C., Wang, Y., Meng, J.-H., Hill, R., Sweeney, K., Kirn, D., et al. (2008). E1A-expressing adenoviral E3B mutants act synergistically with chemotherapeutics in immunocompetent tumor models. *Cancer Gene Therapy, 15*, 40–50.

Chiappori, A. A., Soliman, H., Janssen, W. E., Antonia, S. J., & Gabrilovich, D. I. (2010). INGN-225: A dendritic cell-based p53 vaccine (Ad.p53-DC) in small cell lung cancer: Observed association between immune response and enhanced chemotherapy effect. *Expert Opinion on Biological Therapy, 10*, 983–991.

Chillón, M., Lee, J. H., Fasbender, A., & Welsh, M. J. (1998). Adenovirus complexed with polyethylene glycol and cationic lipid is shielded from neutralizing antibodies in vitro. *Gene Therapy, 5*, 995–1002.

Choi, K.-J., Kim, J.-H., Lee, Y.-S., Kim, J., Suh, B.-S., Kim, H., et al. (2006). Concurrent delivery of GM-CSF and B7-1 using an oncolytic adenovirus elicits potent antitumor effect. *Gene Therapy, 13*, 1010–1020.

Choi, K. J., Zhang, S. N., Choi, I. K., Kim, J. S., & Yun, C. O. (2011). Strengthening of antitumor immune memory and prevention of thymic atrophy mediated by adenovirus expressing IL-12 and GM-CSF. *Gene Therapy*, (epub).

Chu, Y., Heistad, D., Cybulsky, M. I., & Davidson, B. L. (2001). Vascular cell adhesion molecule-1 augments adenovirus-mediated gene transfer. *Arteriosclerosis, Thrombosis, and Vascular Biology, 21*, 238–242.

Churchyard, G. J., Morgan, C., Adams, E., Hural, J., Graham, B. S., Moodie, Z., et al. (2011). A phase IIA randomized clinical trial of a multiclade HIV-1 DNA prime followed by a multiclade rAd5 HIV-1 vaccine boost in healthy adults (HVTN204). *PLoS One, 6*, e21225.

Clive, K. S., Tyler, J. A., Clifton, G. T., Holmes, J. P., Mittendorf, E. A., Ponniah, S., et al. (2010). Use of GM-CSF as an adjuvant with cancer vaccines: Beneficial or detrimental? *Expert Review of Vaccines, 9*, 519–525.

Corjon, S., Gonzalez, G., Henning, P., Grichine, A., Lindholm, L., Boulanger, P., et al. (2011). Cell entry and trafficking of human adenovirus bound to blood factor X is determined by the fiber serotype and not hexon:heparan sulfate interaction. *PLoS One, 6*, e18205.

Crompton, J., Toogood, C. I., Wallis, N., & Hay, R. T. (1994). Expression of a foreign epitope on the surface of the adenovirus hexon. *The Journal of General Virology*, *75* (Pt. 1), 133–139.

Crozat, K., Guiton, R., Contreras, V., Feuillet, V., Dutertre, C.-A., Ventre, E., et al. (2010). The XC chemokine receptor 1 is a conserved selective marker of mammalian cells homologous to mouse CD8alpha+ dendritic cells. *The Journal of Experimental Medicine*, *207*, 1283–1292.

Danthinne, X., & Imperiale, M. J. (2000). Production of first generation adenovirus vectors: A review. *Gene Therapy*, *7*, 1707–1714.

de Gruijl, T. D., Luykx-de Bakker, S. A., Tillman, B. W., van den Eertwegh, A. J. M., Buter, J., Lougheed, S. M., et al. (2002). Prolonged maturation and enhanced transduction of dendritic cells migrated from human skin explants after in situ delivery of CD40-targeted adenoviral vectors. *Journal of Immunology*, *169*, 5322–5331.

de Gruijl, T. D., Ophorst, O. J. A. E., Goudsmit, J., Verhaagh, S., Lougheed, S. M., Radosevic, K., et al. (2006). Intradermal delivery of adenoviral type-35 vectors leads to high efficiency transduction of mature, CD8+ T cell-stimulating skin-emigrated dendritic cells. *Journal of Immunology*, *177*, 2208–2215.

de Gruijl, T. D., Sombroek, C. C., Lougheed, S. M., Oosterhoff, D., Buter, J., van den Eertwegh, A. J. M., et al. (2006). A postmigrational switch among skin-derived dendritic cells to a macrophage-like phenotype is predetermined by the intracutaneous cytokine balance. *Journal of Immunology*, *176*, 7232–7242.

de Jong, J. C., Osterhaus, A. D. M. E., Jones, M. S., & Harrach, B. (2008). Human adenovirus type 52: A type 41 in disguise? *Journal of Virology*, *82*, 3809, author reply 3809–3810.

De Rosa, S. C., Thomas, E. P., Bui, J., Huang, Y., deCamp, A., Morgan, C., et al. (2011). HIV-DNA priming alters T cell responses to HIV-adenovirus vaccine even when responses to DNA are undetectable. *Journal of Immunology*, *187*, 3391–3401.

de Vrij, J., Uil, T. G., van den Hengel, S. K., Cramer, S. J., Koppers-Lalic, D., Verweij, M. C., et al. (2008). Adenovirus targeting to HLA-A1/MAGE-A1-positive tumor cells by fusing a single-chain T-cell receptor with minor capsid protein IX. *Gene Therapy*, *15*, 978–989.

Dechecchi, M. C., Melotti, P., Bonizzato, A., Santacatterina, M., Chilosi, M., & Cabrini, G. (2001). Heparan sulfate glycosaminoglycans are receptors sufficient to mediate the initial binding of adenovirus types 2 and 5. *Journal of Virology*, *75*, 8772–8780.

Dechecchi, M. C., Tamanini, A., Bonizzato, A., & Cabrini, G. (2000). Heparan sulfate glycosaminoglycans are involved in adenovirus type 5 and 2-host cell interactions. *Virology*, *268*, 382–390.

DeFrancesco, L. (2010). Landmark approval for Dendreon's cancer vaccine. *Nature Biotechnology*, *28*, 531–532.

Dehari, H., Ito, Y., Nakamura, T., Kobune, M., Sasaki, K., Yonekura, N., et al. (2003). Enhanced antitumor effect of RGD fiber-modified adenovirus for gene therapy of oral cancer. *Cancer Gene Therapy*, *10*, 75–85.

Diaconu, I., Cerullo, V., Hirvinen, M. L., Escutenaire, S., Ugolini, M., Pesonen, S. K., et al. (2012). Immune response is an important aspect of the antitumor effect produced by a CD40L-encoding oncolytic adenovirus. *Cancer Research*, *72*, 2327–2338.

Dias, J. D., Hemminki, O., Diaconu, I., Hirvinen, M., Bonetti, A., Guse, K., et al. (2011). Targeted cancer immunotherapy with oncolytic adenovirus coding for a fully human monoclonal antibody specific for CTLA-4. *Gene Therapy*, http://dx.doi.org/10.1038/gt.2011.176 (Epub ahead of print).

Dietz, A. B., & Vuk-Pavlović, S. (1998). High efficiency adenovirus-mediated gene transfer to human dendritic cells. *Blood*, *91*, 392–398.

Ding, Z.-Y., Wu, Y., Luo, Y., Su, J.-M., Li, Q., Zhang, X.-W., et al. (2007). Mannan-modified adenovirus as a vaccine to induce antitumor immunity. *Gene Therapy*, *14*, 657–663.

Dmitriev, I. P., Kashentseva, E. A., & Curiel, D. T. (2002). Engineering of adenovirus vectors containing heterologous peptide sequences in the C terminus of capsid protein IX. *Journal of Virology*, *76*, 6893–6899.

Dmitriev, I., Kashentseva, E., Rogers, B. E., Krasnykh, V., & Curiel, D. T. (2000). Ectodomain of coxsackievirus and adenovirus receptor genetically fused to epidermal growth factor mediates adenovirus targeting to epidermal growth factor receptor-positive cells. *Journal of Virology*, *74*, 6875–6884.

Dmitriev, I., Krasnykh, V., Miller, C. R., Wang, M., Kashentseva, E., Mikheeva, G., et al. (1998). An adenovirus vector with genetically modified fibers demonstrates expanded tropism via utilization of a coxsackievirus and adenovirus receptor-independent cell entry mechanism. *Journal of Virology*, *72*, 9706–9713.

Dörig, R. E., Marcil, A., & Richardson, C. D. (1994). CD46, a primate-specific receptor for measles virus. *Trends in Microbiology*, *2*, 312–318.

Dormond, E., & Kamen, A. A. (2011). Manufacturing of adenovirus vectors: Production and purification of helper dependent adenovirus. *Methods in Molecular Biology*, *737*, 139–156.

Dormond, E., Perrier, M., & Kamen, A. (2009). From the first to the third generation adenoviral vector: What parameters are governing the production yield? *Biotechnology Advances*, *27*, 133–144.

Douglas, A. D., de Cassan, S. C., Dicks, M. D. J., Gilbert, S. C., Hill, A. V. S., & Draper, S. J. (2010). Tailoring subunit vaccine immunogenicity: Maximizing antibody and T cell responses by using combinations of adenovirus, poxvirus and protein-adjuvant vaccines against Plasmodium falciparum MSP1. *Vaccine*, *28*, 7167–7178.

Douglas, J. T., Rogers, B. E., Rosenfeld, M. E., Michael, S. I., Feng, M., & Curiel, D. T. (1996). Targeted gene delivery by tropism-modified adenoviral vectors. *Nature Biotechnology*, *14*, 1574–1578.

Dranoff, G. (2003). GM-CSF-secreting melanoma vaccines. *Oncogene*, *22*, 3188–3192.

Draper, S. J., Biswas, S., Spencer, A. J., Remarque, E. J., Capone, S., Naddeo, M., et al. (2010). Enhancing blood-stage malaria subunit vaccine immunogenicity in rhesus macaques by combining adenovirus, poxvirus, and protein-in-adjuvant vaccines. *Journal of Immunology*, *185*, 7583–7595.

Dreier, B., Mikheeva, G., Belousova, N., Parizek, P., Boczek, E., Jelesarov, I., et al. (2011). Her2-specific multivalent adapters confer designed tropism to adenovirus for gene targeting. *Journal of Molecular Biology*, *405*, 410–426.

Dresch, C., Edelmann, S. L., Marconi, P., & Brocker, T. (2008). Lentiviral-mediated transcriptional targeting of dendritic cells for induction of T cell tolerance in vivo. *Journal of Immunology*, *181*, 4495–4506.

Dummer, R., Eichmüller, S., Gellrich, S., Assaf, C., Dreno, B., Schiller, M., et al. (2010). Phase II clinical trial of intratumoral application of TG1042 (adenovirus-interferon-gamma) in patients with advanced cutaneous T-cell lymphomas and multilesional cutaneous B-cell lymphomas. *Molecular Therapy*, *18*, 1244–1247.

Dunn, G. P., Bruce, A. T., Ikeda, H., Old, L. J., & Schreiber, R. D. (2002). Cancer immunoediting: From immunosurveillance to tumor escape. *Nature Immunology*, *3*, 991–998.

Edukulla, R., Ramakrishna, E., Woller, N., Mundt, B., Knocke, S., Gürlevik, E., et al. (2009). Antitumoral immune response by recruitment and expansion of dendritic cells in tumors infected with telomerase-dependent oncolytic viruses. *Cancer Research*, *69*, 1448–1458.

Eiz, B., & Pring-Akerblom, P. (1997). Molecular characterization of the type-specific gamma-determinant located on the adenovirus fiber. *Journal of Virology*, *71*, 6576–6581.

Fallaux, F. J., Bout, A., van der Velde, I., van den Wollenberg, D. J., Hehir, K. M., Keegan, J., et al. (1998). New helper cells and matched early region 1-deleted adenovirus vectors prevent generation of replication-competent adenoviruses. *Human Gene Therapy*, *9*, 1909–1917.

Fallaux, F. J., Kranenburg, O., Cramer, S. J., Houweling, A., Van Ormondt, H., Hoeben, R. C., et al. (1996). Characterization of 911: A new helper cell line for the titration and propagation of early region 1-deleted adenoviral vectors. *Human Gene Therapy*, *7*, 215–222.

Fechner, H., Haack, A., Wang, H., Wang, X., Eizema, K., Pauschinger, M., et al. (1999). Expression of coxsackie adenovirus receptor and alphav-integrin does not correlate with adenovector targeting in vivo indicating anatomical vector barriers. *Gene Therapy*, *6*, 1520–1535.

Fernandes, M. S., Gomes, E. M., Butcher, L. D., Hernandez-Alcoceba, R., Chang, D., Kansopon, J., et al. (2009). Growth inhibition of human multiple myeloma cells by an oncolytic adenovirus carrying the CD40 ligand transgene. *Clinical Cancer Research*, *15*, 4847–4856.

Finnefrock, A. C., Tang, A., Li, F., Freed, D. C., Feng, M., Cox, K. S., et al. (2009). PD-1 blockade in rhesus macaques: Impact on chronic infection and prophylactic vaccination. *Journal of Immunology*, *182*, 980–987.

Fisher, K. D., Stallwood, Y., Green, N. K., Ulbrich, K., Mautner, V., & Seymour, L. W. (2001). Polymer-coated adenovirus permits efficient retargeting and evades neutralising antibodies. *Gene Therapy*, *8*, 341–348.

Flinsenberg, T. W. H., Compeer, E. B., Boelens, J. J., & Boes, M. (2011). Antigen cross-presentation: Extending recent laboratory findings to therapeutic intervention. *Clinical and Experimental Immunology*, *165*, 8–18.

Fox, B. A., Schendel, D. J., Butterfield, L. H., Aamdal, S., Allison, J. P., Ascierto, P. A., et al. (2011). Defining the critical hurdles in cancer immunotherapy. *Journal of Translational Medicine*, *9*, 214.

Fransen, M. F., Sluijter, M., Morreau, H., Arens, R., & Melief, C. J. M. (2011). Local activation of CD8 T cells and systemic tumor eradication without toxicity via slow release and local delivery of agonistic CD40 antibody. *Clinical Cancer Research*, *17*, 2270–2280.

Fukuhara, H., Homma, Y., & Todo, T. (2010). Oncolytic virus therapy for prostate cancer. *International Journal of Urology*, *17*, 20–30.

Fushimi, T., O'Connor, T. P., & Crystal, R. G. (2006). Adenoviral gene transfer of stromal cell-derived factor-1 to murine tumors induces the accumulation of dendritic cells and suppresses tumor growth. *Cancer Research*, *66*, 3513–3522.

Gabitzsch, E. S., Xu, Y., Balcaitis, S., Balint, J. P., & Jones, F. R. (2011). An Ad5[E1-, E2b-]-HER2/neu vector induces immune responses and inhibits HER2/neu expressing tumor progression in Ad5 immune mice. *Cancer Gene Therapy*, *18*, 326–335.

Gabitzsch, E. S., Xu, Y., Balint, J. P., Hartman, Z. C., Lyerly, H. K., & Jones, F. R. (2010). Anti-tumor immunotherapy despite immunity to adenovirus using a novel adenoviral vector Ad5 [E1-, E2b-]-CEA. *Cancer Immunology, Immunotherapy*, *59*, 1131–1135.

Gabitzsch, E. S., Xu, Y., Yoshida, L. H., Balint, J., Gayle, R. B., Amalfitano, A., et al. (2009). A preliminary and comparative evaluation of a novel Ad5 [E1-, E2b-] recombinant-based vaccine used to induce cell mediated immune responses. *Immunology Letters*, *122*, 44–51.

Gabrilovich, D. (2004). Mechanisms and functional significance of tumour-induced dendritic-cell defects. *Nature Reviews. Immunology*, *4*, 941–952.

Gabrilovich, D. I., & Nagaraj, S. (2009). Myeloid-derived suppressor cells as regulators of the immune system. *Nature Reviews. Immunology*, *9*, 162–174.

Gaggar, A., Shayakhmetov, D. M., & Lieber, A. (2003). CD46 is a cellular receptor for group B adenoviruses. *Nature Medicine*, *9*, 1408–1412.

Gajewski, T. F., Fuertes, M., Spaapen, R., Zheng, Y., & Kline, J. (2011). Molecular profiling to identify relevant immune resistance mechanisms in the tumor microenvironment. *Current Opinion in Immunology*, *23*, 286–292.

Gall, J. G., Crystal, R. G., & Falck-Pedersen, E. (1998). Construction and characterization of hexon-chimeric adenoviruses: Specification of adenovirus serotype. *Journal of Virology, 72*, 10260–10264.

Gambotto, A., Tuting, T., McVey, D. L., Kovesdi, I., Tahara, H., Lotze, M. T., et al. (1999). Induction of antitumor immunity by direct intratumoral injection of a recombinant adenovirus vector expressing interleukin-12. *Cancer Gene Therapy, 6*, 45–53.

Gao, J.-Q., Kanagawa, N., Xu, D.-H., Han, M., Sugita, T., Hatanaka, Y., et al. (2008). Combination of two fiber-mutant adenovirus vectors, one encoding the chemokine FKN and another encoding cytokine interleukin 12, elicits notably enhanced anti-tumor responses. *Cancer Immunology, Immunotherapy, 57*, 1657–1664.

Geisbert, T. W., Bailey, M., Hensley, L., Asiedu, C., Geisbert, J., Stanley, D., et al. (2011). Recombinant adenovirus serotype 26 (Ad26) and Ad35 vaccine vectors bypass immunity to Ad5 and protect nonhuman primates against ebolavirus challenge. *Journal of Virology, 85*, 4222–4233.

Gerlini, G., Mariotti, G., Chiarugi, A., Di Gennaro, P., Caporale, R., Parenti, A., et al. (2008). Induction of CD83+CD14+ nondendritic antigen-presenting cells by exposure of monocytes to IFN-alpha. *Journal of Immunology, 181*, 2999–3008.

Gigante, M., Mandic, M., Wesa, A. K., Cavalcanti, E., Dambrosio, M., Mancini, V., et al. (2008). Interferon-alpha (IFN-alpha)-conditioned DC preferentially stimulate type-1 and limit Treg-type in vitro T-cell responses from RCC patients. *Journal of Immunotherapy, 31*, 254–262.

Glasgow, J. N., Mikheeva, G., Krasnykh, V., & Curiel, D. T. (2009). A strategy for adenovirus vector targeting with a secreted single chain antibody. *PLoS One, 4*, e8355.

Glaspy, J., Ribas, A., & Chmielowski, B. (2009). Interferon alfa in the postsurgical management of high-risk melanoma: Is it worth it? *Journal of Clinical Oncology, 27*, 2896–2897.

Gonzalez, R., Vereecque, R., Wickham, T. J., Facon, T., Hetuin, D., Kovesdi, I., et al. (1999). Transduction of bone marrow cells by the AdZ.F(pK7) modified adenovirus demonstrates preferential gene transfer in myeloma cells. *Human Gene Therapy, 10*, 2709–2717.

Gorziglia, M. I., Lapcevich, C., Roy, S., Kang, Q., Kadan, M., Wu, V., et al. (1999). Generation of an adenovirus vector lacking E1, e2a, E3, and all of E4 except open reading frame 3. *Journal of Virology, 73*, 6048–6055.

Grabbe, S., Bruvers, S., & Granstein, R. D. (1992). Effects of immunomodulatory cytokines on the presentation of tumor-associated antigens by epidermal Langerhans cells. *The Journal of Investigative Dermatology, 99*, 66S–68S.

Graham, F. L., Smiley, J., Russell, W. C., & Nairn, R. (1977). Characteristics of a human cell line transformed by DNA from human adenovirus type 5. *The Journal of General Virology, 36*, 59–74.

Gray, G., Buchbinder, S., & Duerr, A. (2010). Overview of STEP and Phambili trial results: Two phase IIb test-of-concept studies investigating the efficacy of MRK adenovirus type 5 gag/pol/nef subtype B HIV vaccine. *Current Opinion in HIV and AIDS, 5*, 357–361.

Grossmann, C., Tenbusch, M., Nchinda, G., Temchura, V., Nabi, G., Stone, G. W., et al. (2009). Enhancement of the priming efficacy of DNA vaccines encoding dendritic cell-targeted antigens by synergistic toll-like receptor ligands. *BMC Immunology, 10*, 43.

Günther, P. S., Mikeler, E., Hamprecht, K., Schneider-Schaulies, J., Jahn, G., & Dennehy, K. M. (2011). CD209/DC-SIGN mediates efficient infection of monocyte-derived dendritic cells by clinical adenovirus 2C isolates in the presence of bovine lactoferrin. *The Journal of General Virology, 92*, 1754–1759.

Gupalo, E., Buriachkovskaia, L., & Othman, M. (2011). Human platelets express CAR with localization at the sites of intercellular interaction. *Virology Journal, 8*, 456.

Haenssle, H., Buhl, T., Knudsen, S., Krueger, U., Rosenberger, A., Reich, K., et al. (2008). CD40 ligation during dendritic cell maturation reduces cell death and prevents interleukin-10-induced regression to macrophage-like monocytes. *Experimental Dermatology, 17*, 177–187.

Halldén, G., Hill, R., Wang, Y., Anand, A., Liu, T.-C., Lemoine, N. R., et al. (2003). Novel immunocompetent murine tumor models for the assessment of replication-competent oncolytic adenovirus efficacy. *Molecular Therapy, 8*, 412–424.

Han, Z., Hong, Z., Chen, C., Gao, Q., Luo, D., Fang, Y., et al. (2009). A novel oncolytic adenovirus selectively silences the expression of tumor-associated STAT3 and exhibits potent antitumoral activity. *Carcinogenesis, 30*, 2014–2022.

Hangalapura, B. N., Oosterhoff, D., Aggarwal, S., Wijnands, P. G. J. T. B., van de Ven, R., Santegoets, S. J. A. M., et al. (2010). Selective transduction of dendritic cells in human lymph nodes and superior induction of high-avidity melanoma-reactive cytotoxic T cells by a CD40-targeted adenovirus. *Journal of Immunotherapy, 33*, 706–715.

Hangalapura, B. N., Oosterhoff, D., de Groot, J., Boon, L., Tüting, T., van den Eertwegh, A. J., et al. (2011). Potent antitumor immunity generated by a CD40-targeted adenoviral vaccine. *Cancer Research, 71*, 5827–5837.

Hangalapura, B. N., Oosterhoff, D., Gupta, T., de Groot, J., Wijnands, P. G., van Beusechem, V. W., et al. (2011). Delivery route, MyD88 signaling and cross-priming events determine the anti-tumor efficacy of an adenovirus based melanoma vaccine. *Vaccine, 29*, 2313–2321.

Hangalapura, B. N., Timares, L., Oosterhoff, D., Scheper, R. J., Curiel, D. T., & de Gruijl, T. D. (2012). CD40-targeted adenoviral cancer vaccines: The long and winding road to the clinic. *The Journal of Gene Medicine*, .

Hara, H., Kobayashi, A., Narumi, K., Kondoh, A., Yoshida, K., Nishimoto, T., et al. (2009). Intratumoral interferon-alpha gene transfer enhances tumor immunity after allogeneic hematopoietic stem cell transplantation. *Cancer Immunology, Immunotherapy, 58*, 1007–1021.

Hara, H., Kobayashi, A., Yoshida, K., Ohashi, M., Ohnami, S., Uchida, E., et al. (2007). Local interferon-alpha gene therapy elicits systemic immunity in a syngeneic pancreatic cancer model in hamster. *Cancer Science, 98*, 455–463.

Harrop, R., John, J., & Carroll, M. W. (2006). Recombinant viral vectors: Cancer vaccines. *Advanced Drug Delivery Reviews, 58*, 931–947.

Hartman, Z. C., Kiang, A., Everett, R. S., Serra, D., Yang, X. Y., Clay, T. M., et al. (2007). Adenovirus infection triggers a rapid, MyD88-regulated transcriptome response critical to acute-phase and adaptive immune responses in vivo. *Journal of Virology, 81*, 1796–1812.

Hartman, Z. C., Osada, T., Glass, O., Yang, X. Y., Lei, G.-J., Lyerly, H. K., et al. (2010). Ligand-independent toll-like receptor signals generated by ectopic overexpression of MyD88 generate local and systemic antitumor immunity. *Cancer Research, 70*, 7209–7220.

Harui, A., Roth, M. D., Sanghvi, M., Vira, D., Mizuguchi, H., & Basak, S. K. (2006). Centrifugation enhances integrin-mediated transduction of dendritic cells by conventional and RGD-modified adenoviral vectors. *Journal of Immunological Methods, 312*, 94–104.

Harui, A., Roth, M. D., Vira, D., Sanghvi, M., Mizuguchi, H., & Basak, S. K. (2006). Adenoviral-encoded antigens are presented efficiently by a subset of dendritic cells expressing high levels of alpha(v)beta3 integrins. *Journal of Leukocyte Biology, 79*, 1271–1278.

Harvey, T. J., Burdon, D., Steele, L., Ingram, N., Hall, G. D., Selby, P. J., et al. (2010). Retargeted adenoviral cancer gene therapy for tumour cells overexpressing epidermal growth factor receptor or urokinase-type plasminogen activator receptor. *Gene Therapy, 17*, 1000–1010.

Heath, W. R., & Carbone, F. R. (2001). Cross-presentation, dendritic cells, tolerance and immunity. *Annual Review of Immunology, 19*, 47–64.

Hensbergen, P. J., Wijnands, P. G. J. T. B., Schreurs, M. W. J., Scheper, R. J., Willemze, R., & Tensen, C. P. (2005). The CXCR3 targeting chemokine CXCL11 has potent antitumor

activity in vivo involving attraction of CD8+ T lymphocytes but not inhibition of angiogenesis. *Journal of Immunotherapy, 28*, 343–351.
Hensley, S. E., & Amalfitano, A. (2007). Toll-like receptors impact on safety and efficacy of gene transfer vectors. *Molecular Therapy, 15*, 1417–1422.
Higham, E. M., Shen, C.-H., Wittrup, K. D., & Chen, J. (2010). Cutting edge: Delay and reversal of T cell tolerance by intratumoral injection of antigen-loaded dendritic cells in an autochthonous tumor model. *Journal of Immunology, 184*, 5954–5958.
Hodi, F. S., O'Day, S. J., McDermott, D. F., Weber, R. W., Sosman, J. A., Haanen, J. B., et al. (2010). Improved survival with ipilimumab in patients with metastatic melanoma. *The New England Journal of Medicine, 363*, 711–723.
Holterman, L., Vogels, R., van der Vlugt, R., Sieuwerts, M., Grimbergen, J., Kaspers, J., et al. (2004). Novel replication-incompetent vector derived from adenovirus type 11 (Ad11) for vaccination and gene therapy: Low seroprevalence and non-cross-reactivity with Ad5. *Journal of Virology, 78*, 13207–13215.
Hong, S. S., Karayan, L., Tournier, J., Curiel, D. T., & Boulanger, P. A. (1997). Adenovirus type 5 fiber knob binds to MHC class I alpha2 domain at the surface of human epithelial and B lymphoblastoid cells. *The EMBO Journal, 16*, 2294–2306.
Hsu, C., Boysen, M., Gritton, L. D., Frosst, P. D., Nemerow, G. R., & Von Seggern, D. J. (2005). In vitro dendritic cell infection by pseudotyped adenoviral vectors does not correlate with their in vivo immunogenicity. *Virology, 332*, 1–7.
Huang, S., Kamata, T., Takada, Y., Ruggeri, Z. M., & Nemerow, G. R. (1996). Adenovirus interaction with distinct integrins mediates separate events in cell entry and gene delivery to hematopoietic cells. *Journal of Virology, 70*, 4502–4508.
Huang, J.-H., Zhang, S.-N., Choi, K.-J., Choi, I.-K., Kim, J.-H., Lee, M.-G., et al. (2010). Therapeutic and tumor-specific immunity induced by combination of dendritic cells and oncolytic adenovirus expressing IL-12 and 4-1BBL. *Molecular Therapy, 18*, 264–274.
Ichikawa, H. T., Williams, L. P., & Segal, B. M. (2002). Activation of APCs through CD40 or Toll-like receptor 9 overcomes tolerance and precipitates autoimmune disease. *Journal of Immunology, 169*, 2781–2787.
Iida, T., Shiba, H., Misawa, T., Ohashi, T., Eto, Y., & Yanaga, K. (2010). Immunogene therapy against colon cancer metastasis using an adenovirus vector expressing CD40 ligand. *Surgery, 148*, 925–935.
Imler, J. L., Chartier, C., Dreyer, D., Dieterle, A., Sainte-Marie, M., Faure, T., et al. (1996). Novel complementation cell lines derived from human lung carcinoma A549 cells support the growth of E1-deleted adenovirus vectors. *Gene Therapy, 3*, 75–84.
Iwata-Kajihara, T., Sumimoto, H., Kawamura, N., Ueda, R., Takahashi, T., Mizuguchi, H., et al. (2011). Enhanced cancer immunotherapy using STAT3-depleted dendritic cells with high Th1-inducing ability and resistance to cancer cell-derived inhibitory factors. *Journal of Immunology, 187*, 27–36.
Jager, L., Hausl, M. A., Rauschhuber, C., Wolf, N. M., Kay, M. A., & Ehrhardt, A. (2009). A rapid protocol for construction and production of high-capacity adenoviral vectors. *Nature Protocols, 4*, 547–564.
Ji, L., Bouvet, M., Price, R. E., Roth, J. A., & Fang, B. (1999). Reduced toxicity, attenuated immunogenicity and efficient mediation of human p53 gene expression in vivo by an adenovirus vector with deleted E1-E3 and inactivated E4 by GAL4-TATA promoter replacement. *Gene Therapy, 6*, 393–402.
Jiang, Y., & Wang, M. (2010). Personalized medicine in oncology: Tailoring the right drug to the right patient. *Biomarkers in Medicine, 4*, 523–533.
Jiang, H., White, E. J., Ríos-Vicil, C. I., Xu, J., Gomez-Manzano, C., & Fueyo, J. (2011). Human adenovirus type 5 induces cell lysis through autophagy and autophagy-triggered caspase activity. *Journal of Virology, 85*, 4720–4729.

Johansson, C., Jonsson, M., Marttila, M., Persson, D., Fan, X.-L., Skog, J., et al. (2007). Adenoviruses use lactoferrin as a bridge for CAR-independent binding to and infection of epithelial cells. *Journal of Virology, 81*, 954–963.
Johnson, D. J. D., Langdown, J., & Huntington, J. A. (2010). Molecular basis of factor IXa recognition by heparin-activated antithrombin revealed by a 1.7-A structure of the ternary complex. *Proceedings of the National Academy of Sciences of the United States of America, 107*, 645–650.
Jones, M. S., Harrach, B., Ganac, R. D., Gozum, M. M. A., Dela Cruz, W. P., Riedel, B., et al. (2007). New adenovirus species found in a patient presenting with gastroenteritis. *Journal of Virology, 81*, 5978–5984.
Jongbloed, S. L., Kassianos, A. J., McDonald, K. J., Clark, G. J., Ju, X., Angel, C. E., et al. (2010). Human CD141+ (BDCA-3)+ dendritic cells (DCs) represent a unique myeloid DC subset that cross-presents necrotic cell antigens. *The Journal of Experimental Medicine, 207*, 1247–1260.
Jonsson, M. I., Lenman, A. E., Frangsmyr, L., Nyberg, C., Abdullahi, M., & Arnberg, N. (2009). Coagulation factors IX and X enhance binding and infection of adenovirus types 5 and 31 in human epithelial cells. *Journal of Virology, 83*, 3816–3825.
Kahl, C. A., Bonnell, J., Hiriyanna, S., Fultz, M., Nyberg-Hoffman, C., Chen, P., et al. (2010). Potent immune responses and in vitro pro-inflammatory cytokine suppression by a novel adenovirus vaccine vector based on rare human serotype 28. *Vaccine, 28*, 5691–5702.
Kaneko, H., Aoki, K., Ishida, S., Ohno, S., Kitaichi, N., Ishiko, H., et al. (2011). Recombination analysis of intermediate human adenovirus type 53 in Japan by complete genome sequence. *The Journal of General Virology, 92*, 1251–1259.
Kaplan, D. H., Shankaran, V., Dighe, A. S., Stockert, E., Aguet, M., Old, L. J., et al. (1998). Demonstration of an interferon gamma-dependent tumor surveillance system in immunocompetent mice. *Proceedings of the National Academy of Sciences of the United States of America, 95*, 7556–7561.
Kar, U. K., Srivastava, M. K., Andersson, A., Baratelli, F., Huang, M., Kickhoefer, V. A., et al. (2011). Novel CCL21-vault nanocapsule intratumoral delivery inhibits lung cancer growth. *PLoS One, 6*, e18758.
Karan, D., Krieg, A. M., & Lubaroff, D. M. (2007). Paradoxical enhancement of CD8 T cell-dependent anti-tumor protection despite reduced CD8 T cell responses with addition of a TLR9 agonist to a tumor vaccine. *International Journal of Cancer, 121*, 1520–1528.
Kasaian, K., & Jones, S. J. (2011). A new frontier in personalized cancer therapy: Mapping molecular changes. *Future Oncology, 7*, 873–894.
Kashentseva, E. A., Seki, T., Curiel, D. T., & Dmitriev, I. P. (2002). Adenovirus targeting to c-erbB-2 oncoprotein by single-chain antibody fused to trimeric form of adenovirus receptor ectodomain. *Cancer Research, 62*, 609–616.
Kass, E., Panicali, D. L., Mazzara, G., Schlom, J., & Greiner, J. W. (2001). Granulocyte/macrophage-colony stimulating factor produced by recombinant avian poxviruses enriches the regional lymph nodes with antigen-presenting cells and acts as an immunoadjuvant. *Cancer Research, 61*, 206–214.
Kemper, C., Leung, M., Stephensen, C. B., Pinkert, C. A., Liszewski, M. K., Cattaneo, R., et al. (2001). Membrane cofactor protein (MCP; CD46) expression in transgenic mice. *Clinical and Experimental Immunology, 124*, 180–189.
Kepp, O., Tesniere, A., Schlemmer, F., Michaud, M., Senovilla, L., Zitvogel, L., et al. (2009). Immunogenic cell death modalities and their impact on cancer treatment. *Apoptosis, 14*, 364–375.
Khare, R., Chen, C. Y., Weaver, E. A., & Barry, M. A. (2011). Advances and future challenges in adenoviral vector pharmacology and targeting. *Current Gene Therapy, 11*, 241–258.

Khare, R., Reddy, V. S., Nemerow, G. R., & Barry, M. A. (2011). Identification of adenovirus serotype 5 hexon regions that interact with scavenger receptors. *Journal of Virology, 86*, 2293–2301.
Kibuuka, H., Kimutai, R., Maboko, L., Sawe, F., Schunk, M. S., Kroidl, A., et al. (2010). A phase 1/2 study of a multiclade HIV-1 DNA plasmid prime and recombinant adenovirus serotype 5 boost vaccine in HIV-Uninfected East Africans (RV 172). *The Journal of Infectious Diseases, 201*, 600–607.
Kikuchi, T., & Crystal, R. G. (1999). Anti-tumor immunity induced by in vivo adenovirus vector-mediated expression of CD40 ligand in tumor cells. *Human Gene Therapy, 10*, 1375–1387.
Kikuchi, T., Moore, M. A., & Crystal, R. G. (2000). Dendritic cells modified to express CD40 ligand elicit therapeutic immunity against preexisting murine tumors. *Blood, 96*, 91–99.
Kim, Y.-S., Kim, Y.-J., Lee, J.-M., Han, S.-H., Ko, H.-J., Park, H.-J., et al. (2010). CD40-targeted recombinant adenovirus significantly enhances the efficacy of antitumor vaccines based on dendritic cells and B cells. *Human Gene Therapy, 21*, 1697–1706.
Kim, Y.-J., Ko, H.-J., Kim, Y.-S., Kim, D.-H., Kang, S., Kim, J.-M., et al. (2008). alpha-Galactosylceramide-loaded, antigen-expressing B cells prime a wide spectrum of antitumor immunity. *International Journal of Cancer, 122*, 2774–2783.
Kim, J., Smith, T., Idamakanti, N., Mulgrew, K., Kaloss, M., Kylefjord, H., et al. (2002). Targeting adenoviral vectors by using the extracellular domain of the coxsackie-adenovirus receptor: Improved potency via trimerization. *Journal of Virology, 76*, 1892–1903.
Kirk, C. J., Hartigan-O'Connor, D., & Mule, J. J. (2001). The dynamics of the T-cell antitumor response: Chemokine-secreting dendritic cells can prime tumor-reactive T cells extranodally. *Cancer Research, 61*, 8794–8802.
Kirk, C. J., Hartigan-O'Connor, D., Nickoloff, B. J., Chamberlain, J. S., Giedlin, M., Aukerman, L., et al. (2001). T cell-dependent antitumor immunity mediated by secondary lymphoid tissue chemokine: Augmentation of dendritic cell-based immunotherapy. *Cancer Research, 61*, 2062–2070.
Knippertz, I., Hesse, A., Schunder, T., Kämpgen, E., Brenner, M. K., Schuler, G., et al. (2009). Generation of human dendritic cells that simultaneously secrete IL-12 and have migratory capacity by adenoviral gene transfer of hCD40L in combination with IFN-gamma. *Journal of Immunotherapy, 32*, 524–538.
Ko, S.-Y., Ko, H.-J., Chang, W.-S., Park, S.-H., Kweon, M.-N., & Kang, C.-Y. (2005). alpha-Galactosylceramide can act as a nasal vaccine adjuvant inducing protective immune responses against viral infection and tumor. *Journal of Immunology, 175*, 3309–3317.
Koizumi, N., Yamaguchi, T., Kawabata, K., Sakurai, F., Sasaki, T., Watanabe, Y., et al. (2007). Fiber-modified adenovirus vectors decrease liver toxicity through reduced IL-6 production. *Journal of Immunology, 178*, 1767–1773.
Korn, T., Nettelbeck, D. M., Völkel, T., Müller, R., & Kontermann, R. E. (2004). Recombinant bispecific antibodies for the targeting of adenoviruses to CEA-expressing tumour cells: A comparative analysis of bacterially expressed single-chain diabody and tandem scFv. *The Journal of Gene Medicine, 6*, 642–651.
Korokhov, N., de Gruijl, T. D., Aldrich, W. A., Triozzi, P. L., Banerjee, P. T., Gillies, S. D., et al. (2005). High efficiency transduction of dendritic cells by adenoviral vectors targeted to DC-SIGN. *Cancer Biology & Therapy, 4*, 289–294.
Korokhov, N., Mikheeva, G., Krendelshchikov, A., Belousova, N., Simonenko, V., Krendelshchikova, V., et al. (2003). Targeting of adenovirus via genetic modification of the viral capsid combined with a protein bridge. *Journal of Virology, 77*, 12931–12940.
Korokhov, N., Noureddini, S. C., Curiel, D. T., Santegoets, S. J. A. M., Scheper, R. J., & de Gruijl, T. D. (2005). A single-component CD40-targeted adenovirus vector displays

highly efficient transduction and activation of dendritic cells in a human skin substrate system. *Molecular Pharmaceutics, 2*, 218–223.
Kortylewski, M., Kujawski, M., Wang, T., Wei, S., Zhang, S., Pilon-Thomas, S., et al. (2005). Inhibiting Stat3 signaling in the hematopoietic system elicits multicomponent antitumor immunity. *Nature Medicine, 11*, 1314–1321.
Koski, A., Kangasniemi, L., Escutenaire, S., Pesonen, S., Cerullo, V., Diaconu, I., et al. (2010). Treatment of cancer patients with a serotype 5/3 chimeric oncolytic adenovirus expressing GMCSF. *Molecular Therapy, 18*, 1874–1884.
Koup, R. A., Roederer, M., Lamoreaux, L., Fischer, J., Novik, L., Nason, M. C., et al. (2010). Priming immunization with DNA augments immunogenicity of recombinant adenoviral vectors for both HIV-1 specific antibody and T-cell responses. *PLoS One, 5*, e9015.
Krasnykh, V., Dmitriev, I., Mikheeva, G., Miller, C. R., Belousova, N., & Curiel, D. T. (1998). Characterization of an adenovirus vector containing a heterologous peptide epitope in the HI loop of the fiber knob. *Journal of Virology, 72*, 1844–1852.
Krasnykh, V. N., Douglas, J. T., & van Beusechem, V. W. (2000). Genetic targeting of adenoviral vectors. *Molecular Therapy, 1*, 391–405.
Krause, A., Joh, J. H., Hackett, N. R., Roelvink, P. W., Bruder, J. T., Wickham, T. J., et al. (2006). Epitopes expressed in different adenovirus capsid proteins induce different levels of epitope-specific immunity. *Journal of Virology, 80*, 5523–5530.
Kresge, K. J. (2007). A STEP back? Newspaper article. *IAVI Report, 11*, 1–5.
Kuball, J., Schuler, M., Antunes Ferreira, E., Herr, W., Neumann, M., Obenauer-Kutner, L., et al. (2002). Generating p53-specific cytotoxic T lymphocytes by recombinant adenoviral vector-based vaccination in mice, but not man. *Gene Therapy, 9*, 833–843.
Kumamoto, T., Huang, E. K., Paek, H. J., Morita, A., Matsue, H., Valentini, R. F., et al. (2002). Induction of tumor-specific protective immunity by in situ Langerhans cell vaccine. *Nature Biotechnology, 20*, 64–69.
Lapenta, C., Santini, S. M., Spada, M., Donati, S., Urbani, F., Accapezzato, D., et al. (2006). IFN-alpha-conditioned dendritic cells are highly efficient in inducing cross-priming CD8(+) T cells against exogenous viral antigens. *European Journal of Immunology, 36*, 2046–2060.
Lapteva, N., Aldrich, M., Rollins, L., Ren, W., Goltsova, T., Chen, S.-Y., et al. (2009). Attraction and activation of dendritic cells at the site of tumor elicits potent antitumor immunity. *Molecular Therapy, 17*, 1626–1636.
Lapteva, N., Aldrich, M., Weksberg, D., Rollins, L., Goltsova, T., Chen, S.-Y., et al. (2009). Targeting the intratumoral dendritic cells by the oncolytic adenoviral vaccine expressing RANTES elicits potent antitumor immunity. *Journal of Immunotherapy, 32*, 145–156.
Lasarte, J. J., Casares, N., Gorraiz, M., Hervás-Stubbs, S., Arribillaga, L., Mansilla, C., et al. (2007). The extra domain A from fibronectin targets antigens to TLR4-expressing cells and induces cytotoxic T cell responses in vivo. *Journal of Immunology, 178*, 748–756.
Ledford, H. (2011). Melanoma drug wins US approval. *Nature, 471*, 561.
Lee, Y.-S., Kim, J.-H., Choi, K.-J., Choi, I.-K., Kim, H., Cho, S., et al. (2006). Enhanced antitumor effect of oncolytic adenovirus expressing interleukin-12 and B7-1 in an immunocompetent murine model. *Clinical Cancer Research, 12*, 5859–5868.
Lee, C. T., Wu, S., Ciernik, I. F., Chen, H., Nadaf-Rahrov, S., Gabrilovich, D., et al. (1997). Genetic immunotherapy of established tumors with adenovirus-murine granulocyte-macrophage colony-stimulating factor. *Human Gene Therapy, 8*, 187–193.
Lemckert, A. A. C., Sumida, S. M., Holterman, L., Vogels, R., Truitt, D. M., Lynch, D. M., et al. (2005). Immunogenicity of heterologous prime-boost regimens involving recombinant adenovirus serotype 11 (Ad11) and Ad35 vaccine vectors in the presence of anti-Ad5 immunity. *Journal of Virology, 79*, 9694–9701.

Lemke, C. D., Graham, J. B., Geary, S. M., Zamba, G., Lubaroff, D. M., & Salem, A. K. (2011). Chitosan is a surprising negative modulator of cytotoxic CD8 + T cell responses elicited by adenovirus cancer vaccines. *Molecular Pharmaceutics*, *8*, 1652–1661.

Li, E., Brown, S. L., Stupack, D. G., Puente, X. S., Cheresh, D. A., & Nemerow, G. R. (2001). Integrin alpha(v)beta1 is an adenovirus coreceptor. *Journal of Virology*, *75*, 5405–5409.

Li, J.-L., Liu, H.-L., Zhang, X.-R., Xu, J.-P., Hu, W.-K., Liang, M., et al. (2009). A phase I trial of intratumoral administration of recombinant oncolytic adenovirus overexpressing HSP70 in advanced solid tumor patients. *Gene Therapy*, *16*, 376–382.

Li, Y., Wang, L.-X., Yang, G., Hao, F., Urba, W. J., & Hu, H.-M. (2008). Efficient cross-presentation depends on autophagy in tumor cells. *Cancer Research*, *68*, 6889–6895.

Lindenberg, J. J., Fehres, C. M., van Cruijsen, H., Oosterhoff, D., & de Gruijl, T. D. (2011). Cross-talk between tumor and myeloid cells: How to tip the balance in favor of anti-tumor immunity. *Immunotherapy*, *3*, 77–96.

Lindsay, R. W. B., Darrah, P. A., Quinn, K. M., Wille-Reece, U., Mattei, L. M., Iwasaki, A., et al. (2010). CD8+ T cell responses following replication-defective adenovirus serotype 5 immunization are dependent on CD11c+ dendritic cells but show redundancy in their requirement of TLR and nucleotide-binding oligomerization domain-like receptor signaling. *Journal of Immunology*, *185*, 1513–1521.

Liu, M. A. (2010). Immunologic basis of vaccine vectors. *Immunity*, *33*, 504–515.

Liu, X., Cao, X., Wei, R., Cai, Y., Li, H., Gui, J., et al. (2011). Gene-viro-therapy targeting liver cancer by a dual-regulated oncolytic adenoviral vector harboring IL-24 and TRAIL. *Cancer Gene Therapy*, *19*, 49–57.

Liu, F., Fang, C., Ye, X., Lu, Q., Yu, D.-H., Zhang, W.-W., et al. (2009). Anti-4-1BB scFv immunogene therapy and low dose cyclophosphamide exhibit a synergistic antitumor effect in established murine lung tumors. *Cancer Biology & Therapy*, *8*, 707–713.

Liu, F., Lu, Q., Ye, X., Fang, C., Zhao, Y., Liang, M., et al. (2008). Cancer gene therapy of adenovirus-mediated anti-4-1BB scFv in immunocompetent mice. *Cancer Biology & Therapy*, *7*, 448–453.

Lopes, L., Dewannieux, M., Gileadi, U., Bailey, R., Ikeda, Y., Whittaker, C., et al. (2008). Immunization with a lentivector that targets tumor antigen expression to dendritic cells induces potent CD8+ and CD4+ T-cell responses. *Journal of Virology*, *82*, 86–95.

Loré, K., Adams, W. C., Havenga, M. J. E., Precopio, M. L., Holterman, L., Goudsmit, J., et al. (2007). Myeloid and plasmacytoid dendritic cells are susceptible to recombinant adenovirus vectors and stimulate polyfunctional memory T cell responses. *Journal of Immunology*, *179*, 1721–1729.

Lotem, M., Zhao, Y., Riley, J., Hwu, P., Morgan, R. A., Rosenberg, S. A., et al. (2006). Presentation of tumor antigens by dendritic cells genetically modified with viral and non-viral vectors. *Journal of Immunotherapy*, *29*, 616–627.

Lubaroff, D. M., Konety, B. R., Link, B., Gerstbrein, J., Madsen, T., Shannon, M., et al. (2009). Phase I clinical trial of an adenovirus/prostate-specific antigen vaccine for prostate cancer: Safety and immunologic results. *Clinical Cancer Research*, *15*, 7375–7380.

Madisch, I., Harste, G., Pommer, H., & Heim, A. (2005). Phylogenetic analysis of the main neutralization and hemagglutination determinants of all human adenovirus prototypes as a basis for molecular classification and taxonomy. *Journal of Virology*, *79*, 15265–15276.

Magnusson, M. K., Kraaij, R., Leadley, R. M., De Ridder, C. M. A., van Weerden, W. M., Van Schie, K. A. J., et al. (2012). A transductionally retargeted adenoviral vector for virotherapy of Her2/neu-expressing prostate cancer. *Human Gene Therapy*, *23*, 70–82.

Malmström, P.-U., Loskog, A. S. I., Lindqvist, C. A., Mangsbo, S. M., Fransson, M., Wanders, A., et al. (2010). AdCD40L immunogene therapy for bladder carcinoma—The first phase I/IIa trial. *Clinical Cancer Research*, *16*, 3279–3287.

Malvicini, M., Ingolotti, M., Piccioni, F., Garcia, M., Bayo, J., Atorrasagasti, C., et al. (2011). Reversal of gastrointestinal carcinoma-induced immunosuppression and induction of antitumoural immunity by a combination of cyclophosphamide and gene transfer of IL-12. *Molecular Oncology*, *5*, 242–255.

Marshall, E. (1999). Gene therapy death prompts review of adenovirus vector. *Science*, *286*, 2244–2245.

Mast, T. C., Kierstead, L., Gupta, S. B., Nikas, A. A., Kallas, E. G., Novitsky, V., et al. (2010). International epidemiology of human pre-existing adenovirus (Ad) type-5, type-6, type-26 and type-36 neutralizing antibodies: Correlates of high Ad5 titers and implications for potential HIV vaccine trials. *Vaccine*, *28*, 950–957.

Matthews, Q. L. (2011). Capsid-incorporation of antigens into adenovirus capsid proteins for a vaccine approach. *Molecular Pharmaceutics*, *8*, 3–11.

Matthews, Q. L., Yang, P., Wu, Q., Belousova, N., Rivera, A. A., Stoff-Khalili, M. A., et al. (2008). Optimization of capsid-incorporated antigens for a novel adenovirus vaccine approach. *Virology Journal*, *5*, 98.

Mazzolini, G., Alfaro, C., Sangro, B., Feijoó, E., Ruiz, J., Benito, A., et al. (2005). Intratumoral injection of dendritic cells engineered to secrete interleukin-12 by recombinant adenovirus in patients with metastatic gastrointestinal carcinomas. *Journal of Clinical Oncology*, *23*, 999–1010.

Mazzolini, G., Prieto, J., & Melero, I. (2003). Gene therapy of cancer with interleukin-12. *Current Pharmaceutical Design*, *9*, 1981–1991.

McConnell, M. J., Danthinne, X., & Imperiale, M. J. (2006). Characterization of a permissive epitope insertion site in adenovirus hexon. *Journal of Virology*, *80*, 5361–5370.

McCoy, K., Tatsis, N., Korioth-Schmitz, B., Lasaro, M. O., Hensley, S. E., Lin, S.-W., et al. (2007). Effect of preexisting immunity to adenovirus human serotype 5 antigens on the immune responses of nonhuman primates to vaccine regimens based on human- or chimpanzee-derived adenovirus vectors. *Journal of Virology*, *81*, 6594–6604.

McGuire, K. A., Barlan, A. U., Griffin, T. M., & Wiethoff, C. M. (2011). Adenovirus type 5 rupture of lysosomes leads to cathepsin B-dependent mitochondrial stress and production of reactive oxygen species. *Journal of Virology*, *85*, 10806–10813.

Melief, C. J. M., van Der Burg, S. H., Toes, R. E. M., Ossendorp, F., & Offringa, R. (2002). Effective therapeutic anticancer vaccines based on precision guiding of cytolytic T lymphocytes. *Immunological Reviews*, *188*, 177–182.

Merad, M., & Manz, M. G. (2009). Dendritic cell homeostasis. *Blood*, *113*, 3418–3427.

Mescher, M. F., Curtsinger, J. M., Agarwal, P., Casey, K. A., Gerner, M., Hammerbeck, C. D., et al. (2006). Signals required for programming effector and memory development by CD8+ T cells. *Immunological Reviews*, *211*, 81–92.

Miller, G., Lahrs, S., Pillarisetty, V. G., Shah, A. B., & DeMatteo, R. P. (2002). Adenovirus infection enhances dendritic cell immunostimulatory properties and induces natural killer and T-cell-mediated tumor protection. *Cancer Research*, *62*, 5260–5266.

Mincheff, M., Tchakarov, S., Zoubak, S., Loukinov, D., Botev, C., Altankova, I., et al. (2000). Naked DNA and adenoviral immunizations for immunotherapy of prostate cancer: A phase I/II clinical trial. *European Urology*, *38*, 208–217.

Mizuguchi, H., Kay, M. A., & Hayakawa, T. (2001). Approaches for generating recombinant adenovirus vectors. *Advanced Drug Delivery Reviews*, *52*, 165–176.

Mok, H., Palmer, D., Ng, P., & Barry, M. (2005). Evaluation of polyethylene glycol modification of first-generation and helper-dependent adenoviral vectors to reduce innate immune responses. *Molecular Therapy*, *11*, 66–79.

Molinier-Frenkel, V., Lengagne, R., Gaden, F., Hong, S.-S., Choppin, J., Gahery-Ségard, H., et al. (2002). Adenovirus hexon protein is a potent adjuvant for activation of a cellular immune response. *Journal of Virology*, *76*, 127–135.

Morita, A., Ariizumi, K., Ritter, R., Jester, J. V., Kumamoto, T., Johnston, S. A., et al. (2001). Development of a Langerhans cell-targeted gene therapy format using a dendritic cell-specific promoter. *Gene Therapy, 8*, 1729–1737.

Muruve, D. A., Cotter, M. J., Zaiss, A. K., White, L. R., Liu, Q., Chan, T., et al. (2004). Helper-dependent adenovirus vectors elicit intact innate but attenuated adaptive host immune responses in vivo. *Journal of Virology, 78*, 5966–5972.

Nefedova, Y., Fishman, M., Sherman, S., Wang, X., Beg, A. A., & Gabrilovich, D. I. (2007). Mechanism of all-trans retinoic acid effect on tumor-associated myeloid-derived suppressor cells. *Cancer Research, 67*, 11021–11028.

Nefedova, Y., Huang, M., Kusmartsev, S., Bhattacharya, R., Cheng, P., Salup, R., et al. (2004). Hyperactivation of STAT3 is involved in abnormal differentiation of dendritic cells in cancer. *Journal of Immunology, 172*, 464–474.

Nefedova, Y., Nagaraj, S., Rosenbauer, A., Muro-Cacho, C., Sebti, S. M., & Gabrilovich, D. I. (2005). Regulation of dendritic cell differentiation and antitumor immune response in cancer by pharmacologic-selective inhibition of the janus-activated kinase 2/signal transducers and activators of transcription 3 pathway. *Cancer Research, 65*, 9525–9535.

Nemerow, G. R., Pache, L., Reddy, V., & Stewart, P. L. (2009). Insights into adenovirus host cell interactions from structural studies. *Virology, 384*, 380–388.

Nemunaitis, J., Meyers, T., Senzer, N., Cunningham, C., West, H., Vallieres, E., et al. (2006). Phase I trial of sequential administration of recombinant DNA and adenovirus expressing L523S protein in early stage non-small-cell lung cancer. *Molecular Therapy, 13*, 1185–1191.

Nettelbeck, D. M., Miller, D. W., Jérôme, V., Zuzarte, M., Watkins, S. J., Hawkins, R. E., et al. (2001). Targeting of adenovirus to endothelial cells by a bispecific single-chain diabody directed against the adenovirus fiber knob domain and human endoglin (CD105). *Molecular Therapy, 3*, 882–891.

Nettelbeck, D. M., Rivera, A. A., Kupsch, J., Dieckmann, D., Douglas, J. T., Kontermann, R. E., et al. (2004). Retargeting of adenoviral infection to melanoma: Combining genetic ablation of native tropism with a recombinant bispecific single-chain diabody (scDb) adapter that binds to fiber knob and HMWMAA. *International Journal of Cancer, 108*, 136–145.

Ni, S., Bernt, K., Gaggar, A., Li, Z.-Y., Kiem, H.-P., & Lieber, A. (2005). Evaluation of biodistribution and safety of adenovirus vectors containing group B fibers after intravenous injection into baboons. *Human Gene Therapy, 16*, 664–677.

Nicklin, S. A., Wu, E., Nemerow, G. R., & Baker, A. H. (2005). The influence of adenovirus fiber structure and function on vector development for gene therapy. *Molecular Therapy, 12*, 384–393.

Nociari, M., Ocheretina, O., Schoggins, J. W., & Falck-Pedersen, E. (2007). Sensing infection by adenovirus: Toll-like receptor-independent viral DNA recognition signals activation of the interferon regulatory factor 3 master regulator. *Journal of Virology, 81*, 4145–4157.

Ogawara, K.-I., Rots, M. G., Kok, R. J., Moorlag, H. E., Van Loenen, A.-M., Meijer, D. K. F., et al. (2004). A novel strategy to modify adenovirus tropism and enhance transgene delivery to activated vascular endothelial cells in vitro and in vivo. *Human Gene Therapy, 15*, 433–443.

Ojima, T., Iwahashi, M., Nakamura, M., Matsuda, K., Nakamori, M., Ueda, K., et al. (2007). Benefits of gene transduction of granulocyte macrophage colony-stimulating factor in cancer vaccine using genetically modified dendritic cells. *International Journal of Oncology, 31*, 931–939.

Okada, N., Sasaki, A., Niwa, M., Okada, Y., Hatanaka, Y., Tani, Y., et al. (2006). Tumor suppressive efficacy through augmentation of tumor-infiltrating immune cells by intratumoral injection of chemokine-expressing adenoviral vector. *Cancer Gene Therapy*, *13*, 393–405.

Olive, M., Eisenlohr, L., Flomenberg, N., Hsu, S., & Flomenberg, P. (2002). The adenovirus capsid protein hexon contains a highly conserved human CD4+ T-cell epitope. *Human Gene Therapy*, *13*, 1167–1178.

Oosterhoff, D., Sluijter, B. J. R., Hangalapura, B. N., & de Gruijl, T. D. (2012). The dermis as a portal for dendritic cell-targeted immunotherapy of cutaneous melanoma. *Current Topics in Microbiology and Immunology*, *351*, 181–220.

Ophorst, O. J. A. E., Kostense, S., Goudsmit, J., de Swart, R. L., Verhaagh, S., Zakhartchouk, A., et al. (2004). An adenoviral type 5 vector carrying a type 35 fiber as a vaccine vehicle: DC targeting, cross neutralization, and immunogenicity. *Vaccine*, *22*, 3035–3044.

Ophorst, O. J. A. E., Radosevic, K., Klap, J. M., Sijtsma, J., Gillissen, G., Mintardjo, R., et al. (2007). Increased immunogenicity of recombinant Ad35-based malaria vaccine through formulation with aluminium phosphate adjuvant. *Vaccine*, *25*, 6501–6510.

O'Riordan, C. R., Lachapelle, A., Delgado, C., Parkes, V., Wadsworth, S. C., Smith, A. E., et al. (1999). PEGylation of adenovirus with retention of infectivity and protection from neutralizing antibody in vitro and in vivo. *Human Gene Therapy*, *10*, 1349–1358.

Othman, M., Labelle, A., Mazzetti, I., Elbatarny, H. S., & Lillicrap, D. (2007). Adenovirus-induced thrombocytopenia: The role of von Willebrand factor and P-selectin in mediating accelerated platelet clearance. *Blood*, *109*, 2832–2839.

Palm, N. W., & Medzhitov, R. (2009). Pattern recognition receptors and control of adaptive immunity. *Immunological Reviews*, *227*, 221–233.

Palucka, K., Ueno, H., Fay, J., & Banchereau, J. (2009). Harnessing dendritic cells to generate cancer vaccines. *Annals of the New York Academy of Sciences*, *1174*, 88–98.

Paoloni, M., & Khanna, C. (2008). Translation of new cancer treatments from pet dogs to humans. *Nature Reviews. Cancer*, *8*, 147–156.

Park, K.-H., Kim, G., Jang, S. H., Kim, C. H., Kwon, S.-Y., Yoo, C.-G., et al. (2003). Gene therapy with GM-CSF, interleukin-4 and herpes simplex virus thymidine kinase shows strong antitumor effect on lung cancer. *Anticancer Research*, *23*, 1559–1564.

Park, J. W., Mok, H., & Park, T. G. (2008). Epidermal growth factor (EGF) receptor targeted delivery of PEGylated adenovirus. *Biochemical and Biophysical Research Communications*, *366*, 769–774.

Parks, R. (2005). Adenovirus protein IX: A new look at an old protein. *Molecular Therapy*, *11*, 19–25.

Patterson, L. J. (2011). The "STEP-wise" future of adenovirus-based HIV vaccines. *Current Medicinal Chemistry*, *18*, 3981–3986.

Pereboev, A. V., Asiedu, C. K., Kawakami, Y., Dong, S. S., Blackwell, J. L., Kashentseva, E. A., et al. (2002). Coxsackievirus-adenovirus receptor genetically fused to anti-human CD40 scFv enhances adenoviral transduction of dendritic cells. *Gene Therapy*, *9*, 1189–1193.

Pereboeva, L., Komarova, S., Roth, J., Ponnazhagan, S., & Curiel, D. T. (2007). Targeting EGFR with metabolically biotinylated fiber-mosaic adenovirus. *Gene Therapy*, *14*, 627–637.

Perkins, S. D., Williams, A. J., O'Brien, L. M., Laws, T. R., & Phillpotts, R. J. (2008). CpG used as an adjuvant for an adenovirus-based Venezuelan equine encephalitis virus vaccine increases the immune response to the vector, but not to the transgene product. *Viral Immunology*, *21*, 451–457.

Pesonen, S., Diaconu, I., Cerullo, V., Escutenaire, S., Raki, M., Kangasniemi, L., et al. (2012). Integrin targeted oncolytic adenoviruses Ad5-D24-RGD and Ad5-RGD-D24-GMCSF for treatment of patients with advanced chemotherapy refractory solid tumors. *International Journal of Cancer*, *130*, 1937–1947.

Pesonen, S., Diaconu, I., Kangasniemi, L., Ranki, T., Kanerva, A., Pesonen, S. K., et al. (2012). Oncolytic immunotherapy of advanced solid tumors with a CD40L-expressing replicating adenovirus: Assessment of safety and immunologic responses in patients. *Cancer Research, 72,* 1621–1631.

Pincha, M., Sundarasetty, B. S., & Stripecke, R. (2010). Lentiviral vectors for immunization: An inflammatory field. *Expert Review of Vaccines, 9,* 309–321.

Pooley, J. L., Heath, W. R., & Shortman, K. (2001). Cutting edge: Intravenous soluble antigen is presented to CD4 T cells by CD8− dendritic cells, but cross-presented to CD8 T cells by CD8+ dendritic cells. *Journal of Immunology, 166,* 5327–5330.

Porgador, A., Irvine, K. R., Iwasaki, A., Barber, B. H., Restifo, N. P., & Germain, R. N. (1998). Predominant role for directly transfected dendritic cells in antigen presentation to CD8+ T cells after gene gun immunization. *The Journal of Experimental Medicine, 188,* 1075–1082.

Poschke, I., Mougiakakos, D., Hansson, J., Masucci, G. V., & Kiessling, R. (2010). Immature immunosuppressive CD14+HLA-DR-/low cells in melanoma patients are Stat3hi and overexpress CD80, CD83, and DC-sign. *Cancer Research, 70,* 4335–4345.

Poulin, L. F., Salio, M., Griessinger, E., Anjos-Afonso, F., Craciun, L., Chen, J.-L., et al. (2010). Characterization of human DNGR-1+ BDCA3+ leukocytes as putative equivalents of mouse CD8alpha+ dendritic cells. *The Journal of Experimental Medicine, 207,* 1261–1271.

Radosevic, K., Rodríguez, A., Lemckert, A. A. C., van der Meer, M., Gillissen, G., Warnar, C., et al. (2010). The Th1 immune response to Plasmodium falciparum circumsporozoite protein is boosted by adenovirus vectors 35 and 26 with a homologous insert. *Clinical and Vaccine Immunology, 17,* 1687–1694.

Raki, M., Sarkioja, M., Escutenaire, S., Kangasniemi, L., Haavisto, E., Kanerva, A., et al. (2011). Switching the fiber knob of oncolytic adenoviruses to avoid neutralizing antibodies in human cancer patients. *The Journal of Gene Medicine, 13,* 253–261.

Rea, D., Havenga, M. J., van Den Assem, M., Sutmuller, R. P., Lemckert, A., Hoeben, R. C., et al. (2001). Highly efficient transduction of human monocyte-derived dendritic cells with subgroup B fiber-modified adenovirus vectors enhances transgene-encoded antigen presentation to cytotoxic T cells. *Journal of Immunology, 166,* 5236–5244.

Rea, D., Schagen, F. H., Hoeben, R. C., Mehtali, M., Havenga, M. J., Toes, R. E., et al. (1999). Adenoviruses activate human dendritic cells without polarization toward a T-helper type 1-inducing subset. *Journal of Virology, 73,* 10245–10253.

Rhee, E. G., Blattman, J. N., Kasturi, S. P., Kelley, R. P., Kaufman, D. R., Lynch, D. M., et al. (2010). Multiple innate immune pathways contribute to the immunogenicity of recombinant adenovirus vaccine vectors. *Journal of Virology, 85,* 315–323.

Ribas, A., Butterfield, L. H., McBride, W. H., Dissette, V. B., Koh, A., Vollmer, C. M., et al. (1999). Characterization of antitumor immunization to a defined melanoma antigen using genetically engineered murine dendritic cells. *Cancer Gene Therapy, 6,* 523–536.

Ribas, A., Butterfield, L. H., McBride, W. H., Jilani, S. M., Bui, L. A., Vollmer, C. M., et al. (1997). Genetic immunization for the melanoma antigen MART-1/Melan-A using recombinant adenovirus-transduced murine dendritic cells. *Cancer Research, 57,* 2865–2869.

Ridge, J. P., Di Rosa, F., & Matzinger, P. (1998). A conditioned dendritic cell can be a temporal bridge between a CD4+ T-helper and a T-killer cell. *Nature, 393,* 474–478.

Rieger, R., & Kipps, T. J. (2003). CpG oligodeoxynucleotides enhance the capacity of adenovirus-mediated CD154 gene transfer to generate effective B-cell lymphoma vaccines. *Cancer Research, 63,* 4128–4135.

Rizza, P., Capone, I., Moretti, F., Proietti, E., & Belardelli, F. (2011). IFN-α as a vaccine adjuvant: Recent insights into the mechanisms and perspectives for its clinical use. *Expert Review of Vaccines, 10,* 487–498.

Robert-Guroff, M. (2007). Replicating and non-replicating viral vectors for vaccine development. *Current Opinion in Biotechnology, 18*, 546–556.
Roberts, D. M., Nanda, A., Havenga, M. J. E., Abbink, P., Lynch, D. M., Ewald, B. A., et al. (2006). Hexon-chimaeric adenovirus serotype 5 vectors circumvent pre-existing anti-vector immunity. *Nature, 441*, 239–243.
Robinson, M., Li, B., Ge, Y., Ko, D., Yendluri, S., Harding, T., et al. (2009). Novel immunocompetent murine tumor model for evaluation of conditionally replication-competent (oncolytic) murine adenoviral vectors. *Journal of Virology, 83*, 3450–3462.
Rodríguez, A., Mintardjo, R., Tax, D., Gillissen, G., Custers, J., Pau, M. G., et al. (2009). Evaluation of a prime-boost vaccine schedule with distinct adenovirus vectors against malaria in rhesus monkeys. *Vaccine, 27*, 6226–6233.
Rodriguez-Rocha, H., Gomez-Gutierrez, J. G., Garcia-Garcia, A., Rao, X.-M., Chen, L., McMasters, K. M., et al. (2011). Adenoviruses induce autophagy to promote virus replication and oncolysis. *Virology, 416*, 9–15.
Roelvink, P. W., Kovesdi, I., & Wickham, T. J. (1996). Comparative analysis of adenovirus fiber-cell interaction: Adenovirus type 2 (Ad2) and Ad9 utilize the same cellular fiber receptor but use different binding strategies for attachment. *Journal of Virology, 70*, 7614–7621.
Roelvink, P. W., Lizonova, A., Lee, J. G., Li, Y., Bergelson, J. M., Finberg, R. W., et al. (1998). The coxsackievirus-adenovirus receptor protein can function as a cellular attachment protein for adenovirus serotypes from subgroups A, C, D, E, and F. *Journal of Virology, 72*, 7909–7915.
Roelvink, P. W., Mi Lee, G., Einfeld, D. A., Kovesdi, I., & Wickham, T. J. (1999). Identification of a conserved receptor-binding site on the fiber proteins of CAR-recognizing adenoviridae. *Science, 286*, 1568–1571.
Rollier, C. S., Reyes-Sandoval, A., Cottingham, M. G., Ewer, K., & Hill, A. V. S. (2011). Viral vectors as vaccine platforms: Deployment in sight. *Current Opinion in Immunology, 23*, 377–382.
Romani, N., Clausen, B. E., & Stoitzner, P. (2010). Langerhans cells and more: Langerin-expressing dendritic cell subsets in the skin. *Immunological Reviews, 234*, 120–141.
Rosa, D. S., Ribeiro, S. P., & Cunha-Neto, E. (2010). CD4+ T cell epitope discovery and rational vaccine design. *Archivum Immunologiae et Therapiae Experimentalis, 58*, 121–130.
Rosario, M., Borthwick, N., Stewart-Jones, G. B., Mbewe-Mvula, A., Bridgeman, A., Colloca, S., et al. (2012). Prime-boost regimens with adjuvanted synthetic long peptides elicit T cells and antibodies to conserved regions of HIV-1 in macaques. *AIDS, 26*, 275–284.
Rosenberg, S. A., Zhai, Y., Yang, J. C., Schwartzentruber, D. J., Hwu, P., Marincola, F. M., et al. (1998). Immunizing patients with metastatic melanoma using recombinant adenoviruses encoding MART-1 or gp100 melanoma antigens. *Journal of the National Cancer Institute, 90*, 1894–1900.
Roy, S., Shirley, P. S., McClelland, A., & Kaleko, M. (1998). Circumvention of immunity to the adenovirus major coat protein hexon. *Journal of Virology, 72*, 6875–6879.
Russell, W. C. (2009). Adenoviruses: Update on structure and function. *The Journal of General Virology, 90*, 1–20.
Sakurai, F., Kawabata, K., & Mizuguchi, H. (2007). Adenovirus vectors composed of subgroup B adenoviruses. *Current Gene Therapy, 7*, 229–238.
Sangro, B., Mazzolini, G., Ruiz, J., Herraiz, M., Quiroga, J., Herrero, I., et al. (2004). Phase I trial of intratumoral injection of an adenovirus encoding interleukin-12 for advanced digestive tumors. *Journal of Clinical Oncology, 22*, 1389–1397.
Sarkar, D., Lebedeva, I. V., Su, Z. Z., Park, E. S., Chatman, L., Vozhilla, N., et al. (2007). Eradication of therapy-resistant human prostate tumors using a cancer terminator virus. *Cancer Research, 67*, 5434–5442.

Sarkar, D., Su, Z. Z., Park, E. S., Vozhilla, N., Dent, P., Curiel, D. T., et al. (2008). A cancer terminator virus eradicates both primary and distant human melanomas. *Cancer Gene Therapy*, *15*, 293–302.

Sarkar, D., Su, Z. Z., Vozhilla, N., Park, E. S., Gupta, P., & Fisher, P. B. (2005). Dual cancer-specific targeting strategy cures primary and distant breast carcinomas in nude mice. *Proceedings of the National Academy of Sciences of the United States of America*, *102*, 14034–14039.

Sarkar, D., Su, Z. Z., Vozhilla, N., Park, E. S., Randolph, A., Valerie, K., et al. (2005). Targeted virus replication plus immunotherapy eradicates primary and distant pancreatic tumors in nude mice. *Cancer Research*, *65*, 9056–9063.

Sas, S., Chan, T., Sami, A., El-Gayed, A., & Xiang, J. (2008). Vaccination of fiber-modified adenovirus-transfected dendritic cells to express HER-2/neu stimulates efficient HER-2/neu-specific humoral and CTL responses and reduces breast carcinogenesis in transgenic mice. *Cancer Gene Therapy*, *15*, 655–666.

Satoh, T., Saika, T., Ebara, S., Kusaka, N., Timme, T. L., Yang, G., et al. (2003). Macrophages transduced with an adenoviral vector expressing interleukin 12 suppress tumor growth and metastasis in a preclinical metastatic prostate cancer model. *Cancer Research*, *63*, 7853–7860.

Sauane, M., Su, Z. Z., Gupta, P., Lebedeva, I. V., Dent, P., Sarkar, D., et al. (2008). Autocrine regulation of mda-7/IL-24 mediates cancer-specific apoptosis. *Proceedings of the National Academy of Sciences of the United States of America*, *105*, 9763–9768.

Schierer, S., Hesse, A., Knippertz, I., Kaempgen, E., Baur, A. S., Schuler, G., et al. (2012). Human dendritic cells efficiently phagocytose adenoviral oncolysate but require additional stimulation to mature. *International Journal of Cancer*, *130*, 1682–1694.

Schierer, S., Hesse, A., Müller, I., Kämpgen, E., Curiel, D. T., Schuler, G., et al. (2007). Modulation of viability and maturation of human monocyte-derived dendritic cells by oncolytic adenoviruses. *International Journal of Cancer*, *122*, 219–229.

Schirmbeck, R., Reimann, J., Kochanek, S., & Kreppel, F. (2008). The immunogenicity of adenovirus vectors limits the multispecificity of CD8 T-cell responses to vector-encoded transgenic antigens. *Molecular Therapy*, *16*, 1609–1616.

Schoenberger, S. P., Toes, R. E., van der Voort, E. I., Offringa, R., & Melief, C. J. (1998). T-cell help for cytotoxic T lymphocytes is mediated by CD40-CD40L interactions. *Nature*, *393*, 480–483.

Schowalter, D. B., Tubb, J. C., Liu, M., Wilson, C. B., & Kay, M. A. (1997). Heterologous expression of adenovirus E3-gp19K in an E1a-deleted adenovirus vector inhibits MHC I expression in vitro, but does not prolong transgene expression in vivo. *Gene Therapy*, *4*, 351–360.

Segerman, A., Atkinson, J. P., Marttila, M., Dennerquist, V., Wadell, G., & Arnberg, N. (2003). Adenovirus type 11 uses CD46 as a cellular receptor. *Journal of Virology*, *77*, 9183–9191.

Segerman, A., Lindman, K., Mei, Y.-F., Allard, A., & Wadell, G. (2006). Adenovirus types 11p and 35 attach to and infect primary lymphocytes and monocytes, but hexon expression in T-cells requires prior activation. *Virology*, *349*, 96–111.

Seiler, M. P., Cerullo, V., & Lee, B. (2007). Immune response to helper dependent adenoviral mediated liver gene therapy: Challenges and prospects. *Current Gene Therapy*, *7*, 297–305.

Seiradake, E., Henaff, D., Wodrich, H., Billet, O., Perreau, M., Hippert, C., et al. (2009). The cell adhesion molecule "CAR" and sialic acid on human erythrocytes influence adenovirus in vivo biodistribution. *PLoS Pathogens*, *5*, e1000277.

Sekaly, R.-P. (2008). The failed HIV Merck vaccine study: A step back or a launching point for future vaccine development? *The Journal of Experimental Medicine*, *205*, 7–12.

Seregin, S. S., & Amalfitano, A. (2010). Improving adenovirus based gene transfer: Strategies to accomplish immune evasion. *Viruses*, *2*, 2013–2036.

Seshidhar Reddy, P., Ganesh, S., Limbach, M. P., Brann, T., Pinkstaff, A., Kaloss, M., et al. (2003). Development of adenovirus serotype 35 as a gene transfer vector. *Virology*, *311*, 384–393.

Seto, D., Chodosh, J., Brister, J. R., & Jones, M. S. Members of the Adenovirus Research Community (2011). Using the whole-genome sequence to characterize and name human adenoviruses. *Journal of Virology*, *85*, 5701–5702.

Seya, T., Hirano, A., Matsumoto, M., Nomura, M., & Ueda, S. (1999). Human membrane cofactor protein (MCP, CD46): Multiple isoforms and functions. *The International Journal of Biochemistry & Cell Biology*, *31*, 1255–1260.

Seya, T., Nomura, M., Murakami, Y., Begum, N. A., Matsumoto, M., & Nagasawa, S. (1998). CD46 (membrane cofactor protein of complement, measles virus receptor): Structural and functional divergence among species (review). *International Journal of Molecular Medicine*, *1*, 809–816.

Sheehy, S. H., Duncan, C. J., Elias, S. C., Collins, K. A., Ewer, K. J., Spencer, A. J., et al. (2011). Phase Ia clinical evaluation of the Plasmodium falciparum blood-stage antigen MSP1 in ChAd63 and MVA vaccine vectors. *Molecular Therapy*, *19*, 2269–2276.

Shimony, N., Elkin, G., Kolodkin-Gal, D., Krasny, L., Urieli-Shoval, S., & Haviv, Y. S. (2009). Analysis of adenoviral attachment to human platelets. *Virology Journal*, *6*, 25.

Shiver, J. W., Fu, T.-M., Chen, L., Casimiro, D. R., Davies, M.-E., Evans, R. K., et al. (2002). Replication-incompetent adenoviral vaccine vector elicits effective anti-immunodeficiency-virus immunity. *Nature*, *415*, 331–335.

Short, J. (2004). Adenovirus serotype 3 utilizes CD80 (B7.1) and CD86 (B7.2) as cellular attachment receptors. *Virology*, *322*, 349–359.

Short, J. J., Vasu, C., Holterman, M. J., Curiel, D. T., & Pereboev, A. (2006). Members of adenovirus species B utilize CD80 and CD86 as cellular attachment receptors. *Virus Research*, *122*, 144–153.

Simmons, A. D., Moskalenko, M., Creson, J., Fang, J., Yi, S., VanRoey, M. J., et al. (2008). Local secretion of anti-CTLA-4 enhances the therapeutic efficacy of a cancer immunotherapy with reduced evidence of systemic autoimmunity. *Cancer Immunology, Immunotherapy*, *57*, 1263–1270.

Sirena, D., Lilienfeld, B., Eisenhut, M., Kälin, S., Boucke, K., Beerli, R. R., et al. (2004). The human membrane cofactor CD46 is a receptor for species B adenovirus serotype 3. *Journal of Virology*, *78*, 4454–4462.

Sloan, J. M., Kershaw, M. H., Touloukian, C. E., Lapointe, R., Robbins, P. F., Restifo, N. P., et al. (2002). MHC class I and class II presentation of tumor antigen in retrovirally and adenovirally transduced dendritic cells. *Cancer Gene Therapy*, *9*, 946–950.

Small, E. J., Tchekmedyian, N. S., Rini, B. I., Fong, L., Lowy, I., & Allison, J. P. (2007). A pilot trial of CTLA-4 blockade with human anti-CTLA-4 in patients with hormone-refractory prostate cancer. *Clinical Cancer Research*, *13*, 1810–1815.

Smith, J. G., Wiethoff, C. M., Stewart, P. L., & Nemerow, G. R. (2010). Adenovirus. *Current Topics in Microbiology and Immunology*, *343*, 195–224.

Smithgall, T. E., Briggs, S. D., Schreiner, S., Lerner, E. C., Cheng, H., & Wilson, M. B. (2000). Control of myeloid differentiation and survival by Stats. *Oncogene*, *19*, 2612–2618.

Smyth, M. J., Thia, K. Y., Street, S. E., MacGregor, D., Godfrey, D. I., & Trapani, J. A. (2000). Perforin-mediated cytotoxicity is critical for surveillance of spontaneous lymphoma. *The Journal of Experimental Medicine*, *192*, 755–760.

Song, M.-Y., Park, S.-H., Nam, H. J., Choi, D.-H., & Sung, Y.-C. (2011). Enhancement of vaccine-induced primary and memory CD8(+) T-cell responses by soluble PD-1. *Journal of Immunotherapy*, *34*, 297–306.

Sorensen, M. R., Holst, P. J., Pircher, H., Christensen, J. P., & Thomsen, A. R. (2009). Vaccination with an adenoviral vector encoding the tumor antigen directly linked to invariant chain induces potent CD4(+) T-cell-independent CD8(+) T-cell-mediated tumor control. *European Journal of Immunology*, *39*, 2725–2736.

Sorensen, M. R., Holst, P. J., Steffensen, M. A., Christensen, J. P., & Thomsen, A. R. (2010). Adenoviral vaccination combined with CD40 stimulation and CTLA-4 blockage can lead to complete tumor regression in a murine melanoma model. *Vaccine*, *28*, 6757–6764.

Steinman, R. M. (2008). Dendritic cells in vivo: A key target for a new vaccine science. *Immunity*, *29*, 319–324.

Steitz, J., Brück, J., Steinbrink, K., Enk, A., Knop, J., & Tuting, T. (2000). Genetic immunization of mice with human tyrosinase-related protein 2: Implications for the immunotherapy of melanoma. *International Journal of Cancer*, *86*, 89–94.

Steitz, J., Tormo, D., Schweichel, D., & Tuting, T. (2006). Comparison of recombinant adenovirus and synthetic peptide for DC-based melanoma vaccination. *Cancer Gene Therapy*, *13*, 318–325.

Sterman, D. H., Recio, A., Haas, A. R., Vachani, A., Katz, S. I., Gillespie, C. T., et al. (2010). A phase I trial of repeated intrapleural adenoviral-mediated interferon-beta gene transfer for mesothelioma and metastatic pleural effusions. *Molecular Therapy*, *18*, 852–860.

Stoff-Khalili, M. A., Rivera, A. A., Glasgow, J. N., Le, L. P., Stoff, A., Everts, M., et al. (2005). A human adenoviral vector with a chimeric fiber from canine adenovirus type 1 results in novel expanded tropism for cancer gene therapy. *Gene Therapy*, *12*, 1696–1706.

Stone, D., Ni, S., Li, Z.-Y., Gaggar, A., DiPaolo, N., Feng, Q., et al. (2005). Development and assessment of human adenovirus type 11 as a gene transfer vector. *Journal of Virology*, *79*, 5090–5104.

Sudowe, S., Ludwig-Portugall, I., Montermann, E., Ross, R., & Reske-Kunz, A. B. (2003). Transcriptional targeting of dendritic cells in gene gun-mediated DNA immunization favors the induction of type 1 immune responses. *Molecular Therapy*, *8*, 567–575.

Sumida, S. M., Truitt, D. M., Lemckert, A. A. C., Vogels, R., Custers, J. H. H. V., Addo, M. M., et al. (2005). Neutralizing antibodies to adenovirus serotype 5 vaccine vectors are directed primarily against the adenovirus hexon protein. *Journal of Immunology*, *174*, 7179–7185.

Sutherland, R. M., Londrigan, S. L., Brady, J. L., Azher, H., Carrington, E. M., Zhan, Y., et al. (2011). Shutdown of immunological priming and presentation after in vivo administration of adenovirus. *Gene Therapy*, http://dx.doi.org/10.1038/gt.2011.187 (Epub ahead of print).

Tacken, P. J., & Figdor, C. G. (2011). Targeted antigen delivery and activation of dendritic cells in vivo: Steps towards cost effective vaccines. *Seminars in Immunology*, *23*, 12–20.

Tacken, P. J., Zeelenberg, I. S., Cruz, L. J., van Hout-Kuijer, M. A., van de Glind, G., Fokkink, R. G., et al. (2011). Targeted delivery of TLR ligands to human and mouse dendritic cells strongly enhances adjuvanticity. *Blood*, *118*, 6836–6844.

Tallone, T., Malin, S., Samuelsson, A., Wilbertz, J., Miyahara, M., Okamoto, K., et al. (2001). A mouse model for adenovirus gene delivery. *Proceedings of the National Academy of Sciences of the United States of America*, *98*, 7910–7915.

Tang, J., Olive, M., Champagne, K., Flomenberg, N., Eisenlohr, L., Hsu, S., et al. (2004). Adenovirus hexon T-cell epitope is recognized by most adults and is restricted by HLA DP4, the most common class II allele. *Gene Therapy*, *11*, 1408–1415.

Tang, J., Olive, M., Pulmanausahakul, R., Schnell, M., Flomenberg, N., Eisenlohr, L., et al. (2006). Human CD8+ cytotoxic T cell responses to adenovirus capsid proteins. *Virology*, *350*, 312–322.

Tatsis, N., Blejer, A., Lasaro, M. O., Hensley, S. E., Cun, A., Tesema, L., et al. (2007). A CD46-binding chimpanzee adenovirus vector as a vaccine carrier. *Molecular Therapy*, *15*, 608–617.

Tatsis, N., Lasaro, M. O., Lin, S.-W., Haut, L. H., Xiang, Z. Q., Zhou, D., et al. (2009). Adenovirus vector-induced immune responses in nonhuman primates: Responses to prime boost regimens. *Journal of Immunology, 182*, 6587–6599.

Tatsis, N., Tesema, L., Robinson, E. R., Giles-Davis, W., McCoy, K., Gao, G. P., et al. (2006). Chimpanzee-origin adenovirus vectors as vaccine carriers. *Gene Therapy, 13*, 421–429.

Tatsumi, T., Takehara, T., Yamaguchi, S., Sasakawa, A., Miyagi, T., Jinushi, M., et al. (2007). Injection of IL-12 gene-transduced dendritic cells into mouse liver tumor lesions activates both innate and acquired immunity. *Gene Therapy, 14*, 863–871.

Tenbusch, M., Kuate, S., Tippler, B., Gerlach, N., Schimmer, S., Dittmer, U., et al. (2008). Coexpression of GM-CSF and antigen in DNA prime-adenoviral vector boost immunization enhances polyfunctional CD8+ T cell responses, whereas expression of GM-CSF antigen fusion protein induces autoimmunity. *BMC Immunology, 9*, 13.

Terme, M., Ullrich, E., Delahaye, N. F., Chaput, N., & Zitvogel, L. (2008). Natural killer cell-directed therapies: Moving from unexpected results to successful strategies. *Nature Immunology, 9*, 486–494.

Tesniere, A., Panaretakis, T., Kepp, O., Apetoh, L., Ghiringhelli, F., Zitvogel, L., et al. (2008). Molecular characteristics of immunogenic cancer cell death. *Cell Death and Differentiation, 15*, 3–12.

Thaci, B., Ulasov, I. V., Wainwright, D. A., & Lesniak, M. S. (2011). The challenge for gene therapy: Innate immune response to adenoviruses. *Oncotarget, 2*, 113–121.

Thacker, E. E., Nakayama, M., Smith, B. F., Bird, R. C., Muminova, Z., Strong, T. V., et al. (2009). A genetically engineered adenovirus vector targeted to CD40 mediates transduction of canine dendritic cells and promotes antigen-specific immune responses in vivo. *Vaccine, 27*, 7116–7124.

Thorner, A. R., Lemckert, A. A. C., Goudsmit, J., Lynch, D. M., Ewald, B. A., Denholtz, M., et al. (2006). Immunogenicity of heterologous recombinant adenovirus prime-boost vaccine regimens is enhanced by circumventing vector cross-reactivity. *Journal of Virology, 80*, 12009–12016.

Tian, S., Liu, Z., Donahue, C., Falo, L. D., & You, Z. (2011). Genetic targeting of the active transcription factor XBP1s to dendritic cells potentiates vaccine-induced prophylactic and therapeutic antitumor immunity. *Molecular Therapy, 20*, 432–442.

Tian, S., Liu, Z., Donahue, C., Noh, H. S., Falo, L. D., & You, Z. (2009). Transcriptional IL-15-directed in vivo DC targeting DNA vaccine. *Gene Therapy, 16*, 1260–1270.

Tillman, B. W., de Gruijl, T. D., Luykx-de Bakker, S. A., Scheper, R. J., Pinedo, H. M., Curiel, T. J., et al. (1999). Maturation of dendritic cells accompanies high-efficiency gene transfer by a CD40-targeted adenoviral vector. *Journal of Immunology, 162*, 6378–6383.

Todorova, K., Ignatova, I., Tchakarov, S., Altankova, I., Zoubak, S., Kyurkchiev, S., et al. (2005). Humoral immune response in prostate cancer patients after immunization with gene-based vaccines that encode for a protein that is proteasomally degraded. *Cancer Immunity, 5*, 1.

Toka, F. N., Gieryńska, M., Suvas, S., Schoenberger, S. P., & Rouse, B. T. (2005). Rescue of memory CD8+T cell reactivity in peptide/TLR9 ligand immunization by codelivery of cytokines or CD40 ligation. *Virology, 331*, 151–158.

Tomihara, K., Kato, K., Masuta, Y., Nakamura, K., Uchida, H., Sasaki, K., et al. (2007). Gene transfer of CD40-ligand to dendritic cells stimulates interferon-γ production to induce growth arrest and apoptosis of tumor cells. *Gene Therapy, 15*, 203–213.

Tomko, R. P., Xu, R., & Philipson, L. (1997). HCAR and MCAR: The human and mouse cellular receptors for subgroup C adenoviruses and group B coxsackieviruses. *Proceedings of the National Academy of Sciences of the United States of America, 94*, 3352–3356.

Toth, K., & Wold, W. S. M. (2010). Increasing the efficacy of oncolytic adenovirus vectors. *Viruses, 2*, 1844–1866.

Trudel, S., Trachtenberg, J., Toi, A., Sweet, J., Li, Z. H., Jewett, M., et al. (2003). A phase I trial of adenovector-mediated delivery of interleukin-2 (AdIL-2) in high-risk localized prostate cancer. *Cancer Gene Therapy, 10*, 755–763.

Tsao, H., Millman, P., Linette, G. P., Hodi, F. S., Sober, A. J., Goldberg, M. A., et al. (2002). Hypopigmentation associated with an adenovirus-mediated gp100/MART-1-transduced dendritic cell vaccine for metastatic melanoma. *Archives of Dermatology, 138*, 799–802.

Tsujimura, A., Shida, K., Kitamura, M., Nomura, M., Takeda, J., Tanaka, H., et al. (1998). Molecular cloning of a murine homologue of membrane cofactor protein (CD46): Preferential expression in testicular germ cells. *The Biochemical Journal, 330*(Pt. 1), 163–168.

Tuettenberg, A., Jonuleit, H., Tuting, T., Brück, J., Knop, J., & Enk, A. H. (2003). Priming of T cells with Ad-transduced DC followed by expansion with peptide-pulsed DC significantly enhances the induction of tumor-specific CD8+ T cells: Implications for an efficient vaccination strategy. *Gene Therapy, 10*, 243–250.

Tuting, T., Steitz, J., Brück, J., Gambotto, A., Steinbrink, K., DeLeo, A. B., et al. (1999). Dendritic cell-based genetic immunization in mice with a recombinant adenovirus encoding murine TRP2 induces effective anti-melanoma immunity. *The Journal of Gene Medicine, 1*, 400–406.

Tuve, S., Chen, B.-M., Liu, Y., Cheng, T.-L., Touré, P., Sow, P. S., et al. (2007). Combination of tumor site-located CTL-associated antigen-4 blockade and systemic regulatory T-cell depletion induces tumor-destructive immune responses. *Cancer Research, 67*, 5929–5939.

Tuve, S., Liu, Y., Tragoolpua, K., Jacobs, J. D., Yumul, R. C., Li, Z.-Y., et al. (2009). In situ adenovirus vaccination engages T effector cells against cancer. *Vaccine, 27*, 4225–4239.

Tuve, S., Wang, H., Jacobs, J. D., Yumul, R. C., Smith, D. F., & Lieber, A. (2008). Role of cellular heparan sulfate proteoglycans in infection of human adenovirus serotype 3 and 35. *PLoS Pathogens, 4*, e1000189.

Ueno, H., Palucka, A. K., & Banchereau, J. (2010). The expanding family of dendritic cell subsets. *Nature Biotechnology, 28*, 813–815.

van Beusechem, V. W., Grill, J., Mastenbroek, D. C. J., Wickham, T. J., Roelvink, P. W., Haisma, H. J., et al. (2002). Efficient and selective gene transfer into primary human brain tumors by using single-chain antibody-targeted adenoviral vectors with native tropism abolished. *Journal of Virology, 76*, 2753–2762.

van Cruijsen, H., van der Veldt, A. A. M., Vroling, L., Oosterhoff, D., Broxterman, H. J., Scheper, R. J., et al. (2008). Sunitinib-induced myeloid lineage redistribution in renal cell cancer patients: CD1c+ dendritic cell frequency predicts progression-free survival. *Clinical Cancer Research, 14*, 5884–5892.

van de Ven, R., Lindenberg, J. J., Oosterhoff, D., van den Tol, M. P., Rosalia, R. A., Murakami, M., et al. (2009). Selective transduction of mature DC in human skin and lymph nodes by CD80/CD86-targeted fiber-modified adenovirus-5/3. *Journal of Immunotherapy, 32*, 895–906.

van de Ven, R., Scheffer, G. L., Reurs, A. W., Lindenberg, J. J., Oerlemans, R., Jansen, G., et al. (2008). A role for multidrug resistance protein 4 (MRP4; ABCC4) in human dendritic cell migration. *Blood, 112*, 2353–2359.

Vardouli, L., Lindqvist, C., Vlahou, K., Loskog, A. S. I., & Eliopoulos, A. G. (2009). Adenovirus delivery of human CD40 ligand gene confers direct therapeutic effects on carcinomas. *Cancer Gene Therapy, 16*, 848–860.

Verhaagh, S., de Jong, E., Goudsmit, J., Lecollinet, S., Gillissen, G., de Vries, M., et al. (2006). Human CD46-transgenic mice in studies involving replication-incompetent adenoviral type 35 vectors. *The Journal of General Virology, 87*, 255–265.

Vesely, M. D., Kershaw, M. H., Schreiber, R. D., & Smyth, M. J. (2011). Natural innate and adaptive immunity to cancer. *Annual Review of Immunology, 29*, 235–271.

Vigne, E., Mahfouz, I., Dedieu, J. F., Brie, A., Perricaudet, M., & Yeh, P. (1999). RGD inclusion in the hexon monomer provides adenovirus type 5-based vectors with a fiber knob-independent pathway for infection. *Journal of Virology, 73*, 5156–5161.

Villadangos, J. A., & Shortman, K. (2010). Found in translation: The human equivalent of mouse CD8+ dendritic cells. *The Journal of Experimental Medicine, 207*, 1131–1134.

von Bubnoff, A. (2010). Phase I trial of adenovirus-based prime-boost regimen begins in Boston. *IAVI Report, 14*, 18–19.

Vonderheide, R. H., Flaherty, K. T., Khalil, M., Stumacher, M. S., Bajor, D. L., Hutnick, N. A., et al. (2007). Clinical activity and immune modulation in cancer patients treated with CP-870,893, a novel CD40 agonist monoclonal antibody. *Journal of Clinical Oncology, 25*, 876–883.

Vuylsteke, R. J. C. L. M., Molenkamp, B. G., Gietema, H. A., van Leeuwen, P. A. M., Wijnands, P. G. J. T. B., Vos, W., et al. (2004). Local administration of granulocyte/macrophage colony-stimulating factor increases the number and activation state of dendritic cells in the sentinel lymph node of early-stage melanoma. *Cancer Research, 64*, 8456–8460.

Vuylsteke, R. J. C. L. M., Molenkamp, B. G., van Leeuwen, P. A. M., Meijer, S., Wijnands, P. G. J. T. B., Haanen, J. B. A. G., et al. (2006). Tumor-specific CD8+ T cell reactivity in the sentinel lymph node of GM-CSF-treated stage I melanoma patients is associated with high myeloid dendritic cell content. *Clinical Cancer Research, 12*, 2826–2833.

Waddington, S. N., McVey, J. H., Bhella, D., Parker, A. L., Barker, K., Atoda, H., et al. (2008). Adenovirus serotype 5 hexon mediates liver gene transfer. *Cell, 132*, 397–409.

Waller, E. K. (2007). The role of sargramostim (rhGM-CSF) as immunotherapy. *The Oncologist, 12*(Suppl 2), 22–26.

Walsh, M. P., Seto, J., Liu, E. B., Dehghan, S., Hudson, N. R., Lukashev, A. N., et al. (2011). Computational analysis of two species C human adenoviruses provides evidence of a novel virus. *Journal of Clinical Microbiology, 49*, 3482–3490.

Wang, W., Edington, H. D., Rao, U. N. M., Jukic, D. M., Land, S. R., Ferrone, S., et al. (2007). Modulation of signal transducers and activators of transcription 1 and 3 signaling in melanoma by high-dose IFNalpha2b. *Clinical Cancer Research, 13*, 1523–1531.

Wang, K., Huang, S., Kapoor-Munshi, A., & Nemerow, G. (1998). Adenovirus internalization and infection require dynamin. *Journal of Virology, 72*, 3455–3458.

Wang, H., Li, Z.-Y., Liu, Y., Persson, J., Beyer, I., Möller, T., et al. (2011). Desmoglein 2 is a receptor for adenovirus serotypes 3, 7, 11 and 14. *Nature Medicine, 17*, 96–104.

Wang, T., Niu, G., Kortylewski, M., Burdelya, L., Shain, K., Zhang, S., et al. (2004). Regulation of the innate and adaptive immune responses by Stat-3 signaling in tumor cells. *Nature Medicine, 10*, 48–54.

Wang, L., Qi, X., Sun, Y., Liang, L., & Ju, D. (2002). Adenovirus-mediated combined P16 gene and GM-CSF gene therapy for the treatment of established tumor and induction of antitumor immunity. *Cancer Gene Therapy, 9*, 819–824.

Wargo, J. A., Schumacher, L. Y., Comin-Anduix, B., Dissette, V. B., Glaspy, J. A., McBride, W. H., et al. (2005). Natural killer cells play a critical role in the immune response following immunization with melanoma-antigen-engineered dendritic cells. *Cancer Gene Therapy, 12*, 516–527.

Weber, J. (2009). Ipilimumab: Controversies in its development, utility and autoimmune adverse events. *Cancer Immunology, Immunotherapy, 58*, 823–830.

Weber, J. (2010). Immune checkpoint proteins: A new therapeutic paradigm for cancer—Preclinical background: CTLA-4 and PD-1 blockade. *Seminars in Oncology*, *37*, 430–439.
Wickham, T. J. (2000). Targeting adenovirus. *Gene Therapy*, 7, 110–114.
Wickham, T. J., Mathias, P., Cheresh, D. A., & Nemerow, G. R. (1993). Integrins alpha v beta 3 and alpha v beta 5 promote adenovirus internalization but not virus attachment. *Cell*, *73*, 309–319.
Wickham, T. J., Tzeng, E., Shears, L. L., Roelvink, P. W., Li, Y., Lee, G. M., et al. (1997). Increased in vitro and in vivo gene transfer by adenovirus vectors containing chimeric fiber proteins. *Journal of Virology*, *71*, 8221–8229.
Worgall, S., Busch, A., Rivara, M., Bonnyay, D., Leopold, P. L., Merritt, R., et al. (2004). Modification to the capsid of the adenovirus vector that enhances dendritic cell infection and transgene-specific cellular immune responses. *Journal of Virology*, *78*, 2572–2580.
Worgall, S., Krause, A., Rivara, M., Hee, K.-K., Vintayen, E. V., Hackett, N. R., et al. (2005). Protection against P. aeruginosa with an adenovirus vector containing an OprF epitope in the capsid. *The Journal of Clinical Investigation*, *115*, 1281–1289.
Wu, H., & Curiel, D. T. (2008). Fiber-modified adenoviruses for targeted gene therapy. *Methods in Molecular Biology*, *434*, 113–132.
Wu, H., Han, T., Belousova, N., Krasnykh, V., Kashentseva, E., Dmitriev, I., et al. (2005). Identification of sites in adenovirus hexon for foreign peptide incorporation. *Journal of Virology*, *79*, 3382–3390.
Wu, H., Seki, T., Dmitriev, I., Uil, T., Kashentseva, E., Han, T., et al. (2002). Double modification of adenovirus fiber with RGD and polylysine motifs improves coxsackievirus-adenovirus receptor-independent gene transfer efficiency. *Human Gene Therapy*, *13*, 1647–1653.
Wu, E., Trauger, S. A., Pache, L., Mullen, T.-M., von Seggern, D. J., Siuzdak, G., et al. (2004). Membrane cofactor protein is a receptor for adenoviruses associated with epidemic keratoconjunctivitis. *Journal of Virology*, *78*, 3897–3905.
Xia, D., Henry, L. J., Gerard, R. D., & Deisenhofer, J. (1994). Crystal structure of the receptor-binding domain of adenovirus type 5 fiber protein at 1.7 A resolution. *Structure*, *2*, 1259–1270.
Xiong, J., Sun, W. J., Wang, W. F., Liao, Z. K., Zhou, F. X., Kong, H. Y., et al. (2012). Novel, chimeric, cancer-specific, and radiation-inducible gene promoters for suicide gene therapy of cancer. *Cancer*, *118*, 536–548.
Xu, Y., Liu, Z., Kong, H., Sun, W., Liao, Z., Zhou, F., et al. (2011). Co-expression of interleukin 12 enhances antitumor effects of a novel chimeric promoter-mediated suicide gene therapy in an immunocompetent mouse model. *Biochemical and Biophysical Research Communications*, *412*, 763–768.
Yamaguchi, T., Kawabata, K., Koizumi, N., Sakurai, F., Nakashima, K., Sakurai, H., et al. (2007). Role of MyD88 and TLR9 in the innate immune response elicited by serotype 5 adenoviral vectors. *Human Gene Therapy*, *18*, 753–762.
Yamamoto, M., & Curiel, D. T. (2010). Current issues and future directions of oncolytic adenoviruses. *Molecular Therapy*, *18*, 243–250.
Yang, S.-C., Hillinger, S., Riedl, K., Zhang, L., Zhu, L., Huang, M., et al. (2004). Intratumoral administration of dendritic cells overexpressing CCL21 generates systemic antitumor responses and confers tumor immunity. *Clinical Cancer Research*, *10*, 2891–2901.
Ye, Z., Shi, M., Chan, T., Sas, S., Xu, S., & Xiang, J. (2007). Engineered CD8+ cytotoxic T cells with fiber-modified adenovirus-mediated TNF-alpha gene transfection counteract immunosuppressive interleukin-10-secreting lung metastasis and solid tumors. *Cancer Gene Therapy*, *14*, 661–675.

Youil, R., Toner, T. J., Su, Q., Chen, M., Tang, A., Bett, A. J., et al. (2002). Hexon gene switch strategy for the generation of chimeric recombinant adenovirus. *Human Gene Therapy, 13*, 311–320.

Yu, H., Kortylewski, M., & Pardoll, D. (2007). Crosstalk between cancer and immune cells: Role of STAT3 in the tumour microenvironment. *Nature Reviews. Immunology, 7*, 41–51.

Yu, H., Pardoll, D., & Jove, R. (2009). STATs in cancer inflammation and immunity: A leading role for STAT3. *Nature Reviews. Cancer, 9*, 798–809.

Zamai, L., Ponti, C., Mirandola, P., Gobbi, G., Papa, S., Galeotti, L., et al. (2007). NK cells and cancer. *Journal of Immunology, 178*, 4011–4016.

Zhai, Y., Yang, J. C., Kawakami, Y., Spiess, P., Wadsworth, S. C., Cardoza, L. M., et al. (1996). Antigen-specific tumor vaccines. Development and characterization of recombinant adenoviruses encoding MART1 or gp100 for cancer therapy. *Journal of Immunology, 156*, 700–710.

Zhang, Y., & Bergelson, J. M. (2005). Adenovirus receptors. *Journal of Virology, 79*, 12125–12131.

Zhang, S.-N., Choi, I.-K., Huang, J.-H., Yoo, J.-Y., Choi, K.-J., & Yun, C.-O. (2011). Optimizing DC vaccination by combination with oncolytic adenovirus coexpressing IL-12 and GM-CSF. *Molecular Therapy, 19*, 1558–1568.

Zhang, X., Kedl, R. M., & Xiang, J. (2009). CD40 ligation converts TGF-beta-secreting tolerogenic CD4-8- dendritic cells into IL-12-secreting immunogenic ones. *Biochemical and Biophysical Research Communications, 379*, 954–958.

Zhang, L., Tang, Y., Akbulut, H., Zelterman, D., Linton, P.-J., & Deisseroth, A. B. (2003). An adenoviral vector cancer vaccine that delivers a tumor-associated antigen/CD40-ligand fusion protein to dendritic cells. *Proceedings of the National Academy of Sciences of the United States of America, 100*, 15101–15106.

Zhao, X., Bose, A., Komita, H., Taylor, J. L., Kawabe, M., Chi, N., et al. (2011). Intratumoral IL-12 gene therapy results in the crosspriming of Tc1 cells reactive against tumor-associated stromal antigens. *Molecular Therapy, 19*, 805–814.

Zhu, J., Huang, X., & Yang, Y. (2007). Innate immune response to adenoviral vectors is mediated by both toll-like receptor-dependent and -independent pathways. *Journal of Virology, 81*, 3170–3180.

Zhu, J., Huang, X., & Yang, Y. (2008). A critical role for type I IFN-dependent NK cell activation in innate immune elimination of adenoviral vectors in vivo. *Molecular Therapy, 16*, 1300–1307.

Zhu, J., Huang, X., & Yang, Y. (2010). NKG2D is required for NK cell activation and function in response to E1-deleted adenovirus. *Journal of Immunology, 185*, 7480–7486.

Zitvogel, L., Kepp, O., & Kroemer, G. (2011). Immune parameters affecting the efficacy of chemotherapeutic regimens. *Nature Reviews. Clinical Oncology, 8*, 151–160.

CHAPTER SEVEN

# Cancer Treatment with Gene Therapy and Radiation Therapy

**Sergey A. Kaliberov*, Donald J. Buchsbaum[†,1]**
*Department of Radiation Oncology, Washington University in St. Louis, St. Louis, Missouri, USA
[†]Department of Radiation Oncology, University of Alabama at Birmingham, Birmingham, Alabama, USA
[1]Corresponding author: e-mail address: donald.buchsbaum@ccc.uab.edu; djb@uab.edu

## Contents



*Advances in Cancer Research*, Volume 115
ISSN 0065-230X
http://dx.doi.org/10.1016/B978-0-12-398342-8.00007-0

## Abstract

Radiation therapy methods have evolved remarkably in recent years which have resulted in more effective local tumor control with negligible toxicity of surrounding normal tissues. However, local recurrence and distant metastasis often occur following radiation therapy mostly due to the development of radioresistance through the deregulation of the cell cycle, apoptosis, and inhibition of DNA damage repair mechanisms. Over the last decade, extensive progress in radiotherapy and gene therapy combinatorial approaches has been achieved to overcome resistance of tumor cells to radiation. In this review, we summarize the results from experimental cancer therapy studies on the combination of radiation therapy and gene therapy.

## 1. INTRODUCTION

Approximately half of all cancer patients worldwide are treated with radiation therapy alone or in combination with chemotherapy or surgery. Radiation therapy approaches employ ionizing radiation delivered either externally by linear accelerators or X-rays and $\gamma$-rays or internally with the use of radioisotopes to destroy cancer cells. Radiation therapy is, in general, localized, noninvasive and does not produce systemic toxicity after treatment in comparison with chemotherapy. However, despite progress in radiation therapy, development of resistance to therapeutic doses of irradiation still remains a major obstacle for anticancer therapy. There are generally two major forms of resistance encountered in clinical practice. One is intrinsic radiation resistance, which is an innate property of the cancer cells. The other is acquired resistance, which develops subsequent to radiotherapy (Shimura, 2011). It is becoming more evident from clinical studies of conventional radiotherapy that multimodality treatment strategies have a higher chance of success, presumably because combined therapies simultaneously utilize diverse target signaling pathways in tumor cells and may have nonoverlapping toxicities. In order to overcome mechanisms of radiation resistance, resulting in recurrence of disease, strategies involving a combination of radiation therapy and gene therapy have been extensively investigated. This review focuses on cancer gene therapy strategies that are currently being evaluated in conjunction with radiation therapy, including the employment of radiation-inducible promoters in combination with therapeutic genes that ultimately lead to radiosensitization of tumor cells. Furthermore, in some instances, radiation treatment has resulted in increased proliferation and intratumor distribution and tumor cell killing of replicative viral vectors.

## 2. MOLECULAR MECHANISMS OF TUMOR RESPONSE TO RADIATION THERAPY

Significant progress has been made in recent years in understanding the molecular mechanisms of cellular responses to radiation therapy (summarized in Fig. 7.1). Exposure of tumor cells to clinically relevant doses of ionizing radiation produces a wide range of DNA damages that include DNA single-strand breaks, DNA double-strand breaks (DSBs), DNA base damage and apyrimidinic/apurinic (AP) sites, as well as DNA–protein cross-links. In addition to direct DNA damage, irradiation triggers a variety of lesions in cellular macromolecules (e.g., lipid peroxidation) in part due to mitochondria-dependent generation of reactive oxygen species (ROS). This complex of DNA and non-DNA lesions can stimulate various signal transduction pathways (such as protein kinase C (PKC), JNK, ceramide, and MAPK activation). Thus, the cellular responses to radiation-mediated DNA damage are characterized by coordinated processes involving DNA repair mechanisms engaged with other events that are occurring in the cell, including cell cycle progression or apoptosis. Initially, the damage to DNA initiates signaling pathways that halt cell cycle progression through G1, intra-S, and G2/M to prevent transfer of DNA damage to progeny cells.

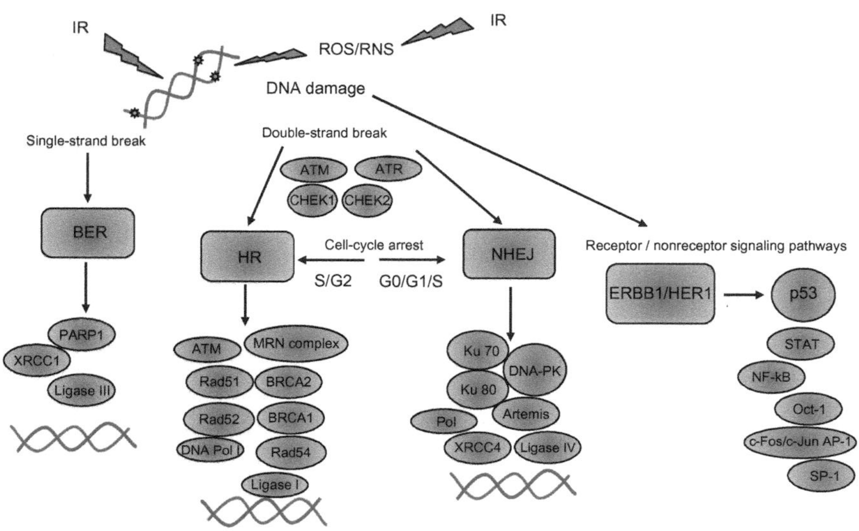


**Figure 7.1** Molecular mechanisms of cell response to radiotherapy. (For color version of this figure, the reader is referred to the online version of this chapter.)

Activation of the DNA repair machinery is facilitated by increased accessibility of DNA damage sites. Concurrently with the arrest of cell cycle progression, malignant cells mobilize mechanisms preparing the cell to undergo apoptosis when DNA cannot be successfully repaired. Also, inactivation of phosphatases by mitochondria-dependent generation of ROS induces signaling at the receptor level that consequently causes ligand-independent activation of cell-surface receptors with intrinsic tyrosine kinase activity.

## 2.1. Molecular mechanisms of DNA damage repair

A growing body of evidence suggests that DNA DSBs are the main cause of cell death induced by ionizing radiation. The two major pathways involved in DSB repair are homologous recombination (HR) and non-homologous end joining (NHEJ). The primary function of HR is to repair DSBs at the replication fork, whereas NHEJ is the key mechanism of DSB repair during G0, G1, and early S phase, and NHEJ continues to repair a minority of breaks during late S and G2 phases of the cell cycle (Fig. 7.1).

The HR pathway employs numerous proteins (RPA, Rad51, Rad52, Rad54, BRCA1, BRCA2, and Mre11–Rad50–Nbs1 (MRN) complex). Ataxia-telangiectasia mutated (ATM), RAD3-related protein kinases (ATR), and RAD3-related protein kinases from the PI3K-related kinase family are central mediators of cellular responses to DNA DSBs in cells. ATM/ATR and Rad3-related protein kinases phosphorylate numerous substrates in the cell that modulate the cell response to DNA damage via activation of receptors/intracellular signaling pathways as well as stimulating cell cycle checkpoints, p53 activity, and DNA repair complex function. ATM is activated when it is recruited to sites of DSB damage by the MRN complex (Bakkenist & Kastan, 2004; Falck, Coates, & Jackson, 2005; Jazayeri et al., 2006; Lee & Paull, 2005). Once activated, ATM and ATR protein kinases phosphorylate numerous substrates in the cell and induce the cell response to DNA damage via facilitating DNA repair and modulation of cell cycle arrest (Gatei et al., 2003). The best characterized among the ATM-activated substrates include p53 (Canman et al., 1998; Khanna & Lavin, 1993), the kinases CHK1 and CHK2 (Zhou et al., 2000), and the histone H2AX and the MRN complex (Burma, Chen, Murphy, Kurimasa, & Chen, 2001). Similarly, ATM and ATR phosphorylate CHK1 and DNA helicase BLM1. Thus, ATM/ATR protein kinases are involved in the modulation of apoptosis via p53 and CHK2 expression, DNA repair (H2AX and MRN), and cell cycle arrest using p53, CHK1, and CHK2 signaling pathways (Andreassen, Ho, & D'Andrea, 2006). Finally, the core DNA repair machinery is activated, as a result of CHK1/CHK2-mediated cell cycle arrest (mostly at intra-S or G2/M phase) (Zhou & Bartek, 2004).

NHEJ predominantly repairs DSBs that have been generated by ionizing radiation during the cell cycle. The main NHEJ pathway involves binding of the DNA-PK complex composed of Ku70, Ku80/Ku86, and the DNA-dependent protein kinase catalytic subunit (DNA-PKcs) kinase to the ends at a DSB. In this process, the Ku70/Ku80 heterodimers load onto a DNA terminus during the initial "synapsis" phase. The Ku70/Ku80–DNA complex binds to the DNA-PKcs serine/threonine kinase. The activated protein kinase complex is involved in the phosphorylation of the DNA DSB repair/*V(D)J* recombination protein Artemis. Subsequently, Artemis–DNA-$PK_{cs}$ complex with endonuclease activity processes DNA ends before XRCC4 and DNA ligase IV to facilitate the final ligation step (Burma, Chen, & Chen, 2006; Lieber, Ma, Pannicke, & Schwarz, 2003). In contrast, HR is generally restricted to late S and G2 phase and rejoins DSBs using a sister homologue as a template. An early step in HR involves the generation of a single-stranded region of DNA, followed by invasion of the template strand, which creates a Holliday junction. DNA synthesis using the sister strand as a template is followed by branch migration and subsequent resolution of the heteroduplex. Rad51, a central player in HR, is loaded onto ssDNA and promotes strand invasion, with BRCA2 having a role in delivering Rad51 to the DNA (O'Driscoll & Jeggo, 2006; Pellegrini et al., 2002).

Additionally, clinically relevant low-dose ionizing radiation can generate ROS through radiolysis of water that results in a multitude of oxidized base lesions, DNA strand breaks, and base loss, largely caused by hydroxyl radicals. Inhibition of protein tyrosine phosphatases (PTPases) by ROS and reactive nitrogen species (RNS) leads to activation of receptor and nonreceptor tyrosine kinases and activation of downstream signaling pathways (Chiarugi & Cirri, 2003). Radiation dose-escalation studies revealed that high doses of ionizing irradiation increase the levels of acidic sphingomyelinase and production of ceramide and promote membrane-associated receptor activation by facilitating receptor clustering within the lipid rafts (Galabova-Kovacs et al., 2006). The activation of cell-surface RTK receptors initiates a cascade of proliferative signaling events involving several members of the epidermal growth factor receptor (ERBB1/HER1) family (Goldkorn, Balaban, Shannon, & Matsukuma, 1997; Gulbins &

Kolesnick, 2003; Kavanagh, Dent, Schmidt-Ullrich, Chen, & Mikkelsen, 1998; Kavanagh, Lin, Chen, & Schmidt-Ullrich, 1995; Schlessinger & Lemmon, 2006). It was shown that heterodimerization of ERBB1 with other members of the ERBB1 receptor family (ERBB2, ERBB3, and ERBB4) is linked to downstream activation of intracellular signaling pathways, including the RAF-1–MEK1/2–ERK1/2 and the PI3K–phosphoinositide-dependent kinase-1–AKT pathways (Hynes & Lane, 2005; Nyati, Morgan, Feng, & Lawrence, 2006; Schlessinger, 2004). The initial radiation-induced activation of ERK1/2 can temporally delay reactivation of receptors as well as intracellular signaling pathways by facilitating cleavage and release of paracrine ligands in irradiated cells. Several vital signaling pathways are also activated by the receptors including the c-Src tyrosine kinase, the signal transducer and activator of transcription proteins (STATs), and mammalian target of rapamycin (Bjornsti & Houghton, 2004; Ishizawar & Parsons, 2004; Yu & Jove, 2004). Perturbations of such signaling pathways can lead to radiation resistance in cancer cells.

## 2.2. Downstream signaling networks

Signaling pathway networks activated through both the cell-surface receptors and damaged DNA play an important role in the regulation of the downstream pathways. These signaling pathways are ultimately involved in the activation of a variety of transcription factors (Criswell, Leskov, Miyamoto, Luo, & Boothman, 2003). Radiation-activated transcription factors include tumor suppressor p53 (Lu & Lane, 1993), proinflammatory transcription factor NF-κB (Prasad, Ravindran, & Aggarwal, 2010), signal transducer and activator of transcription-3 (STAT3) (Aggarwal et al., 2009), the octamer-binding protein-1 (Oct-1) transcription factor (Jin et al., 2001), the c-Fos/c-Jun AP-1 complex (Hallahan, Virudachalam, et al., 1991; Sherman, Datta, Hallahan, Weichselbaum, & Kufe, 1990), and the early growth response 1 SP1-like transcription factor (Datta et al., 1993), which are important in the regulation of gene expression in major pathways of inflammation, tumor cell proliferation, as well as invasion and metastasis. Radiation-mediated activation of intracellular signaling pathways can modulate levels of transcription of proteins involved in proliferative and inflammatory responses of tumor cells to radiation therapy. Thus, the balance between DNA damage and DNA repair signaling pathways determines the radiosensitivity of tumor cells.

# 3. GENE TRANSFER SYSTEMS

Radiation treatment is a well-established modality for anticancer therapy. A growing body of evidence suggests that a manipulation of genes involved in tumor progression and radiation resistance represents a promising approach for therapeutic intervention in combination with radiation therapy. Cancer gene therapy is the selective transfer of recombinant DNA/RNA using viral and nonviral gene delivery vectors, into malignant cells as well as tumor-associated cells for therapeutic purposes. Generally, gene therapy approaches are based on the employment of genetically engineered vectors that promote the transfer of the transgene expression unit into the tumor cells and allow transient or stable expression of the therapeutic gene. The search for methods to introduce genes into cancer cells led to many advances in viral and nonviral vector development. Vectors derived from adenoviruses, adeno-associated viruses, retroviruses, herpesviruses, and poxviruses are the most commonly used gene carriers in cancer gene therapy. Nonviral gene transfer systems include DNA transfection using chemical methods such as utilization of a variety of liposomes, cationic polymers, or peptide delivery systems, as well as disruption of the cell membrane by physical methods (such as electroporation or ultrasonography). Many of these gene delivery systems have been tested for cancer therapy in combination with radiation therapy. Each group of vectors demonstrates a number of advantages and limitations. Most commonly used in cancer gene therapy, recombinant viral vectors are biological systems derived from naturally evolved pathogens which as part of the infection process transfer their genomes into the target tumor cells. The common restrictions associated with viral vectors are related to their safety, immunogenicity, preexisting immune response, and limited capacity of transgene insert.

## 3.1. Adenoviruses

Adenoviruses (Ads) are nonenveloped linear double-stranded DNA viruses. At the present time, human and nonhuman Ad vectors have been modified extensively and are commonly used vectors in clinical trials of cancer gene therapy. Ads can transduce both dividing and quiescent cells. In contrast with parvoviruses and retroviruses, Ads do not integrate into the cell genome and therefore pose no risk of insertional mutagenesis. Helper-dependent high-capacity Ad vectors are completely devoid of all viral coding sequences but retain the viral ITRs and the packaging signal and were demonstrated to be significantly safer in comparison with the first generation of Ad vectors. The Ad vectors can accommodate up to a 30-kb DNA insert and demonstrate a high transduction efficiency and relatively simple production and manipulation. These vectors have been used in gene transfer for basic studies of gene function and for gene therapy of genetic diseases and cancer. Conditionally replicating Ads (CRAds) with *E1A* driven by tumor-specific promoters have often been explored as oncolytic viruses. A substantial problem associated with Ad vectors is high immunogenicity in combination with proinflammatory effects that limit transgene expression and hamper systemic readministration (Bauerschmitz, Barker, & Hemminki, 2002). However, the Ad immunogenicity is applicable to treatment of chronic diseases, and an appropriate immune response induced by Ad vectors may be beneficial in some cancer gene therapy approaches related to the regulation of the host immune system.

### 3.2. Adeno-associated viruses

Adeno-associated viral (AAV) vectors are replication-defective, single-stranded DNA parvoviruses that require a helper Ad for their replication. Site-specific or random AAV vector integration into the host cell genome, in the absence of a helper virus, results in long-term transgene expression. Also, AAV vectors demonstrate high transduction efficiency in a broad range of quiescent and dividing target cells. In contrast to Ads, AAV vectors do not produce a significant immune response (Zaiss et al., 2002). A major limitation of using AAV vector is a relatively small transgene size (4 kb), a necessity of helper Ad or herpesviruses for successful transduction, and integration into the genome of target tumor cells as well as host cells that could lead to mutagenesis. In the absence of a helper virus, AAV vectors establish latency by preferential integration into a specific site on human chromosome 19. AAV vectors are frequently used for gene therapy for genetic diseases and cancer.

### 3.3. Retroviruses/lentiviruses

Retroviruses are single-stranded RNA viruses that can integrate into the genome of cells, which results in stable replication and transmission to all the progeny of these cells. Retroviruses continue to be employed as gene delivery vehicles, although recent adverse events following retroviral gene therapy have raised concerns about potential insertional mutagenesis (Hacein-Bey-Abina et al., 2003). Lentiviral vectors can transduce both proliferating and quiescent cells. Advantages of retroviral vectors for gene therapy include potential long-term transgene expression due to integration into the target tumor cells or host genome and low immunogenicity. However, most retroviruses demonstrate low levels of transduction efficiency and only infect dividing cells during mitosis (Adamina, Daetwiler, Rosenthal, & Zajac, 2005).

## 3.4. Herpesviruses

The herpesviruses, including herpes simplex virus HSV-1, are receiving increasing attention because of their ability to replicate in the nucleus as latent episomes and kill dividing and nondividing tumor cells. Attenuated and replication-competent HSV-1 mutants offer the capability to insert up to almost a 40-kb transgenic sequence. However, low and transient transfection efficiency as well as elicited host immune response and severe symptoms associated with wild-type viruses are serious obstacles for HSV-1-based gene therapy (Marconi et al., 1996).

## 3.5. Poxviruses

Similar to HSV-1, the vaccinia virus has a large genome (about 186 kb) and can accommodate up to 25 kb foreign DNA. Recombinant vaccinia viral vectors have been used mainly for delivery of cancer-specific antigens and immunoregulatory genes for vaccine-based gene therapy, which is based on the long history of utilization of these viruses for vaccination. However, the strong immune response as a result of preexisting anti-vaccinia immunity following childhood smallpox immunization limits the use of recombinant vaccinia viral vectors for gene therapy (Arlen, Gulley, Madan, Hodge, & Schlom, 2007).

## 3.6. Nonviral vectors

In an attempt to improve both the efficacy and safety of virus-mediated gene therapy, many vectors have been modified. However, several obstacles including limitation of the size of the inserted genetic material, difficulty of propagation and purification, and toxic adverse events (in particular immunogenic and oncogenic activity) have encouraged researchers to focus on nonbiological gene transfer systems as an alternative to viral vectors. Progress has been made using a variety of liposomes, cationic polymers, and peptides. In this approach, recombinant DNA is associated with positively charged lipid bilayers, cationic polymers, or peptides and can enter cells either by endocytosis or by fusion with the cell membrane. DNA–protein complexes have been developed employing natural or synthetic peptides that allow the use of these gene delivery vehicles in cell-targeting strategies. In contrast with recombinant viruses, these nonviral gene transfer systems are characterized by low cost and simple production, large gene insert size without toxic effects, and lack of an immune response after readministration. However, nonbiological gene transfer systems produce short-term expression, and their efficiency and specificity of gene delivery is less compared to viral vectors (Glover, Lipps, & Jans, 2005).

## 4. RADIATION-RESPONSIVE PROMOTERS

Gene therapy strategies offer the potential for specific targeting at several levels including transcriptional targeting. Transcriptional regulatory elements are composed of both *cis*-acting and *trans*-acting elements. The *cis*-acting components include promoters as well as enhancer regions that also regulate expression of a particular gene. The *trans*-acting elements belong to a growing group of transcription factors that bind to specific sites within promoter and/or enhancer regions. Transcriptional control of therapeutic gene expression using promoters that are responsive to ionizing radiation provides an attractive approach for the combination of radiation therapy and gene therapy (summarized in Table 7.1). It has been known for many years that ionizing radiation can modulate gene transcription as well as protein expression of a large number of cellular proteins involved in the early response of tumor cells to ionizing radiation-mediated damage. This includes genes involved in cell cycle checkpoints, cellular stress, DNA repair, and apoptosis. It was shown that expression of radiation-inducible genes such as cytokines and growth factors (e.g., early growth factor-1) (Hallahan, Sukhatme, et al., 1991) as well as immediate-early genes encoding transcription factors (e.g., Fos, Jun, AP-1, NF-κB) (Weichselbaum et al., 1994) is variable and cell dependent (Hamdi et al., 2005). Activated immediate-early genes interact with the promoter region of so-called late-responding genes, coding for cytokines, involved in the inflammation process (e.g., IL-6) and growth factors, involved in cell survival and tissue recovery (e.g., TGFβ1, TNF-α) (Beetz et al., 2000; Blalock et al., 1999; Martin et al., 1997). Importantly, upregulation of therapeutic genes is achievable at clinically relevant doses of radiation both *in vitro* and *in vivo*. The molecular mechanisms of radiation-inducible gene expression remain

**Table 7.1** Summary of radiation-responsive promoters employed in combination with irradiation

| Promoter (putative radiation response elements) | Gene/assay | References |
|---|---|---|
| Egr-1 (CArG, Sp-1, MAP) | *CAT* | Datta et al. (1992) |
| | *TNF-α* | Mezhir et al. (2006), Meyer, Kupper, Kandolf, and Rodemann (2002), Park et al. (2002), Hallahan et al. (1995), Senzer et al. (2004), McLoughlin et al. (2005), Bickenbach et al. (2008) |
| | *CYP4B1* | Hsu et al. (2003) |
| | *HSV-TK* | Joki, Nakamura, and Ohno (1995) |
| tPA (AP-2, AP-2-like, NFκB) | *tPA* | Boothman, Lee, and Sahijdak (1994) |
| p21/WAF1 (p53) | *iNOS* | Worthington et al. (2004), McCarthy et al. (2007) |
| | *HSV-TK* | Nenoi, Daino, Ichimura, Takahash, and Akuta (2006) |
| GADD45α (AP-1, p53, Oct-1, NF-κB, HNF, NF-YA, KLF) | Electrophoretic mobility shift analysis (EMSA), chromatin immunoprecipitation (ChIP) | Daino, Ichimura, and Nenoi (2003), Takahashi, Saito, Ohtani, and Sakai (2001), Daino, Ichimura, and Nenoi (2006) |
| Synthetic promoter containing isolated CArG elements | *GFP, HSV-TK* | Greco, Powell, Marples, Joiner, and Scott (2005) |
| | *GFP* | Scott, Joiner, and Marples (2002) |
| | *iNOS* | Coulter et al. (2008) |

unclear but involve putative radiation-responsive elements in gene promoter binding sites for selected transcription factors.

## 4.1. Early growth response-1 gene

The early growth response-1 (*Egr-1*) promoter, first described more than two decades ago, is one of the most widely investigated radiation-inducible promoters. The *Egr-1* gene encodes a nuclear protein with a Cys2-His2 zinc-finger motif that is partially homologous to the corresponding domain in the Wilms tumor susceptibility gene (Gessler et al., 1990). Egr-1 is an immediate-early protein that is involved in signal transduction during cellular proliferation and differentiation. It was shown that induction of Egr-1 expression in the transition of quiescent cells from the G0 to G1 phase of the cell cycle is regulated by efficient *cis-* and *trans-*regulatory mechanisms. The mechanism of regulation of Egr-1 expression by ionizing radiation is incompletely defined. However, functional analysis of *Egr-1* revealed a number of DNA sequence motifs including GCGGGGGCG consensus sequence identified in the promoters of many regulatory cytokines including *bFGF, IL-1, PDGF-α,* and *TNF-α*, as well as putative AP-1 sites located in the enhancer region of the *Egr-1* promoter that can modulate the radiation-mediated response (Datta et al., 1992). Ionizing radiation as well as DNA-damaging chemotherapy agents can readily induce the *Egr-1* promoter via a 10-nucleotide motif of consensus sequence in the CArG (CC(A+T rich) 6GG or serum response element) (Datta et al., 1992, 1993; Hsu et al., 2003; Mezhir et al., 2006; Tsai-Morris, Cao, & Sukhatme, 1988). The comparative analysis of a series deletion in the *Egr-1* promoter region using constructs with the chloramphenicol acetyltransferase (*CAT*) reporter gene demonstrated that the three 5′ distal CArG boxes were essential to the radiation response. It has been shown that CArG elements are responsive to hydrogen peroxide treatment, and this could play a role in radiation inducibility via ROS produced following irradiation (Datta et al., 1992; Hsu et al., 2003). CArG elements were shown to regulate the expression of other immediate-early genes including several members of the actin family and c-fos via the binding of specific accessory proteins such as Elk-1 and/or SAP-1. The presence of the putative binding site moieties in the *Egr-1* promoter also suggests potential involvement of the Sp-1 transcription factor as well as the Fos-jun heterodimer activator protein (AP)-1. Also, the phosphorylation by mitogen-activated protein (MAP) kinases involves the *Egr-1* promoter response to cellular stress following irradiation of cancer cells (Christy & Nathans, 1989; Datta et al., 1992; Treisman, 1990; Tsai-Morris et al., 1988). A number of gene therapy vectors encoding the therapeutic transgene under the control of the radiation-responsive elements of *Egr-1* promoter were extensively tested in combination with ionizing radiation. The early gene therapy studies of this system utilized an Ad.Egr.TNF adenoviral vector encoding the proapoptotic cytokine TNF-α under the control of CArG elements. It was shown that radiation and chemotherapy increase TNF-α expression (Meyer et al., 2002; Mezhir et al., 2006; Park et al., 2002). Ad.Egr.TNF demonstrated increased antitumor efficacy in combination with radiation therapy in phase I and phase II trials (Bickenbach et al., 2008; Hallahan et al., 1995; McLoughlin et al., 2005; Senzer et al., 2004). Currently, several suicide gene/prodrug systems using radiation-inducible promoters and gene delivery systems are under intense investigation. Recently, *Egr-1*-driven molecular chemotherapy using the herpes simplex virus-thymidine kinase (HSV-TK) with the antiherpetic drug ganciclovir (GCV) (Joki et al., 1995) or cytochrome p450 isozyme (CYP4B1) in combination with 4-ipomeanol (4-IM) (Hsu et al., 2003) systems was tested in combination with radiation *in vitro* and *in vivo*.

## 4.2. Other radiation-inducible promoters

The promoters from many stress response genes related to cellular fate following DNA damage are potential candidates for cancer gene therapy. In the last two decades, several radiation-inducible promoters including tissue plasminogen activator (*tPA*), *p21/WAF-1*, and *GADD45α* were proposed for gene therapy in combination with radiation therapy strategies.

### *4.2.1 Tissue plasminogen activator gene*

Human tPA is a glycoprotein produced mainly by vascular endothelial cells. tPA activates clot dissolution in the presence of fibrin by conversion of plasminogen to plasmin, thereby cleaving cross-linked fibrin to D-dimer and other degradation products. The *tPA* promoter comprises several putative binding sites for the AP-2, NFκB-like and AP-2-like regulatory factors. The mutagenesis of the AP-2 and NFκB binding sequences resulted in the reduction of basal activity and UV-induced tPA expression. Additionally, mutations in the AP-2-like sequence resulted in complete loss of the UV response and produced negligible modification of basal expression. tPA was induced in radioresistant human melanoma cells over 50-fold following X-irradiation at 4.5 Gy (Boothman et al., 1994).

#### 4.2.2 Wild-type p53-activated fragment (p21/WAF1)

WAF1, also known as p21 or CIP1, is a cyclin-dependent kinase inhibitor, suppressing cell growth. It has also been shown to be involved in the immediate-early response to a range of stresses (Ishida, Sasaguri, Kosaka, Nojima, & Ogata, 1997; Steinman et al., 1994). The *p21/WAF1* promoter region has a p53 binding site that has been shown to confer p53-dependent WAF1 expression. These p53 recognition elements comprise four tandem PuPuPuC(A/T) pentamers and located 1.3, 1.7, and 2.2 kb upstream of the first exon of human *p21/WAF1* (Ishida et al., 1997). It was shown *in vitro* that the radiation inducibility of the *p21/WAF1* promoter is p53 dependent (el-Deiry et al., 1993). However, there is also evidence of p53-independent activation pathways and that the *p53* status did not alter significantly the radiation inducibility of *p21/WAF1 in vivo* (Blundell, Kaminski, & Harrison, 2004; Rimner et al., 2001; Sourvinos & Spandidos, 1998). The expression of the inducible nitric oxide synthetase (*iNOS*) gene under the control of the *p21/WAF1* promoter has been employed to improve the efficacy of conventional radiation therapy. It was shown that a *p21/WAF1* promoter-driven *iNOS* gene therapy significantly sensitized both p53 wild-type RIF-1 tumors and p53 mutant HT29 tumors to fractionated radiation therapy (McCarthy et al., 2007; Worthington et al., 2004). Molecular chemotherapy using a recombinant AAV vector expressing the *HSV-TK* gene under the control of the *p21/WAF1* promoter significantly increased killing of MCF7 human breast cancer cells in the presence of the prodrug GCV in combination with fractionated radiation treatment at 1 Gy (Nenoi et al., 2006).

#### 4.2.3 Growth arrest and DNA damage-inducible protein 45 alpha

Growth arrest and DNA damage-inducible protein 45 alpha (GADD45α) plays an important role in controlling cell cycle checkpoints, apoptosis, and DNA repair after DNA damage. A number of studies have reported that GADD45α expression is modulated by a variety of DNA-damaging agents, including ionizing radiation (Fornace, Alamo, & Alamo, 1988; Fornace et al., 1989; Papathanasiou et al., 1991). Several X-ray-responsive elements were found including the p53 recognition site in *GADD45α*. Detailed analysis of the upstream region (−2244 bp/+89 bp) and the third intron (+1389 bp/+2488 bp) of *GADD45α* revealed binding sites for Oct-1, NF-κB, HNF, NF-AT, and KLF family transcription factors. Radiation treatment stimulated transcription factor binding to *GADD45α* gene promoter after a single dose of 0.5 Gy with expression levels reaching a maximum at 1 h after irradiation (Daino et al., 2003). It was shown that induction of *GADD45α* gene expression following ionizing irradiation is regulated in a p53-dependent manner (Daino et al., 2006; Jin et al., 2001; Takahashi et al., 2001).

### 4.3. Radiation-inducible hybrid/synthetic enhancer–promoter elements

Synthetic or hybrid enhancer and/or promoters have been designed to improve both the strength and tumor specificity for gene expression. In early experiments using synthetic promoters for radiation-mediated gene therapy, increased levels of expression of *GFP* and *HSV-TK* genes under the control of the CArG elements from the *Egr-1* promoter were detected after single dose and multiple fractions of irradiation using human breast cancer and glioma cell models (Greco et al., 2005). An optimization study of radiation-responsive gene promoters comprised 4 (E4), 6 (E6), 9 (E9) or 12 (E12) copies of the CArG element sequences in tandem was carried out. Radiation treatment of MCF7 human breast cancer cells at 3 Gy increased expression of a reporter gene. The maximal levels of GFP expression were reached using a promoter with nine copies of the CArG element (Scott et al., 2002). Another synthetic promoter containing CArG element (pE9) was made to drive production of iNOS, which radiosensitizes hypoxic tumor cells to radiation treatment (Coulter et al., 2008).

## 5. THERAPEUTIC TRANSGENES

Over the last decades, substantial progress has been made in radiation therapy methods. However, often tumor cells escape irradiation induced cell death either by deregulation of the cell cycle, suppressing apoptosis or by an increased ability of cancer cells to repair the DNA damage. Manipulation of genes involved in radiation resistance represents an important approach for therapeutic intervention in cancer. Combination of radiation therapy with gene therapy provides a promising strategy to modify the radiation response and overcome radioresistance of tumor cells. In the last decade, significant progress in cancer therapy using a combination of gene therapy and radiation treatment has been demonstrated. A broad range of different gene therapy approaches, including cytolytic and corrective gene therapy, have been successfully applied for experimental therapy in combination with radiation therapy in animal models, with translation into early clinical trials (summarized in Table 7.2).

**Table 7.2** Summary of gene therapy strategies to enhance radiation therapy efficacy

| Target | Signaling pathways/ mechanisms | References |
|---|---|---|
| *Suppression of radiation-induced DNA damage repair* | | |
| BRCA2 | HR | Yu, Sekine, Fujimori, Ochiya, and Okayasu (2008) |
| Rad51 | | Vernole et al. (2009) |
| Ku70 | NHEJ | Li et al. (2003) |
| Ku80/Ku86 | | Nimura et al. (2007) |
| APE1 | BER | Xiang et al. (2008) |
| Poly(ADP-ribose) | | Ratnam and Low (2007), Trofimova et al. (2002) |
| *Radiosensitization by tumor targeting* | | |
| hSSTr2 | Radiolabeled ligand | Rogers, Zinn, Lin, Chaudhuri, and Buchsbaum (2002), Cotugno et al. (2011) |
| GRPR | | Stackhouse et al. (1999) |
| NIS | Radionuclide uptake | Spitzweg, Baker, Bergert, O'Connor, and Morris (2007) |
| NAT | | Jia et al. (2011) |
| *Facilitation of apoptosis* | | |
| p53 | Apoptosis induction | Gallardo, Drazan, and McBride (1996), Spitz et al. (1996), Li, Lax, Kim, Klamut, and Liu (1999), Xu, Pirollo, Tang, Rait, and Chang (1999), Wakasa et al. (2002), Higuchi et al. (2003) |
| p73, p51/p63 | | Sasaki et al. (2001) |
| Bax | | Kaliberov et al. (2002) |
| Caspases | Apoptosis induction/ execution | Shariat et al. (2001), Uchida et al. (2003), Winter, Rhee, and Kyprianou (2004), Yanamandra et al. (2004) |
| TRAIL | Multiple pathways | Kaliberov, Kaliberova, Stockard, Grizzle, and Buchsbaum (2004) |
| Fas ligand | | Shinoura, Yamamoto, Asai, Kirino, and Hamada (2000) |

**Table 7.2** Summary of gene therapy strategies to enhance radiation therapy efficacy—cont'd

| Target | Signaling pathways/ mechanisms | References |
|---|---|---|
| IL-24/mda-7 | | Nishikawa, Ramesh, Munshi, Chada, and Meyn (2004) |
| Bcl-2 | Inhibition of antiapoptotic pathways | Hara et al. (2005) |
| Bcl-XL | | Mercatante, Mohler, and Kole (2002) |
| XIAP, survivin | | Cao, Mu, Hallahan, and Lu (2004) |
| *Modulation of tumor-associated microenvironment* | | |
| Endostatin | Inhibition of angiogenesis | Shi, Teschendorf, Muzyczka, and Siemann (2003) |
| Soluble VEGF receptor 2 | | Kaliberov, Kaliberova, and Buchsbaum (2005) |
| TNF-α | Multiple pathways | Hallahan, Beckett, Kufe, and Weichselbaum (1990), Gridley, Glisson, and Uhm (1994), Azria et al. (2003), Weichselbaum and Kufe (2009), Sartor (2000) |
| EGFR | Inhibition of proliferation | Gokhale et al. (1999), Dritschilo et al. (2006) |
| c-raf-1 | | Lammering et al. (2003), Lai et al. (2009) |
| *Immunomodulation* | | |
| IL-2 | Facilitating the antitumor immune response | Bray et al. (2003), Hillman et al. (2004), Formenti (2010) |
| IL-3 | | Chiang, Hong, Wu, McBride, and Dougherty (2000), Oh, Chen, Dougherty, and McBride (2004) |
| IL-4 | | Iwadate, Yamaura, Sakiyama, Sato, and Tagawa (2003) |
| IL-12 | | Lohr et al. (2000), Nishitani et al. (2002), Yang, Liu, and Fu (2004) |

*Continued*

**Table 7.2** Summary of gene therapy strategies to enhance radiation therapy efficacy—cont'd

| Target | Signaling pathways/ mechanisms | References |
|---|---|---|
| IL-24/mda-7 | | Emdad et al. (2006), Yacoub et al. (2008), Yang et al. (2009) |
| TNF-α | | Weichselbaum et al. (1994, 2002), Gridley et al. (2004) |
| *Radiosensitization by molecular chemotherapy* | | |
| CD | Disruption of DNA synthesis | Pederson et al. (1997), Stackhouse et al. (2000), Kaliberova et al. (2008), Kaliberov et al. (2007) |
| HSV-TK | | Kim, Kim, Kolozsvary, Brown, and Freytag (1997), Atkinson and Hall (1999), Blackburn, Galoforo, Corry, and Lee (1998), Cheon et al. (2000), Chen et al. (2001), Teh et al. (2001), Tetzlaff et al. (2006) |
| CD:HSV-TK fusion | | Freytag et al. (2002, 2007) |
| NTR | | White et al. (2008) |
| HRP | | Greco, Tozer, and Dachs (2002) |

## 5.1. DNA repair pathways

Exposure to ionizing radiation initiates a cascade of multiple signaling pathways including a number of DNA repair pathways which are essential for the intrinsic resistance of the tumor cell to radiotherapy (Chen, Sin, Leung, Ng, & Poon, 2005; Fischer & Meese, 2007; Kauffmann et al., 2008; Munshi et al., 2005; Rass & Reichrath, 2008). It is well known that the HR and NHEJ pathways are involved in radiation-induced repair. Extensive efforts have been made to develop gene therapy strategies to interfere with the main DSB repair pathways. In the last decade, gene therapy approaches were tested for downregulation of protein expression involved in HR or NHEJ by using short-interfering RNA (siRNA) against BRCA2 and Rad51 (Vernole et al., 2009; Yu et al., 2008) or Ku70 and Ku80 (Li et al., 2003; Nimura et al., 2007), respectively. Recently, a strategy for sensitizing tumors to radiotherapy employing short modified double-stranded DNA molecules (siDNA) has been explored. Use of this signal-interfering DNA, mimicking DSB damage, recruits the Ku70/Ku80/DNA-PKcs DNA damage repair complex. Activation of the DNA-PK kinase phosphorylates a large number of proteins in the DNA damage response cascade including H2AX, Rpa32, Chk1, Chk2, Nbs1, and p53 that results in the inhibition of DNA repair (Berthault et al., 2011; Quanz et al., 2009). Gene correction therapies in combination with radiation have also been evaluated for inhibition of the base excision repair (BER) pathways that are responsible for the repair of oxidative and alkylation DNA damage. Combination of intratumoral (i.t.) injection of Ad5/F35 vector expressing fiber from Ad serotype 35 and siRNA against the human apurinic/apyrimidinic endonuclease (APE1), the important enzyme in the BER pathway, with irradiation at 6 Gy produced delay in LOVO colon tumor growth in comparison with radiation therapy alone (Xiang et al., 2008). It was shown that inhibition of the poly(ADP-ribose) polymerase-1 enzyme, which plays a vital role in DNA damage signaling and regulation of the BER pathways, significantly improved survival of patients with BRCA1 or BRCA2 mutation in the tumor cells (Ratnam & Low, 2007). The efficacy of cancer gene therapy by targeting the poly (ADP-ribose) polymerase dominant negative gene in combination with single dose radiation at 5 Gy was evaluated using a prostate tumor xenograft model (Trofimova et al., 2002).

## 5.2. Radiotargeted gene therapy

DNA damage mediated by radiotargeted gene therapy provides an additional approach to sensitize tumor cells to irradiation. The utilization of tumor-targeted gene transfer for the expression of high-affinity membrane receptors can enhance the specificity as well as extent of localization of radiolabeled ligand within the tumor mass and decrease normal tissue toxicity. Using a genetic radioisotope targeting strategy, induction of high levels of human somatostatin receptor subtype 2 (hSSTr2) or gastrin-releasing peptide receptor (GRPR) expression resulted in selective tumor uptake of radiolabeled peptides and significant inhibition of tumor growth of human tumor xenografts was achieved (Cotugno et al., 2011; Rogers et al., 2002; Stackhouse et al., 1999). A second strategy involves transduction of the sodium iodide symporter (*NIS*) or the noradrenaline transporter (*NAT*) genes in combination with free radionuclide therapy. *NIS* and *NAT* are responsible for the active uptake of $^{131}$I-iodine (Na$^{131}$I) and $^{131}$I-meta-iodobenzylguanidine ($^{131}$I-MIBG), respectively. Therapy studies have been performed on several tumor xenograft models using *NIS/NAT* gene transfer followed by the administration of targeted radionuclides $Na^{131}I$ for NIS and $^{131}I$-MIBG. There is considerable clinical experience in the use of $^{131}I$-MIBG and $Na^{131}I$ in the treatment of medullary thyroid and hepatocellular carcinoma (Jia et al., 2011; Spitzweg et al., 2007). The use of gene transfer technology to induce expression of high-affinity membrane receptors or transporters can enhance the specificity and extent of radioligand or radionuclide localization in tumors, and the use of radionuclides with appropriate emissions can deliver radiation-absorbed cytotoxic doses across several cell diameters to compensate for limited transduction efficiency.

## 5.3. Modulation of the apoptosis pathways

It is now well known that exposure of tumor cells to clinically relevant doses of radiation generates DNA damage in the nucleus as well as triggers a large network of intracellular signaling pathways that collectively produce a multilayered biological response. Apoptosis is recognized as a fundamental mechanism of cellular response to ionizing radiation. Suppressed apoptosis is an important distinctive feature that contributes to the malignant phenotype of cancer. Also, the ability of a cancer cell to respond to radiotherapy is believed to be due, in part, to its apoptotic capacity. Thus, combination of radiation and gene correction strategies is a promising approach to increase apoptosis in tumor cells. Apoptosis is controlled by multiple genes, many of which are mutated in human cancers. *p53* is one of the genes most widely used for gene therapy in combination with radiation. The p53 tumor suppressor protein exerts its growth inhibitory activity by interacting with diverse signaling pathways which leads to many outcomes in cells, including cell cycle arrest and apoptosis. The mutation/deletion of *p53* impairs cellular response to DNA damage, prevents apoptosis, and remains a strong predictor of radioresistance (Lowe et al., 1994). Restoration of p53 activity by i.t. injection of Ad vectors encoding *p53* has been shown to significantly increase tumor cell sensitivity to irradiation. Combination of radiation with wild-type *p53* gene therapy has a synergistic suppressive effect on tumor growth via increasing apoptosis in a wide variety of tumor xenograft models including colorectal, ovarian, nasopharyngeal, head and neck, and squamous cell carcinomas (Gallardo et al., 1996; Higuchi et al., 2003; Li et al., 1999; Spitz et al., 1996; Wakasa et al., 2002; Xu et al., 1999). As a result, advanced clinical studies are now underway which demonstrate promising results in patients with SCCHN, thyroid tumors, non-small cell lung cancer (NSCLC), prostate cancer, or pancreatic cancer with Ad-p53 i.t.

injections (bronchoalveolar lavage for NSCLC) combined with fractionated radiation therapy. Ad-mediated overexpression of the *p53* family of genes, *p73* and *p51/p63*, produces cell cycle arrest and apoptosis in colorectal cancer cells resistant to *p53*-mediated apoptosis (Sasaki et al., 2001). A number of gene therapy approaches were evaluated for the modulation of apoptosis response to radiation through overexpression of proapoptotic genes, including *bax* (Kaliberov et al., 2002); family of cysteine proteases—caspases 1, 3, 8, and 9 (Shariat et al., 2001; Uchida et al., 2003; Winter et al., 2004; Yanamandra et al., 2004); the tumor necrosis factor-related apoptosis-inducing ligand (TRAIL/Apo2L) (Kaliberov et al., 2004); Fas ligand (CD95L/Apo1L) (Shinoura et al., 2000); melanoma differentiation-associated gene-7/interleukin-24 (mda-7/IL-24) (Su et al., 2006); and/or downregulation of antiapoptotic genes by siRNA, antisense oligos, or catalytic RNA (ribozymes), including *Bcl-2* (Hara et al., 2005), *Bcl-XL* (Mercatante et al., 2002), and *XIAP* and *survivin* (Cao et al., 2004).

## 5.4. Targeting of tumor microenvironment

Data from experimental and clinical studies indicate that the tumor–microenvironment interaction have an essential role in tumor growth. Development of radiation resistance is accompanied by remarkable changes in the gene expression profile of surrounding normal cells. Cross talk between malignant cells and other tumor-associated cells represents a new target for anticancer therapy, which has been explored through different forms of intervention. Strategies to inhibit tumor–microenvironment interaction include inhibition of expression of growth factors and their receptors, inhibition of growth factor signaling pathways, abrogation of the binding between ligands and their receptors, and prevention of intracellular transduction of the growth factor signal. The formation of new blood vessels is an important step in cancer progression. Tumor angiogenesis is a complex of coordinated interactions between numerous proteins, involved in different signaling pathways. Each step provides an opportunity for therapeutic intervention. Angiogenesis mediated by the vascular endothelial growth factor (VEGF) constitutes a new target for anticancer therapy, which has been explored through different forms of intervention aimed at blocking of tumor neovascularization. Strategies to inhibit tumor angiogenesis include inhibition of angiogenic factor production and their receptors, inhibition of the VEGF signaling pathway, inhibition of the binding between VEGF and its receptors, and inhibition of intracellular transduction of the VEGF signal (Kaliberov et al., 2005; Shi et al., 2003; Yang, Elliott, & Head, 2002). The importance of signal transduction through the VEGF receptor 2 is illustrated by use of soluble KDR. Treatment with AdVEGF-sKDR, encoding *sKDR* under the control of the human *VEGF* promoter, significantly inhibited the proliferation of human vascular endothelial cells and cancer cells. Ionizing radiation upregulated VEGF promoter activity in prostate carcinoma and endothelial cells. *In vivo* tumor therapy studies demonstrated significant inhibition of DU145 tumor growth in mice that received combined AdVEGF-sKDR infection and ionizing radiation versus AdVEGF-sKDR alone or radiation therapy alone (Kaliberov et al., 2005).

TNF-α has a broad range of potent anticancer properties including activation and recruitment of immune cells, and induction of hemorrhagic necrosis by disruption of the tumor vasculature. The oxidative damage produced by TNF-α might enhance cellular damage produced by ionizing radiation (Zimmerman, Marafino, Chan, Landre, & Winkelhake, 1989). Combination of *TNF-α* gene therapy with radiation produced an additive or a supra-additive increase in antitumor efficacy of a range of tumor cells in comparison with these modalities alone (Azria et al., 2003; Gridley et al., 1994; Hallahan et al., 1990). Furthermore, the combination of radiation with *TNF-α* gene delivery using Ad vectors encoding *TNF-α* under the control of the radiation-inducible CArG elements from the *Egr-1* gene increased radiation targeting specificity of cancer gene therapy (Weichselbaum & Kufe, 2009; Weichselbaum et al., 1994).

Cell growth is controlled by a balance of growth-promoting and growth-inhibiting factors. It is well known that a disturbed balance of growth factors in tumor cells results in uncontrolled cellular proliferation. Recent evidence suggests that radioresistance is due to, in part, tumor cell repopulation and involves several signaling pathways, including of the epidermal growth factor receptor (EGFR) pathway. Members of the EGFR family are overexpressed in many epithelial tumors. Cancer cells have been shown to upregulate the tyrosine kinase activity of EGFR in response to radiation therapy via ROS-mediated inhibition of protein tyrosine phosphatases, the enzymes responsible for dephosphorylating activated tyrosine kinases. Signaling by EGFR can also be upregulated in a delayed manner through the induction of paracrine growth factors, such as EGFR ligand transforming growth factor-α. EGFR overexpression has been linked to poor clinical outcomes with radiation therapy (Barrett et al., 2005; Harari & Huang, 2001; Sartor, 2000). Based on these findings, a variety of EGFR signaling-targeted therapeutic strategies have been developed including monoclonal antibodies (cetuximab), tyrosine kinase inhibitors (erlotinib) as well as antisense oligonucleotides to EGFR or c-raf-1, serine/threonine protein kinase (Dritschilo et al., 2006; Gokhale et al., 1999; Lai et al., 2009; Lammering et al., 2003).

## 5.5. Immunomodulation

Immunotherapy is based on activation of a host immune response against the tumor cells using cytokines and activated immune cells. Immunomodulation represents an approach to cancer therapy, in which induction of an antitumor immune response leads to tumor cell killing. Cancer cells express several unique tissue-specific proteins that can potentially function as target antigens for genetic immunotherapy. Although a large number of tumor-associated antigens that can be recognized by the immune system are expressed on the cell surface, tumor cells demonstrate relatively low immunogenicity. A large body of evidence demonstrates the ability of radiation to induce inflammation and expression of host-derived proinflammatory cytokines, such as IL-1β and TNF-α via modification of the tumor microenvironment (Apetoh et al., 2007; Hallahan, Spriggs, Beckett, & Kufe, 1989; Ishihara, Tsuneoka, Dimchev, & Shikita, 1993). Also, it was shown that different immunotherapeutic strategies can be combined with radiation therapy to enhance antitumor effects as well as employing radiation therapy to promote tumor antigen presentation by dendritic cells (Demaria, Bhardwaj, McBride, & Formenti, 2005). Recently developed approaches in this field are strategies based on dendritic cells (Chakravarty et al., 1999; Demaria et al., 2004; Nikitina & Gabrilovich, 2001; Teitz-Tennenbaum et al., 2003), targeting of toll-like transmembrane receptors with CpG oligodeoxynucleotides (Jahrsdorfer & Weiner, 2008; Krieg, 2004; Mason et al., 2005; Yuan, Qiao, & Chen, 2011), vaccination with modified autologous tumor cells (Chakraborty et al., 2004; Lumniczky et al., 2002), and cytokines. Cytokines are proteins produced following activation of lymphocytes, macrophages, or other cells and are employed in immunotherapy to enhance immune mechanisms directed against tumors. Cytokines can induce a cascade of activation, proliferation, and differentiation of immune cells and increase an antitumor immune response. Interleukin 2 (IL-2), produced by activated T lymphocytes and critical for immune responsiveness, is one of the first cytokines/lymphokines that have been extensively explored in combination with radiation (Bray et al., 2003; Formenti, 2010; Hillman et al., 2004). Promising results have also emerged from studies using a wide range of immunomodulators in combination with radiation, including IL-3 (Chiang et al., 2000; Oh et al., 2004), IL-4 (Iwadate et al., 2003), IL-12 (Lohr et al., 2000; Nishitani et al., 2002; Yang et al., 2004), mda-7/IL-24 (Emdad et al., 2006; Yacoub et al., 2008; Yang et al., 2009), and *TNF*-α (Gridley et al., 2004; Weichselbaum et al., 1994, 2002).

## 5.6. Oncolytic virotherapy

At the present time, several strategies of cytoreductive gene therapy such as oncolytic virotherapy have been used for cancer treatment in combination with radiation. Conditional oncolytic virotherapy has employed replication-competent viral vectors that specifically kill tumor cells and represents a promising tool for cancer treatment. A major advantage of replicative oncolytic viral vectors is that their number should increase via viral replication within infected tumor cells, and the resulting viral progeny can infect and kill neighboring cancer cells within the tumor mass. G207 is an attenuated/replication-competent HSV-1 mutant that lacks both copies of the *ICP34.5* (*RL1*) gene and contains an insertion of *lacZ* in the *ICP6* (*UL39*) gene. These gene modifications offer a new modality in cancer therapy through the ability of G207 to selectively replicate within and kill malignant cells with minimal toxicity to normal tissues. G207 induced a significant inhibition of tumor growth alone and in combination with radiation in human head and neck cancer and prostate tumor xenograft models (Kim et al., 2005; Walker et al., 1999). Significant tumor growth delays were noted in *p53* mutant tumors receiving virotherapy using ONYX-015 (*dl*1520) CRAd vector containing a deletion in E1B-55kD protein in conjunction with irradiation at 5 Gy in comparison with CRAd or radiation alone (Geoerger et al., 2003; Rogulski et al., 2000). An EBV-dependent transcriptional targeting strategy was used to construct a novel conditionally replicating Ad, ad.oriP.E1A. The experimental group treated with ad.oriP.E1A in combination with ionizing radiation showed complete regressions of established tumors for at least 2 weeks in two distinct EBV-positive nasopharyngeal carcinoma xenograft models (Chia et al., 2004). CV706 is an oncolytic CRAd vector, in which the essential *E1A* viral replication gene is under the control of an exogenous minimal enhancer/promoter construct derived from the 5′ flank of the human prostate-specific antigen (*PSA*) gene promoter that is preferentially active in prostate cancer cells (Rodriguez et al., 1997). CV706 has been shown to selectively kill human prostate cancer xenografts in preclinical models (DeWeese et al., 2001). KD1 and KD3 are replication-competent infectivity-enhanced Ads that contain mutations in the *E1A* gene that knock out binding of the E1A protein to cellular p300/CBP and pRB; these mutations facilitate KD1 and KD3 proliferation selectively in tumor cells, but not in normal cells. It was shown that the combination of irradiation with CRAd increased suppression of the growth of human tumor xenografts in nude mice in comparison with these treatments alone (Toth et al., 2003). Enhanced antitumor efficacy was also observed in radiation combination studies with Ad5-Delta24RGD (Lamfers et al., 2002) and the probasin promoter-driven CRAd expressing the *NIS* gene (Trujillo, Oneal, McDonough, Qin, & Morris, 2010). In an attempt to improve both the efficacy and safety of oncolytic virotherapy, a replication-competent Ad vector encoding a *CD/HSV-1 TK* fusion gene was constructed. Phase I studies demonstrated the safety of intraprostatic administration of this Ad in combination with conventional-dose three-dimensional conformal radiation therapy in patients with intermediate to high-risk prostate cancer (Freytag et al., 2003).

## 6. MOLECULAR CHEMOTHERAPY AND RADIATION THERAPY

One of the most widely investigated radiation therapy and gene therapy combination approaches utilizes molecular chemotherapy or suicide gene-directed enzyme/prodrug therapy (GDEPT). This strategy involves delivery of a specific enzyme that can produce cell death through the conversion of an inactive nontoxic prodrug into a cytotoxic drug metabolite. Specifically, targeted expression of the prodrug-activating enzyme avoids systemic toxicity and results in a high drug concentration in the tumor mass and an improved therapeutic index compared to systemic drug administration. The key element of a molecular chemotherapy system is a gene that encodes an enzyme, which converts a prodrug to an active cytotoxic drug. Importantly, prodrug-activating enzymes are normally absent or poorly expressed in mammalian cells. This means that tumor targeting of gene therapy, using specific gene delivery vehicles, can restrict enzyme expression to the transduced tumor cells and adjacent surrounding tumor cells through diffusion of the drug metabolite to generate a bystander effect.

### 6.1. Cytosine deaminase/5-fluorocytosine

One of the most promising molecular chemotherapy systems is cytosine deaminase/5-fluorocytosine (CD/5-FC) that has been studied intensively during the last decade. CD is a bacterial or yeast enzyme that can convert the antifungal agent 5-FC into the chemotherapy agent 5-fluorouracil (5-FU). Importantly, CD is present in prokaryotes and fungi but not in higher eukaryotes, where it is an important member of the pyrimidine salvage pathway. The antifungal agent 5-FC has been extensively used in the past as an antifungal agent for therapy of *Cryptococcus* and *Candida* infection and requires activation by the fungal CD. Therefore, its toxicological properties in humans are well established. 5-FU is a pyrimidine antagonist in which the hydrogen atom at the C-5 position of uracil is substituted by fluorine. Since the structure of 5-FU is similar to that of uracil, 5-FU is converted into biologically active metabolites, 5-fluoro-2′-deoxyuridine-5′-monophosphate (FdUMP) or 5-fluorouridine-5′-triphosphate as well as being metabolized to an inactive form, 5-fluoro-β-alanine, ammonia, and $CO_2$, by the same anabolic and catabolic pathways as that of uracil. 5-Fluorouracil triphosphate is incorporated into RNA and interferes with RNA processing, while FdUMP with the coenzyme *N*5,*N*10-methylene tetrahydrofolate forms a covalent ternary complex with thymidylate synthase (TS), an essential DNA *de novo* synthetic enzyme that catalyzes the methylation of deoxyuridine monophosphate (dUMP) to deoxythymidine monophosphate (dTMP). This irreversibly inhibits the conversion of dUMP to dTMP and hence the disruption of nucleotide pools of dTMP, a thymidine precursor in DNA synthesis. The *codA* gene encoding bacterial CD (bCD) has been cloned from *Escherichia coli* into eukaryotic expression vectors and was first used for GDEPT of mouse fibroblasts *in vitro* and in murine tumor models almost two decades ago (Mullen, Coale, Lowe, & Blaese, 1994).

One of the major advantages of the CD/5-FC-mediated molecular chemotherapy system is a strong bystander effect that does not require gap junctions in contrast to other suicide gene therapy systems such as the HSV-TK/GCV, since 5-FU is a small, uncharged molecule capable of nonfacilitated diffusion through cellular membranes (Hoganson, Batra, Olsen, & Boucher, 1996; Nishihara et al., 1998; Trinh, Austin, Murray, Knick, & Huber, 1995). The bystander effect can be defined as an enlargement of cancer cell cytotoxicity in which nontransfected neighboring cells are killed through the transfer of metabolites of 5-FU from CD-expressing cells in close proximity. This means that even if only a small number of cancer cells express the *CD* gene, tumor eradication may still be achieved by using CD/5-FC molecular chemotherapy. Thus, the bystander activity is important for successful GDEPT due to the low transfection efficiency of gene delivery vectors currently employed for cancer gene therapy. Generally, molecular mechanisms of therapeutic agent-mediated cytotoxicity can be divided into two groups that involve development of a local immune-mediated effect and distant bystander activity. The bystander effect may be produced via the diffusion of a soluble chemotherapy drug such as 5-FU to adjacent surrounding cells (Huber, Austin, Richards, Davis, & Good, 1994; Kuriyama et al., 1998), via transfer of toxic products through gap junctions (Dilber et al., 1997; Elshami et al., 1996), or via apoptotic vesicles (Freeman et al., 1993). It was shown that gap junctions produce restricted transfer due to the requirement of cell-to-cell contact, and the levels of intercellular gap junction communication are often decreased between tumor cells (Holder, Elmore, & Barrett, 1993; Mesnil, Piccoli, Tiraby, Willecke, & Yamasaki, 1996).

Another advantage of the CD/5-FC system is based on the extensive pre-clinical and clinical experience with 5-FU as both a chemotherapeutic agent and a strong radiation sensitizer. Several groups have reported *in vivo* results with CD to have a significant bystander effect at clinically relevant 5-FC doses and radiation regimens (Kambara et al., 2002; Kievit et al., 2000; Kim et al., 1998). Intrinsic differences in sensitivity to 5-FU and ionizing radiation may contribute to overall efficacy of molecular chemotherapy in combination with radiation therapy. The precise mechanisms underlying the interaction of radiation and 5-FU are not clear. However, there are a number of potential pathways by which 5-FU could increase radiation sensitivity at the cellular level. Although the disruption of either RNA or DNA synthesis can produce cytotoxicity, a substantial body of evidence suggests that radiosensitization is a result of TS inhibition (Lawrence, Tepper, & Blackstock, 1997; Miller & Kinsella, 1992). It was shown that 5-FU produces tumor cell killing particularly during S phase, when cells are relatively radioresistant (McGinn & Lawrence, 2001). Also, the combination of 5-FU and radiation sensitization probably results in cells that are progressing inappropriately through S phase due to misrepair of DNA damage imposed by ionizing radiation. This loss of the S-phase checkpoint in cancer cells may provide the molecular basis for selective killing of tumors compared with normal tissues (Davis, Tang, Maybaum, & Lawrence, 1995; Lawrence, Davis, & Loney, 1996). The hydrolytic deamination of cytosine that produces a miscoding uracil residue is a frequent DNA lesion induced by oxidative damage, through exposure to ROS-producing agents such as ionizing radiation. Cytostatic agents interfering with pyrimidine metabolism may enhance uracil incorporation by increasing the dUTP:dTTP ratio. Thus, cytostatic drugs inhibiting TS, for example, 5-FU may enhance uracil incorporation and thus contribute to the cytotoxicity of combined CD/5-FC GDEPT with radiation therapy. Analysis of the linear quadratic or single-hit-multi-target parameters of survival curves indicated a significant reduction in cell survival at both low and high doses of radiation therapy (Pederson et al., 1997; Stackhouse et al., 2000). The combination of CD/5-FC and ionizing radiation produced a measurable antitumor effect using 2 or 5 Gy per fraction over 1 week which is a clinically relevant dose regimen of radiation therapy (Stackhouse et al., 2000). Additional preclinical studies have been carried out using yCD (Hamstra et al., 1999), and clinical trials have been conducted in patients with prostate cancer administered a replicative Ad expressing *CD* in combination with radiotherapy with promising results (Freytag et al., 2002, 2007).

A major problem associated with this GDEPT approach is the low affinity displayed by the *CD* gene product toward 5-FC in comparison with cytosine. Recent studies have demonstrated that substitution of an alanine (A) for the aspartic acid (D) at position 314 of bCD increased relative specificity of the mutant bCD-D314A enzyme to 5-FC in comparison with bCDwt and may be a superior suicide gene (Mahan, Ireton, Knoeber, Stoddard, & Black, 2004; Mahan, Ireton, Stoddard, & Black, 2004). It was shown that infection of AdbCD-D314A vector encoding a mutant *bCD* gene resulted in increased 5-FC-mediated cell killing, compared with AdbCDwt. Furthermore, a significantly greater inhibition of growth of human pancreatic tumor xenografts was produced by the combination of AdbCD-D314A/5-FC with radiation as compared with molecular chemotherapy or radiation therapy alone (Kaliberova et al., 2008). Also, animal studies showed significant inhibition of subcutaneous or intracranial tumor growth of human glioma xenografts by the combination of AdbCD-D314A/5-FC with ionizing radiation as compared with either agent alone or AdbCDwt/5-FC plus radiation. The results suggest that the combination of CD/5-FC with radiation produces markedly increased cytotoxic effects in cancer cells *in vitro* and *in vivo* (Kaliberov et al., 2007).

## 6.2. Herpes simplex virus-thymidine kinase/ganciclovir

Another well-studied suicide gene/prodrug system for cancer therapy employs a combination of HSV-TK with antiherpetic drugs such as GCV or its analogues (acyclovir and valacyclovir) (Eastham et al., 1996). The HSV-TK/GCV system is characterized by the effective phosphorylation of GCV into a toxic compound that is incorporated into DNA during its replication. This incorporation into guanine sites in newly synthesized DNA chains causes termination of synthesis and the selective killing of dividing cells by activation of apoptosis pathways. It was shown that HSV-TK/GCV molecular chemotherapy sensitized cancer cells to ionizing radiation (Atkinson & Hall, 1999; Kim et al., 1997). A number of Ad vectors encoding the *HSV-TK* gene were employed in combination with ionizing radiation and have shown therapeutic effects in clinical trials without substantial toxic effects when used alone or in combination with radiotherapy (Blackburn et al., 1998; Cheon et al., 2000; Eastham et al., 1996). In an extensive series of preclinical studies, a combination of radiation therapy and molecular chemotherapy using Ad vectors encoding *HSV-TK* or *CD:HSV-TK* fusion gene demonstrated potent tumor cell radiosensitization activity (Chen et al., 2001; Teh et al., 2001; Tetzlaff et al., 2006).

Currently, a number of alternative suicide GDEPT systems using various gene delivery systems are under intense investigation. A series of clinical trials using a replication-defective Ad vector expressing *E. coli* nitroreductase (*NTR*) gene were initiated in patients with primary or secondary liver cancer, head and neck cancer, or prostate cancer (Searle et al., 2004). It was shown that the prodrug CB1954 (5-(aziridin-1-yl)-2,4-dinitrobenzamide) is activated by NTR to a potent DNA-cross-linking agent. Adenoviral delivery of *NTR* gene in combination with CB1954 administration in nude mice enhanced the effect of fractionated radiotherapy using four fractions at 3 Gy in SW480 xenograft tumors (White et al., 2008). Another example of a novel molecular chemotherapy system is the combination of gene delivery of horseradish peroxidase (HRP) with indole-3-acetic (IAA) as a prodrug was developed and its antitumor effect was evaluated *in vitro* and *in vivo* (Xu et al., 2011). Metabolism of IAA by HRP produces a wide range of cytotoxic species, including superoxide and peroxyl radicals (de Melo, de Lima, Pithon-Curi, & Curi, 2004) and increased cytotoxicity in combination with irradiation (Greco et al., 2002).

## 7. CONCLUSION

Although in recent years significant progress in radiation therapy methods has been achieved, development of resistance to irradiation remains a major problem for successful control of cancer in patients. The combination of gene therapy and radiation therapy could improve the outcome of cancer treatment. Important advances in the combination of irradiation and gene therapy over the last decade are a result of improved understanding of the molecular mechanisms that mediate cancer progression and resistance to radiation treatment. Ionizing radiation produces DNA damage and activates multiple intracellular signaling pathways. The ability of radiation to activate pathways depends on the levels of DNA damage, generation of ROS/RNS, activation of a variety of transcription factors, as well as alterations in the expression of growth factor receptors and their cognate ligands. A growing body of evidence suggests that interaction between tumor cells and surrounding normal cells controls local tumor development and involves the development of radiation resistance. Intervention against the interaction between tumor cells and their microenvironment is a promising area for the development of combination of gene therapy and radiation therapy modalities. This would provide the possibility to utilize multiple targeting approaches for combined gene therapy and radiotherapy.

Modifications of gene delivery systems, such as using tumor-specific promoters for selective expression of therapeutic genes, or development of new oncolytic virotherapy vectors that are designed to selectively replicate in tumor cells or tumor-associated cells, should significantly increase the efficacy and safety of cancer gene therapy. These efforts for development of cancer gene therapy vectors in combination with radiation can be illustrated by using transductional and transcriptional approaches for targeting of Ad vectors. Importantly, irradiation at therapeutic doses can increase the level of viral uptake, replication, and gene expression. Manipulation of the Ad capsid can modify viral tropism by binding of Ad to receptors other than the coxsackie and adenovirus receptor (CAR)—the primary receptor for Ad5. This can be achieved through genetic modification of Ad5 fiber knob, employment of adaptor molecules (i.e., fusion protein comprising extracellular domain of CAR and tumor-specific single-chain variable fragment), or chemical modification such as PEGylation. The modification of the fiber or fiber knob in Ad serotype 5 with an alternative fiber knob can increase CAR-independent binding of Ad5-based vectors. Additionally, the use of specific promoters allows for Ad vector-mediated tumor targeting of therapeutic gene expression. These modifications increase specificity and reduce the risk of toxic side effects with the combination of gene therapy and radiation therapy. Challenges regarding current approaches and future directions include uptake of vectors in metastatic tumor sites after systemic administration, and adequate expression of the transgene to produce radiosensitization. Importantly, the combination of such targeted cancer therapies can contribute to the growing field of personalized cancer medicine that is based on the unique genetic characteristics of the tumor as well as the patient.

## ACKNOWLEDGMENT

This work was supported in part by NIH grant P50101955.

## REFERENCES

Adamina, M., Daetwiler, S., Rosenthal, R., & Zajac, P. (2005). Clinical applications of recombinant virus-based cancer immunotherapy. *Expert Opinion on Biological Therapy, 5*, 1211–1224.

Aggarwal, B. B., Kunnumakkara, A. B., Harikumar, K. B., Gupta, S. R., Tharakan, S. T., Koca, C., et al. (2009). Signal transducer and activator of transcription-3, inflammation, and cancer: How intimate is the relationship? *Annals of the New York Academy of Sciences, 1171*, 59–76.

Andreassen, P. R., Ho, G. P., & D'Andrea, A. D. (2006). DNA damage responses and their many interactions with the replication fork. *Carcinogenesis, 27*, 883–892.

Apetoh, L., Ghiringhelli, F., Tesniere, A., Obeid, M., Ortiz, C., Criollo, A., et al. (2007). Toll-like receptor 4-dependent contribution of the immune system to anticancer chemotherapy and radiotherapy. *Nature Medicine, 13*, 1050–1059.

Arlen, P. M., Gulley, J. L., Madan, R. A., Hodge, J. W., & Schlom, J. (2007). Preclinical and clinical studies of recombinant poxvirus vaccines for carcinoma therapy. *Critical Reviews in Immunology, 27*, 451–462.

Atkinson, G., & Hall, S. J. (1999). Prodrug activation gene therapy and external beam irradiation in the treatment of prostate cancer. *Urology, 54*, 1098–1104.

Azria, D., Dorvillius, M., Gourgou, S., Martineau, P., Robert, B., Pugniere, M., et al. (2003). Enhancement of radiation therapy by tumor necrosis factor alpha in human colon cancer using a bispecific antibody. *International Journal of Radiation Oncology, Biology, Physics, 55*, 1363–1373.

Bakkenist, C. J., & Kastan, M. B. (2004). Phosphatases join kinases in DNA-damage response pathways. *Trends in Cell Biology, 14*, 339–341.

Barrett, D. M., Black, S. M., Todor, H., Schmidt-Ullrich, R. K., Dawson, K. S., & Mikkelsen, R. B. (2005). Inhibition of protein-tyrosine phosphatases by mild oxidative stresses is dependent on S-nitrosylation. *The Journal of Biological Chemistry, 280*, 14453–14461.

Bauerschmitz, G. J., Barker, S. D., & Hemminki, A. (2002). Adenoviral gene therapy for cancer: From vectors to targeted and replication competent agents (review). *International Journal of Oncology, 21*, 1161–1174.

Beetz, A., Peter, R. U., Oppel, T., Kaffenberger, W., Rupec, R. A., Meyer, M., et al. (2000). NF-kappaB and AP-1 are responsible for inducibility of the IL-6 promoter by ionizing radiation in HeLa cells. *International Journal of Radiation Biology, 76*, 1443–1453.

Berthault, N., Maury, B., Agrario, C., Herbette, A., Sun, J. S., Peyrieras, N., et al. (2011). Comparison of distribution and activity of nanoparticles with short interfering DNA (Dbait) in various living systems. *Cancer Gene Therapy, 18*, 695–706.

Bickenbach, K. A., Veerapong, J., Shao, M. Y., Mauceri, H. J., Posner, M. C., Kron, S. J., et al. (2008). Resveratrol is an effective inducer of CArG-driven TNF-alpha gene therapy. *Cancer Gene Therapy, 15*, 133–139.

Bjornsti, M. A., & Houghton, P. J. (2004). The TOR pathway: A target for cancer therapy. *Nature Reviews. Cancer, 4*, 335–348.

Blackburn, R. V., Galoforo, S. S., Corry, P. M., & Lee, Y. J. (1998). Adenoviral-mediated transfer of a heat-inducible double suicide gene into prostate carcinoma cells. *Cancer Research, 58*, 1358–1362.

Blalock, W. L., Weinstein-Oppenheimer, C., Chang, F., Hoyle, P. E., Wang, X. Y., Algate, P. A., et al. (1999). Signal transduction, cell cycle regulatory, and anti-apoptotic pathways regulated by IL-3 in hematopoietic cells: Possible sites for intervention with anti-neoplastic drugs. *Leukemia, 13,* 1109–1166.

Blundell, R., Kaminski, N., & Harrison, D. (2004). Increase in p21 expression independent of the p53 pathway in bleomycin-induced lung fibrosis. *Experimental and Molecular Pathology, 77,* 231–237.

Boothman, D. A., Lee, I. W., & Sahijdak, W. M. (1994). Isolation of an X-ray-responsive element in the promoter region of tissue-type plasminogen activator: Potential uses of X-ray-responsive elements for gene therapy. *Radiation Research, 138,* S68–S71.

Bray, D., Yu, S. Z., Koprowski, H., 2nd, Rhee, J., Kumar, S., Pericle, F., et al. (2003). Combination nonviral interleukin 2 gene therapy and external-beam radiation therapy for head and neck cancer. *Archives of Otolaryngology—Head & Neck Surgery, 129,* 618–622.

Burma, S., Chen, B. P., & Chen, D. J. (2006). Role of non-homologous end joining (NHEJ) in maintaining genomic integrity. *DNA Repair, 5,* 1042–1048.

Burma, S., Chen, B. P., Murphy, M., Kurimasa, A., & Chen, D. J. (2001). ATM phosphorylates histone H2AX in response to DNA double-strand breaks. *The Journal of Biological Chemistry, 276,* 42462–42467.

Canman, C. E., Lim, D. S., Cimprich, K. A., Taya, Y., Tamai, K., Sakaguchi, K., et al. (1998). Activation of the ATM kinase by ionizing radiation and phosphorylation of p53. *Science, 281,* 1677–1679.

Cao, C., Mu, Y., Hallahan, D. E., & Lu, B. (2004). XIAP and survivin as therapeutic targets for radiation sensitization in preclinical models of lung cancer. *Oncogene, 23,* 7047–7052.

Chakraborty, M., Abrams, S. I., Coleman, C. N., Camphausen, K., Schlom, J., & Hodge, J. W. (2004). External beam radiation of tumors alters phenotype of tumor cells to render them susceptible to vaccine-mediated T-cell killing. *Cancer Research, 64,* 4328–4337.

Chakravarty, P. K., Alfieri, A., Thomas, E. K., Beri, V., Tanaka, K. E., Vikram, B., et al. (1999). Flt3-ligand administration after radiation therapy prolongs survival in a murine model of metastatic lung cancer. *Cancer Research, 59,* 6028–6032.

Chen, Y., DeWeese, T., Dilley, J., Zhang, Y., Li, Y., Ramesh, N., et al. (2001). CV706, a prostate cancer-specific adenovirus variant, in combination with radiotherapy produces synergistic antitumor efficacy without increasing toxicity. *Cancer Research, 61,* 5453–5460.

Chen, G. G., Sin, F. L., Leung, B. C., Ng, H. K., & Poon, W. S. (2005). Glioblastoma cells deficient in DNA-dependent protein kinase are resistant to cell death. *Journal of Cellular Physiology, 203,* 127–132.

Cheon, J., Kim, H. K., Moon, D. G., Yoon, D. K., Cho, J. H., & Koh, S. K. (2000). Adenovirus-mediated suicide-gene therapy using the herpes simplex virus thymidine kinase gene in cell and animal models of human prostate cancer: Changes in tumour cell proliferative activity. *BJU International, 85,* 759–766.

Chia, M. C., Shi, W., Li, J. H., Sanchez, O., Strathdee, C. A., Huang, D., et al. (2004). A conditionally replicating adenovirus for nasopharyngeal carcinoma gene therapy. *Molecular Therapy, 9,* 804–817.

Chiang, C.-S., Hong, J.-H., Wu, Y. C., McBride, W. H., & Dougherty, G. J. (2000). Combining radiation therapy with interleukin-3 gene immunotherapy. *Cancer Gene Therapy, 7,* 1172–1178.

Chiarugi, P., & Cirri, P. (2003). Redox regulation of protein tyrosine phosphatases during receptor tyrosine kinase signal transduction. *Trends in Biochemical Sciences, 28,* 509–514.

Christy, B., & Nathans, D. (1989). Functional serum response elements upstream of the growth factor-inducible gene zif268. *Molecular and Cellular Biology, 9,* 4889–4895.

Cotugno, G., Aurilio, M., Annunziata, P., Capalbo, A., Faella, A., Rinaldi, V., et al. (2011). Noninvasive repetitive imaging of somatostatin receptor 2 gene transfer with positron emission tomography. *Human Gene Therapy, 22*, 189–196.

Coulter, J. A., McCarthy, H. O., Worthington, J., Robson, T., Scott, S., & Hirst, D. G. (2008). The radiation-inducible pE9 promoter driving inducible nitric oxide synthase radiosensitizes hypoxic tumour cells to radiation. *Gene Therapy, 15*, 495–503.

Criswell, T., Leskov, K., Miyamoto, S., Luo, G., & Boothman, D. A. (2003). Transcription factors activated in mammalian cells after clinically relevant doses of ionizing radiation. *Oncogene, 22*, 5813–5827.

Daino, K., Ichimura, S., & Nenoi, M. (2003). Comprehensive search for X-ray-responsive elements and binding factors in the regulatory region of the GADD45a gene. *Journal of Radiation Research (Tokyo), 44*, 311–318.

Daino, K., Ichimura, S., & Nenoi, M. (2006). Both the basal transcriptional activity of the GADD45A gene and its enhancement after ionizing irradiation are mediated by AP-1 element. *Biochimica et Biophysica Acta, 1759*, 458–469.

Datta, R., Rubin, E., Sukhatme, V., Qureshi, S., Hallahan, D., Weichselbaum, R. R., et al. (1992). Ionizing radiation activates transcription of the EGR1 gene via CArG elements. *Proceedings of the National Academy of Sciences of the United States of America, 89*, 10149–10153.

Datta, R., Taneja, N., Sukhatme, V. P., Qureshi, S. A., Weichselbaum, R., & Kufe, D. W. (1993). Reactive oxygen intermediates target CC(A/T)6GG sequences to mediate activation of the early growth response 1 transcription factor gene by ionizing radiation. *Proceedings of the National Academy of Sciences of the United States of America, 90*, 2419–2422.

Davis, M. A., Tang, H. Y., Maybaum, J., & Lawrence, T. S. (1995). Dependence of fluorodeoxyuridine-mediated radiosensitization on S phase progression. *International Journal of Radiation Biology, 67*, 509–517.

de Melo, M. P., de Lima, T. M., Pithon-Curi, T. C., & Curi, R. (2004). The mechanism of indole acetic acid cytotoxicity. *Toxicology Letters, 148*, 103–111.

Demaria, S., Bhardwaj, N., McBride, W. H., & Formenti, S. C. (2005). Combining radiotherapy and immunotherapy: A revived partnership. *International Journal of Radiation Oncology, Biology, Physics, 63*, 655–666.

Demaria, S., Ng, B., Devitt, M. L., Babb, J. S., Kawashima, N., Liebes, L., et al. (2004). Ionizing radiation inhibition of distant untreated tumors (abscopal effect) is immune mediated. *International Journal of Radiation Oncology, Biology, Physics, 58*, 862–870.

DeWeese, T. L., van der Poel, H., Li, S., Mikhak, B., Drew, R., Goemann, M., et al. (2001). A phase I trial of CV706, a replication-competent, PSA selective oncolytic adenovirus, for the treatment of locally recurrent prostate cancer following radiation therapy. *Cancer Research, 61*, 7464–7472.

Dilber, M. S., Abedi, M. R., Christensson, B., Bjorkstrand, B., Kidder, G. M., Naus, C. C., et al. (1997). Gap junctions promote the bystander effect of herpes simplex virus thymidine kinase in vivo. *Cancer Research, 57*, 1523–1528.

Dritschilo, A., Huang, C. H., Rudin, C. M., Marshall, J., Collins, B., Dul, J. L., et al. (2006). Phase I study of liposome-encapsulated c-raf antisense oligodeoxyribonucleotide infusion in combination with radiation therapy in patients with advanced malignancies. *Clinical Cancer Research, 12*, 1251–1259.

Eastham, J. A., Chen, S. H., Sehgal, I., Yang, G., Timme, T. L., Hall, S. J., et al. (1996). Prostate cancer gene therapy: Herpes simplex virus thymidine kinase gene transduction followed by ganciclovir in mouse and human prostate cancer models. *Human Gene Therapy, 7*, 515–523.

el-Deiry, W. S., Tokino, T., Velculescu, V. E., Levy, D. B., Parsons, R., Trent, J. M., et al. (1993). WAF1, a potential mediator of p53 tumor suppression. *Cell, 75*, 817–825.

Elshami, A. A., Saavedra, A., Zhang, H., Kucharczuk, J. C., Spray, D. C., Fishman, G. I., et al. (1996). Gap junctions play a role in the 'bystander effect' of the herpes simplex virus thymidine kinase/ganciclovir system *in vitro*. *Gene Therapy, 3*, 85–92.

Emdad, L., Sarkar, D., Lebedeva, I. V., Su, Z. Z., Gupta, P., Mahasreshti, P. J., et al. (2006). Ionizing radiation enhances adenoviral vector expressing mda-7/IL-24-mediated apoptosis in human ovarian cancer. *Journal of Cellular Physiology, 208*, 298–306.

Falck, J., Coates, J., & Jackson, S. P. (2005). Conserved modes of recruitment of ATM, ATR and DNA-PKcs to sites of DNA damage. *Nature, 434*, 605–611.

Fischer, U., & Meese, E. (2007). Glioblastoma multiforme: The role of DSB repair between genotype and phenotype. *Oncogene, 26*, 7809–7815.

Formenti, S. C. (2010). Immunological aspects of local radiotherapy: Clinical relevance. *Discovery Medicine, 9*, 119–124.

Fornace, A. J., Jr., Alamo, I., Jr., & Hollander, M. C. (1988). DNA damage-inducible transcripts in mammalian cells. *Proceedings of the National Academy of Sciences of the United States of America, 85*, 8800–8804.

Fornace, A. J., Jr., Nebert, D. W., Hollander, M. C., Luethy, J. D., Papathanasiou, M., Fargnoli, J., et al. (1989). Mammalian genes coordinately regulated by growth arrest signals and DNA-damaging agents. *Molecular and Cellular Biology, 9*, 4196–4203.

Freeman, S. M., Abboud, C. N., Whartenby, K. A., Packman, C. H., Koeplin, D. S., Moolten, F. L., et al. (1993). The "bystander effect": Tumor regression when a fraction of the tumor mass is genetically modified. *Cancer Research, 53*, 5274–5283.

Freytag, S. O., Khil, M., Stricker, H., Peabody, J., Menon, M., DePeralta-Venturina, M., et al. (2002). Phase I study of replication-competent adenovirus-mediated double suicide gene therapy for the treatment of locally recurrent prostate cancer. *Cancer Research, 62*, 4968–4976.

Freytag, S. O., Stricker, H., Peabody, J., Pegg, J., Paielli, D., Movsas, B., et al. (2007). Five-year follow-up of trial of replication-competent adenovirus-mediated suicide gene therapy for treatment of prostate cancer. *Molecular Therapy, 15*, 636–642.

Freytag, S. O., Stricker, H., Pegg, J., Paielli, D., Pradhan, D. G., Peabody, J., et al. (2003). Phase I study of replication-competent adenovirus-mediated double-suicide gene therapy in combination with conventional-dose three-dimensional conformal radiation therapy for the treatment of newly diagnosed, intermediate- to high-risk prostate cancer. *Cancer Research, 63*, 7497–7506.

Galabova-Kovacs, G., Kolbus, A., Matzen, D., Meissl, K., Piazzolla, D., Rubiolo, C., et al. (2006). ERK and beyond: Insights from B-Raf and Raf-1 conditional knockouts. *Cell Cycle, 5*, 1514–1518.

Gallardo, D., Drazan, K. E., & McBride, W. H. (1996). Adenovirus-based transfer of wild-type *p53* gene increases ovarian tumor radiosensitivity. *Cancer Research, 56*, 4891–4893.

Gatei, M., Sloper, K., Sorensen, C., Syljuasen, R., Falck, J., Hobson, K., et al. (2003). Ataxia-telangiectasia-mutated (ATM) and NBS1-dependent phosphorylation of Chk1 on Ser-317 in response to ionizing radiation. *The Journal of Biological Chemistry, 278*, 14806–14811.

Geoerger, B., Grill, J., Opolon, P., Morizet, J., Aubert, G., Lecluse, Y., et al. (2003). Potentiation of radiation therapy by the oncolytic adenovirus dl1520 (ONYX-015) in human malignant glioma xenografts. *British Journal of Cancer, 89*, 577–584.

Gessler, M., Poustka, A., Cavenee, W., Neve, R. L., Orkin, S. H., & Bruns, G. A. (1990). Homozygous deletion in Wilms tumours of a zinc-finger gene identified by chromosome jumping. *Nature, 343*, 774–778.

Glover, D. J., Lipps, H. J., & Jans, D. A. (2005). Towards safe, non-viral therapeutic gene expression in humans. *Nature Reviews. Genetics, 6*, 299–310.

Gokhale, P. C., McRae, D., Monia, B. P., Bagg, A., Rahman, A., Dritschilo, A., et al. (1999). Antisense raf oligodeoxyribonucleotide is a radiosensitizer in vivo. *Antisense & Nucleic Acid Drug Development*, *9*, 191–201.
Goldkorn, T., Balaban, N., Shannon, M., & Matsukuma, K. (1997). EGF receptor phosphorylation is affected by ionizing radiation. *Biochimica et Biophysica Acta*, *1358*, 289–299.
Greco, O., Powell, T. M., Marples, B., Joiner, M. C., & Scott, S. D. (2005). Gene therapy vectors containing CArG elements from the Egr1 gene are activated by neutron irradiation, cisplatin and doxorubicin. *Cancer Gene Therapy*, *12*, 655–662.
Greco, O., Tozer, G. M., & Dachs, G. U. (2002). Oxic and anoxic enhancement of radiation-mediated toxicity by horseradish peroxidase/indole-3-acetic acid gene therapy. *International Journal of Radiation Biology*, *78*, 173–181.
Gridley, D. S., Glisson, W. C., & Uhm, J. R. (1994). Interaction of tumour necrosis factor-alpha and radiation against human colon tumour cells. *Therapeutic Immunology*, *1*, 25–31.
Gridley, D. S., Miller, G. M., Luo, X., Cao, J. D., Timiryasova, T. M., Fodor, I., et al. (2004). Proton radiation and TNF-alpha/Bax gene therapy for orthotopic C6 brain tumor in Wistar rats. *Technology in Cancer Research & Treatment*, *3*, 217–227.
Gulbins, E., & Kolesnick, R. (2003). Raft ceramide in molecular medicine. *Oncogene*, *22*, 7070–7077.
Hacein-Bey-Abina, S., von Kalle, C., Schmidt, M., Le Deist, F., Wulffraat, N., McIntyre, E., et al. (2003). A serious adverse event after successful gene therapy for X-linked severe combined immunodeficiency. *The New England Journal of Medicine*, *348*, 255–256.
Hallahan, D. E., Beckett, M. A., Kufe, D., & Weichselbaum, R. R. (1990). The interaction between recombinant human tumor necrosis factor and radiation in 13 human tumor cell lines. *International Journal of Radiation Oncology, Biology, Physics*, *19*, 69–74.
Hallahan, D. E., Mauceri, H. J., Seung, L. P., Dunphy, E. J., Wayne, J. D., Hanna, N. N., et al. (1995). Spatial and temporal control of gene therapy using ionizing radiation. *Nature Medicine*, *1*, 786–791.
Hallahan, D. E., Spriggs, D. R., Beckett, M. A., & Kufe, D. W. (1989). Increased tumor necrosis factor alpha mRNA after cellular exposure to ionizing radiation. *Proceedings of the National Academy of Sciences of the United States of America*, *86*, 10104–10107.
Hallahan, D. E., Sukhatme, V. P., Sherman, M. L., Virudachalam, S., Kufe, D., & Weichselbaum, R. R. (1991). Protein kinase C mediates x-ray inducibility of nuclear signal transducers EGR1 and JUN. *Proceedings of the National Academy of Sciences of the United States of America*, *88*, 2156–2160.
Hallahan, D. E., Virudachalam, S., Beckett, M., Sherman, M. L., Kufe, D., & Weichselbaum, R. R. (1991). Mechanisms of X-ray-mediated protooncogene c-jun expression in radiation-induced human sarcoma cell lines. *International Journal of Radiation Oncology, Biology, Physics*, *21*, 1677–1681.
Hamdi, M., Kool, J., Cornelissen-Steijger, P., Carlotti, F., Popeijus, H. E., van der Burgt, C., et al. (2005). DNA damage in transcribed genes induces apoptosis via the JNK pathway and the JNK-phosphatase MKP-1. *Oncogene*, *24*, 7135–7144.
Hamstra, D. A., Rice, D. J., Pu, A., Oyedijo, D., Ross, B. D., & Rehemtulla, A. (1999). Combined radiation and enzyme/prodrug treatment for head and neck cancer in an orthotopic animal model. *Radiation Research*, *152*, 499–507.
Hara, T., Omura-Minamisawa, M., Chao, C., Nakagami, Y., Ito, M., & Inoue, T. (2005). Bcl-2 inhibitors potentiate the cytotoxic effects of radiation in Bcl-2 overexpressing radioresistant tumor cells. *International Journal of Radiation Oncology, Biology, Physics*, *61*, 517–528.
Harari, P. M., & Huang, S. M. (2001). Radiation response modification following molecular inhibition of epidermal growth factor receptor signaling. *Seminars in Radiation Oncology*, *11*, 281–289.

Higuchi, Y., Asaumi, J., Murakami, J., Matsuzaki, H., Wakasa, T., Inoue, T., et al. (2003). Thermoradiotherapy combined with adenoviral p53 gene therapy in head and neck squamous cell carcinoma cell lines. *Oncology Reports*, *10*, 415–420.

Hillman, G. G., Slos, P., Wang, Y., Wright, J. L., Layer, A., De Meyer, M., et al. (2004). Tumor irradiation followed by intratumoral cytokine gene therapy for murine renal adenocarcinoma. *Cancer Gene Therapy*, *11*, 61–72.

Hoganson, D. K., Batra, R. K., Olsen, J. C., & Boucher, R. C. (1996). Comparison of the effects of three different toxin genes and their levels of expression on cell growth and bystander effect in lung adenocarcinoma. *Cancer Research*, *56*, 1315–1323.

Holder, J. W., Elmore, E., & Barrett, J. C. (1993). Gap junction function and cancer. *Cancer Research*, *53*, 3475–3485.

Hsu, H., Rainov, N. G., Quinones, A., Eling, D. J., Sakamoto, K. M., & Spear, M. A. (2003). Combined radiation and cytochrome CYP4B1/4-ipomeanol gene therapy using the EGR1 promoter. *Anticancer Research*, *23*, 2723–2728.

Huber, B. E., Austin, E. A., Richards, C. A., Davis, S. T., & Good, S. S. (1994). Metabolism of 5-fluorocytosine to 5-fluorouracil in human colorectal tumor cells transduced with cytosine deaminase gene: Significant antitumor effects when only a small percentage of tumor cells express cytosine deaminase. *Proceedings of the National Academy of Sciences of the United States of America*, *91*, 8302–8306.

Hynes, N. E., & Lane, H. A. (2005). ERBB receptors and cancer: The complexity of targeted inhibitors. *Nature Reviews. Cancer*, *5*, 341–354.

Ishida, A., Sasaguri, T., Kosaka, C., Nojima, H., & Ogata, J. (1997). Induction of the cyclin-dependent kinase inhibitor p21(Sdi1/Cip1/Waf1) by nitric oxide-generating vasodilator in vascular smooth muscle cells. *The Journal of Biological Chemistry*, *272*, 10050–10057.

Ishihara, H., Tsuneoka, K., Dimchev, A. B., & Shikita, M. (1993). Induction of the expression of the interleukin-1 beta gene in mouse spleen by ionizing radiation. *Radiation Research*, *133*, 321–326.

Ishizawar, R., & Parsons, S. J. (2004). c-Src and cooperating partners in human cancer. *Cancer Cell*, *6*, 209–214.

Iwadate, Y., Yamaura, A., Sakiyama, S., Sato, Y., & Tagawa, M. (2003). Glioma-specific cytotoxic T cells can be effectively induced by subcutaneous vaccination of irradiated wild-type tumor cells without artificial cytokine production. *International Journal of Oncology*, *23*, 483–488.

Jahrsdorfer, B., & Weiner, G. J. (2008). CpG oligodeoxynucleotides as immunotherapy in cancer. *Update on Cancer Therapeutics*, *3*, 27–32.

Jazayeri, A., Falck, J., Lukas, C., Bartek, J., Smith, G. C., Lukas, J., et al. (2006). ATM- and cell cycle-dependent regulation of ATR in response to DNA double-strand breaks. *Nature Cell Biology*, *8*, 37–45.

Jia, Z. Y., Deng, H. F., Huang, R., Yang, Y. Y., Yang, X. C., Qi, Z. Z., et al. (2011). In vitro and in vivo studies of adenovirus-mediated human norepinephrine transporter gene transduction to hepatocellular carcinoma. *Cancer Gene Therapy*, *18*, 196–205.

Jin, S., Fan, F., Fan, W., Zhao, H., Tong, T., Blanck, P., et al. (2001). Transcription factors Oct-1 and NF-YA regulate the p53-independent induction of the GADD45 following DNA damage. *Oncogene*, *20*, 2683–2690.

Joki, T., Nakamura, M., & Ohno, T. (1995). Activation of the radiosensitive EGR-1 promoter induces expression of the herpes simplex virus thymidine kinase gene and sensitivity of human glioma cells to ganciclovir. *Human Gene Therapy*, *6*, 1507–1513.

Kaliberov, S., Buchsbaum, D. J., Gillespie, G. Y., Curiel, D. T., Carpenter, M., Arafat, W. O., et al. (2002). Adenovirus-mediated transfer of Bax driven by the vascular endothelial growth factor promoter induces apoptosis in lung cancer cells. *Molecular Therapy*, *6*, 190–198.

Kaliberov, S. A., Kaliberova, L. N., & Buchsbaum, D. J. (2005). Combined ionizing radiation and *sKDR* gene delivery for treatment of prostate carcinomas. *Gene Therapy, 12*, 407–417.
Kaliberov, S. A., Kaliberova, L., Stockard, C. R., Grizzle, W. E., & Buchsbaum, D. J. (2004). Adenovirus-mediated flt-1 targeted proapoptotic gene therapy of human prostate cancer. *Molecular Therapy, 10*, 1059–1070.
Kaliberov, S. A., Markert, J. M., Gillespie, G. Y., Krendelchtchikova, V., Della Manna, D., Sellers, J. C., et al. (2007). Mutation of *Escherichia coli* cytosine deaminase significantly enhances molecular chemotherapy of human glioma. *Gene Therapy, 14*, 1111–1119.
Kaliberova, L. N., Della Manna, D. L., Krendelchtchikova, V., Black, M. E., Buchsbaum, D. J., & Kaliberov, S. A. (2008). Molecular chemotherapy of pancreatic cancer using novel mutant bacterial cytosine deaminase gene. *Molecular Cancer Therapeutics, 7*, 2845–2854.
Kambara, H., Tamiya, T., Ono, Y., Ohtsuka, S., Terada, K., Adachi, Y., et al. (2002). Combined radiation and gene therapy for brain tumors with adenovirus-mediated transfer of cytosine deaminase and uracil phosphoribosyltransferase genes. *Cancer Gene Therapy, 9*, 840–845.
Kauffmann, A., Rosselli, F., Lazar, V., Winnepenninckx, V., Mansuet-Lupo, A., Dessen, P., et al. (2008). High expression of DNA repair pathways is associated with metastasis in melanoma patients. *Oncogene, 27*, 565–573.
Kavanagh, B. D., Dent, P., Schmidt-Ullrich, R. K., Chen, P., & Mikkelsen, R. B. (1998). Calcium-dependent stimulation of mitogen-activated protein kinase activity in A431 cells by low doses of ionizing radiation. *Radiation Research, 149*, 579–587.
Kavanagh, B. D., Lin, P.-S., Chen, P., & Schmidt-Ullrich, R. K. (1995). Radiation-induced enhanced proliferation of human squamous cancer cells *in vitro*: A release from inhibition by epidermal growth factor. *Clinical Cancer Research, 1*, 1557–1562.
Khanna, K. K., & Lavin, M. F. (1993). Ionizing radiation and UV induction of p53 protein by different pathways in ataxia-telangiectasia cells. *Oncogene, 8*, 3307–3312.
Kievit, E., Nyati, M. K., Ng, E., Stegman, L. D., Parsels, J., Ross, B. D., et al. (2000). Yeast cytosine deaminase improves radiosensitization and bystander effect by 5-fluorocytosine of human colorectal cancer xenografts. *Cancer Research, 60*, 6649–6655.
Kim, S. H., Kim, J. H., Kolozsvary, A., Brown, S. L., & Freytag, S. O. (1997). Preferential radiosensitization of 9L glioma cells transduced with HSV-tk gene by acyclovir. *Journal of Neuro-Oncology, 33*, 189–194.
Kim, J. H., Kolozsvary, A., Rogulski, K., Khil, M. S., Brown, S. L., & Freytag, S. O. (1998). Selective radiosensitization of 9L glioma in the brain transduced with double suicide fusion gene. *The Cancer Journal from Scientific American, 4*, 364–369.
Kim, S. H., Wong, R. J., Kooby, D. A., Carew, J. F., Adusumilli, P. S., Patel, S. G., et al. (2005). Combination of mutated herpes simplex virus type 1 (G207 virus) with radiation for the treatment of squamous cell carcinoma of the head and neck. *European Journal of Cancer, 41*, 313–322.
Krieg, A. M. (2004). Antitumor applications of stimulating toll-like receptor 9 with CpG oligodeoxynucleotides. *Current Oncology Reports, 6*, 88–95.
Kuriyama, S., Masui, K., Sakamoto, T., Nakatani, T., Kikukawa, M., Tsujinoue, H., et al. (1998). Bystander effect caused by cytosine deaminase gene and 5-fluorocytosine in vitro is substantially mediated by generated 5-fluorouracil. *Anticancer Research, 18*, 3399–3406.
Lai, S. Y., Koppikar, P., Thomas, S. M., Childs, E. E., Egloff, A. M., Seethala, R. R., et al. (2009). Intratumoral epidermal growth factor receptor antisense DNA therapy in head and neck cancer: First human application and potential antitumor mechanisms. *Journal of Clinical Oncology, 27*, 1235–1242.
Lamfers, M. L., Grill, J., Dirven, C. M., Van Beusechem, V. W., Geoerger, B., Van Den Berg, J., et al. (2002). Potential of the conditionally replicative adenovirus

Ad5-Delta24RGD in the treatment of malignant gliomas and its enhanced effect with radiotherapy. *Cancer Research, 62*, 5736–5742.

Lammering, G., Hewit, T. H., Valerie, K., Lin, P. S., Contessa, J. N., & Schmidt-Ullrich, R. K. (2003). Anti-erbB receptor strategy as a gene therapeutic intervention to improve radiotherapy in malignant human tumours. *International Journal of Radiation Biology, 79*, 561–568.

Lawrence, T. S., Davis, M. A., & Loney, T. L. (1996). Fluoropyrimidine-mediated radiosensitization depends on cyclin E-dependent kinase activation. *Cancer Research, 56*, 3203–3206.

Lawrence, T. S., Tepper, J. E., & Blackstock, A. W. (1997). Fluoropyrimidine-radiation Interactions in cells and tumors. *Seminars in Radiation Oncology, 7*, 260–266.

Lee, J. H., & Paull, T. T. (2005). ATM activation by DNA double-strand breaks through the Mre11-Rad50-Nbs1 complex. *Science, 308*, 551–554.

Li, G. C., He, F., Shao, X., Urano, M., Shen, L., Kim, D., et al. (2003). Adenovirus-mediated heat-activated antisense Ku70 expression radiosensitizes tumor cells in vitro and in vivo. *Cancer Research, 63*, 3268–3274.

Li, J. H., Lax, S. A., Kim, J., Klamut, H., & Liu, F. F. (1999). The effects of combining ionizing radiation and adenoviral p53 therapy in nasopharyngeal carcinoma. *International Journal of Radiation Oncology, Biology, Physics, 43*, 607–616.

Lieber, M. R., Ma, Y., Pannicke, U., & Schwarz, K. (2003). Mechanism and regulation of human non-homologous DNA end-joining. *Nature Reviews. Molecular Cell Biology, 4*, 712–720.

Lohr, F., Hu, K., Haroon, Z., Samulski, T. V., Huang, Q., Beaty, J., et al. (2000). Combination treatment of murine tumors by adenovirus-mediated local B7/IL12 immunotherapy and radiotherapy. *Molecular Therapy, 2*, 195–203.

Lowe, S. W., Bodis, S., McClatchey, A., Remington, L., Ruley, H. E., Fisher, D. E., et al. (1994). *p53* status and the efficacy of cancer therapy in vivo. *Science, 266*, 807–810.

Lu, X., & Lane, D. P. (1993). Differential induction of transcriptionally active p53 following UV or ionizing radiation: Defects in chromosome instability syndromes? *Cell, 75*, 765–778.

Lumniczky, K., Desaknai, S., Mangel, L., Szende, B., Hamada, H., Hidvegi, E. J., et al. (2002). Local tumor irradiation augments the antitumor effect of cytokine-producing autologous cancer cell vaccines in a murine glioma model. *Cancer Gene Therapy, 9*, 44–52.

Mahan, S. D., Ireton, G. C., Knoeber, C., Stoddard, B. L., & Black, M. E. (2004). Random mutagenesis and selection of Escherichia coli cytosine deaminase for cancer gene therapy. *Protein Engineering, Design & Selection, 17*, 625–633.

Mahan, S. D., Ireton, G. C., Stoddard, B. L., & Black, M. E. (2004). Alanine-scanning mutagenesis reveals a cytosine deaminase mutant with altered substrate preference. *Biochemistry, 43*, 8957–8964.

Marconi, P., Krisky, D., Oligino, T., Poliani, P. L., Ramakrishnan, R., Goins, W. F., et al. (1996). Replication-defective herpes simplex virus vectors for gene transfer in vivo. *Proceedings of the National Academy of Sciences of the United States of America, 93*, 11319–11320.

Martin, M., Vozenin, M. C., Gault, N., Crechet, F., Pfarr, C. M., & Lefaix, J. L. (1997). Coactivation of AP-1 activity and TGF-beta1 gene expression in the stress response of normal skin cells to ionizing radiation. *Oncogene, 15*, 981–989.

Mason, K. A., Ariga, H., Neal, R., Valdecanas, D., Hunter, N., Krieg, A. M., et al. (2005). Targeting toll-like receptor 9 with CpG oligodeoxynucleotides enhances tumor response to fractionated radiotherapy. *Clinical Cancer Research, 11*, 361–369.

McCarthy, H. O., Worthington, J., Barrett, E., Cosimo, E., Boyd, M., Mairs, R. J., et al. (2007). p21(WAF1)-mediated transcriptional targeting of inducible nitric oxide synthase gene therapy sensitizes tumours to fractionated radiotherapy. *Gene Therapy, 14*, 246–255.

McGinn, C. J., & Lawrence, T. S. (2001). Recent advances in the use of radiosensitizing nucleosides. *Seminars in Radiation Oncology*, *11*, 270–280.

McLoughlin, J. M., McCarty, T. M., Cunningham, C., Clark, V., Senzer, N., Nemunaitis, J., et al. (2005). TNFerade, an adenovector carrying the transgene for human tumor necrosis factor alpha, for patients with advanced solid tumors: Surgical experience and long-term follow-up. *Annals of Surgical Oncology*, *12*, 825–830.

Mercatante, D. R., Mohler, J. L., & Kole, R. (2002). Cellular response to an antisense-mediated shift of Bcl-x pre-mRNA splicing and antineoplastic agents. *The Journal of Biological Chemistry*, *277*, 49374–49382.

Mesnil, M., Piccoli, C., Tiraby, G., Willecke, K., & Yamasaki, H. (1996). Bystander killing of cancer cells by herpes simplex virus thymidine kinase gene is mediated by connexins. *Proceedings of the National Academy of Sciences of the United States of America*, *93*, 1831–1835.

Meyer, R. G., Kupper, J. H., Kandolf, R., & Rodemann, H. P. (2002). Early growth response-1 gene (Egr-1) promoter induction by ionizing radiation in U87 malignant glioma cells in vitro. *European Journal of Biochemistry*, *269*, 337–346.

Mezhir, J. J., Smith, K. D., Posner, M. C., Senzer, N., Yamini, B., Kufe, D. W., et al. (2006). Ionizing radiation: A genetic switch for cancer therapy. *Cancer Gene Therapy*, *13*, 1–6.

Miller, E. M., & Kinsella, T. J. (1992). Radiosensitization by fluorodeoxyuridine: Effects of thymidylate synthase inhibition and cell synchronization. *Cancer Research*, *52*, 1687–1694.

Mullen, C. A., Coale, M., Lowe, R., & Blaese, R. M. (1994). Tumors expressing the cytosine deaminase suicide gene can be eliminated in vivo with 5-fluorocytosine and induce protective immunity to wild type tumor. *Cancer Research*, *54*, 1503–1506.

Munshi, A., Kurland, J. F., Nishikawa, T., Tanaka, T., Hobbs, M. L., Tucker, S. L., et al. (2005). Histone deacetylase inhibitors radiosensitize human melanoma cells by suppressing DNA repair activity. *Clinical Cancer Research*, *11*, 4912–4922.

Nenoi, M., Daino, K., Ichimura, S., Takahash, S., & Akuta, T. (2006). Low-dose radiation response of the p21WAF1/CIP1 gene promoter transduced by adeno-associated virus vector. *Experimental & Molecular Medicine*, *38*, 553–564.

Nikitina, E. Y., & Gabrilovich, D. I. (2001). Combination of gamma-irradiation and dendritic cell administration induces a potent antitumor response in tumor-bearing mice: Approach to treatment of advanced stage cancer. *International Journal of Cancer*, *94*, 825–833.

Nimura, Y., Kawata, T., Uzawa, K., Okamura, J., Liu, C., Saito, M., et al. (2007). Silencing Ku80 using small interfering RNA enhanced radiation sensitivity in vitro and in vivo. *International Journal of Oncology*, *30*, 1477–1484.

Nishihara, E., Nagayama, Y., Narimatsu, M., Namba, H., Watanabe, M., Niwa, M., et al. (1998). Treatment of thyroid carcinoma cells with four different suicide gene/prodrug combinations in vitro. *Anticancer Research*, *18*, 1521–1525.

Nishikawa, T., Ramesh, R., Munshi, A., Chada, S., & Meyn, R. E. (2004). Adenovirus-mediated mda-7 (IL24) gene therapy suppresses angiogenesis and sensitizes NSCLC xenograft tumors to radiation. *Molecular Therapy*, *9*, 818–828.

Nishitani, M. A., Sakai, T., Ishii, K., Zhang, M., Nakano, Y., Nitta, Y., et al. (2002). A convenient cancer vaccine therapy with in vivo transfer of interleukin 12 expression plasmid using gene gun technology after priming with irradiated carcinoma cells. *Cancer Gene Therapy*, *9*, 156–163.

Nyati, M. K., Morgan, M. A., Feng, F. Y., & Lawrence, T. S. (2006). Integration of EGFR inhibitors with radiochemotherapy. *Nature Reviews. Cancer*, *6*, 876–885.

O'Driscoll, M., & Jeggo, P. A. (2006). The role of double-strand break repair—Insights from human genetics. *Nature Reviews. Genetics*, *7*, 45–54.

Oh, Y. T., Chen, D. W., Dougherty, G. J., & McBride, W. H. (2004). Adenoviral interleukin-3 gene-radiation therapy for prostate cancer in mouse model. *International Journal of Radiation Oncology, Biology, Physics*, *59*, 579–583.

Papathanasiou, M. A., Kerr, N. C., Robbins, J. H., McBride, O. W., Alamo, I., Jr., Barrett, S. F., et al. (1991). Induction by ionizing radiation of the gadd45 gene in cultured human cells: Lack of mediation by protein kinase C. *Molecular and Cellular Biology*, *11*, 1009–1016.

Park, J. O., Lopez, C. A., Gupta, V. K., Brown, C. K., Mauceri, H. J., Darga, T. E., et al. (2002). Transcriptional control of viral gene therapy by cisplatin. *The Journal of Clinical Investigation*, *110*, 403–410.

Pederson, L. C., Buchsbaum, D. J., Vickers, S. M., Kancharla, S. R., Mayo, M. S., Curiel, D. T., et al. (1997). Molecular chemotherapy combined with radiation therapy enhances killing of cholangiocarcinoma cells *in vitro* and *in vivo*. *Cancer Research*, *57*, 4325–4332.

Pellegrini, L., Yu, D. S., Lo, T., Anand, S., Lee, M., Blundell, T. L., et al. (2002). Insights into DNA recombination from the structure of a RAD51-BRCA2 complex. *Nature*, *420*, 287–293.

Prasad, S., Ravindran, J., & Aggarwal, B. B. (2010). NF-kappaB and cancer: How intimate is this relationship. *Molecular and Cellular Biochemistry*, *336*, 25–37.

Quanz, M., Berthault, N., Roulin, C., Roy, M., Herbette, A., Agrario, C., et al. (2009). Small-molecule drugs mimicking DNA damage: A new strategy for sensitizing tumors to radiotherapy. *Clinical Cancer Research*, *15*, 1308–1316.

Rass, K., & Reichrath, J. (2008). UV damage and DNA repair in malignant melanoma and nonmelanoma skin cancer. *Advances in Experimental Medicine and Biology*, *624*, 162–178.

Ratnam, K., & Low, J. A. (2007). Current development of clinical inhibitors of poly(ADP-ribose) polymerase in oncology. *Clinical Cancer Research*, *13*, 1383–1388.

Rimner, A., Wischhusen, J., Naumann, U., Gleichmann, M., Steinbach, J. P., & Weller, M. (2001). Identification by suppression subtractive hybridization of p21 as a radio-inducible gene in human glioma cells. *Anticancer Research*, *21*, 3505–3508.

Rodriguez, R., Schuur, E. R., Lim, H. Y., Henderson, G. A., Simons, J. W., & Henderson, D. R. (1997). Prostate attenuated replication competent adenovirus (ARCA) CN706: A selective cytotoxic for prostate-specific antigen-positive prostate cancer cells. *Cancer Research*, *57*, 2559–2563.

Rogers, B. E., Zinn, K. R., Lin, C.-Y., Chaudhuri, T. R., & Buchsbaum, D. J. (2002). Targeted radiotherapy with [$^{90}$Y]-SMT 487 in mice bearing human nonsmall cell lung tumor xenografts induced to express human somatostatin receptor subtype 2 with an adenoviral vector. *Cancer*, *94*, 1298–1305.

Rogulski, K. R., Freytag, S. O., Zhang, K., Gilbert, J. D., Paielli, D. L., Kim, J. H., et al. (2000). In vivo antitumor activity of ONYX-015 is influenced by p53 status and is augmented by radiotherapy. *Cancer Research*, *60*, 1193–1196.

Sartor, C. I. (2000). Biological modifiers as potential radiosensitizers: Targeting the epidermal growth factor receptor family. *Seminars in Oncology*, *27*, 15–20 discussion 92–100.

Sasaki, Y., Morimoto, I., Ishida, S., Yamashita, T., Imai, K., & Tokino, T. (2001). Adenovirus-mediated transfer of the p53 family genes, p73 and p51/p63 induces cell cycle arrest and apoptosis in colorectal cancer cell lines: Potential application to gene therapy of colorectal cancer. *Gene Therapy*, *8*, 1401–1408.

Schlessinger, J. (2004). Common and distinct elements in cellular signaling via EGF and FGF receptors. *Science*, *306*, 1506–1507.

Schlessinger, J., & Lemmon, M. A. (2006). Nuclear signaling by receptor tyrosine kinases: The first robin of spring. *Cell*, *127*, 45–48.

Scott, S. D., Joiner, M. C., & Marples, B. (2002). Optimizing radiation-responsive gene promoters for radiogenetic cancer therapy. *Gene Therapy*, *9*, 1396–1402.

Searle, P. F., Chen, M. J., Hu, L., Race, P. R., Lovering, A. L., Grove, J. I., et al. (2004). Nitroreductase: A prodrug-activating enzyme for cancer gene therapy. *Clinical and Experimental Pharmacology & Physiology*, *31*, 811–816.

Senzer, N., Mani, S., Rosemurgy, A., Nemunaitis, J., Cunningham, C., Guha, C., et al. (2004). TNFerade biologic, an adenovector with a radiation-inducible promoter, carrying the human tumor necrosis factor alpha gene: A phase I study in patients with solid tumors. *Journal of Clinical Oncology*, *22*, 592–601.

Shariat, S. F., Desai, S., Song, W., Khan, T., Zhao, J., Nguyen, C., et al. (2001). Adenovirus-mediated transfer of inducible caspases: A novel "death switch" gene therapeutic approach to prostate cancer. *Cancer Research*, *61*, 2562–2571.

Sherman, M. L., Datta, R., Hallahan, D. E., Weichselbaum, R. R., & Kufe, D. W. (1990). Ionizing radiation regulates expression of the *c-jun* protooncogene. *Proceedings of the National Academy of Sciences of the United States of America*, *87*, 5663–5666.

Shi, W., Teschendorf, C., Muzyczka, N., & Siemann, D. W. (2003). Gene therapy delivery of endostatin enhances the treatment efficacy of radiation. *Radiotherapy and Oncology*, *66*, 1–9.

Shimura, T. (2011). Acquired radioresistance of cancer and the AKT/GSK3beta/cyclin D1 overexpression cycle. *Journal of Radiation Research (Tokyo)*, *52*, 539–544.

Shinoura, N., Yamamoto, N., Asai, A., Kirino, T., & Hamada, H. (2000). Adenovirus-mediated transfer of Fas ligand gene augments radiation-induced apoptosis in U-373MG glioma cells. *Japanese Journal of Cancer Research*, *91*, 1044–1050.

Sourvinos, G., & Spandidos, D. A. (1998). Decreased BRCA1 expression levels may arrest the cell cycle through activation of p53 checkpoint in human sporadic breast tumors. *Biochemical and Biophysical Research Communications*, *245*, 75–80.

Spitz, F. R., Nguyen, D., Skibber, J. M., Meyn, R. E., Cristiano, R. J., & Roth, J. A. (1996). Adenoviral-mediated wild-type *p53* gene expression sensitizes colorectal cancer cells to ionizing radiation. *Clinical Cancer Research*, *2*, 1665–1671.

Spitzweg, C., Baker, C. H., Bergert, E. R., O'Connor, M. K., & Morris, J. C. (2007). Image-guided radioiodide therapy of medullary thyroid cancer after carcinoembryonic antigen promoter-targeted sodium iodide symporter gene expression. *Human Gene Therapy*, *18*, 916–924.

Stackhouse, M. A., Buchsbaum, D. J., Kancharla, S. R., Grizzle, W. E., Grimes, C., Laffoon, K., et al. (1999). Specific membrane receptor gene expression targeted with radiolabeled peptide employing the erbB-2 and DF3 promoter elements in adenoviral vectors. *Cancer Gene Therapy*, *6*, 209–219.

Stackhouse, M. A., Pederson, L. C., Grizzle, W. E., Curiel, D. T., Gebert, J., Haack, K., et al. (2000). Fractionated radiation therapy in combination with adenoviral delivery of the cytosine deaminase gene and 5-fluorocytosine enhances cytotoxic and antitumor effects in human colorectal and cholangiocarcinoma models. *Gene Therapy*, *7*, 1019–1026.

Steinman, R. A., Hoffman, B., Iro, A., Guillouf, C., Liebermann, D. A., & el-Houseini, M. E. (1994). Induction of p21 (WAF-1/CIP1) during differentiation. *Oncogene*, *9*, 3389–3396.

Su, Z. Z., Lebedeva, I. V., Sarkar, D., Emdad, L., Gupta, P., Kitada, S., et al. (2006). Ionizing radiation enhances therapeutic activity of mda-7/IL-24: Overcoming radiation- and mda-7/IL-24-resistance in prostate cancer cells overexpressing the antiapoptotic proteins bcl-xL or bcl-2. *Oncogene*, *25*, 2339–2348.

Takahashi, S., Saito, S., Ohtani, N., & Sakai, T. (2001). Involvement of the Oct-1 regulatory element of the gadd45 promoter in the p53-independent response to ultraviolet irradiation. *Cancer Research*, *61*, 1187–1195.

Teh, B. S., Aguilar-Cordova, E., Kernen, K., Chou, C. C., Shalev, M., Vlachaki, M. T., et al. (2001). Phase I/II trial evaluating combined radiotherapy and in situ gene therapy with or without hormonal therapy in the treatment of prostate cancer—A preliminary report. *International Journal of Radiation Oncology, Biology, Physics, 51*, 605–613.

Teitz-Tennenbaum, S., Li, Q., Rynkiewicz, S., Ito, F., Davis, M. A., McGinn, C. J., et al. (2003). Radiotherapy potentiates the therapeutic efficacy of intratumoral dendritic cell administration. *Cancer Research, 63*, 8466–8475.

Tetzlaff, M. T., Teh, B. S., Timme, T. L., Fujita, T., Satoh, T., Tabata, K., et al. (2006). Expanding the therapeutic index of radiation therapy by combining in situ gene therapy in the treatment of prostate cancer. *Technology in Cancer Research & Treatment, 5*, 23–36.

Toth, K., Tarakanova, V., Doronin, K., Ward, P., Kuppuswamy, M., Locke, J. E., et al. (2003). Radiation increases the activity of oncolytic adenovirus cancer gene therapy vectors that overexpress the ADP (E3-11.6K) protein. *Cancer Gene Therapy, 10*, 193–200.

Treisman, R. (1990). The SRE: A growth factor responsive transcriptional regulator. *Seminars in Cancer Biology, 1*, 47–58.

Trinh, Q. T., Austin, E. A., Murray, D. M., Knick, V. C., & Huber, B. E. (1995). Enzyme/prodrug gene therapy: Comparison of cytosine deaminase/5-fluorocytosine versus thymidine kinase/ganciclovir enzyme/prodrug systems in a human colorectal carcinoma cell line. *Cancer Research, 55*, 4808–4812.

Trofimova, I., Dimtchev, A., Jung, M., Rosenthal, D., Smulson, M., Dritschilo, A., et al. (2002). Gene therapy for prostate cancer by targeting poly(ADP-ribose) polymerase. *Cancer Research, 62*, 6879–6883.

Trujillo, M. A., Oneal, M. J., McDonough, S., Qin, R., & Morris, J. C. (2010). A probasin promoter, conditionally replicating adenovirus that expresses the sodium iodide symporter (NIS) for radiovirotherapy of prostate cancer. *Gene Therapy, 17*, 1325–1332.

Tsai-Morris, C. H., Cao, X. M., & Sukhatme, V. P. (1988). 5' flanking sequence and genomic structure of Egr-1, a murine mitogen inducible zinc finger encoding gene. *Nucleic Acids Research, 16*, 8835–8846.

Uchida, H., Shinoura, N., Kitayama, J., Watanabe, T., Nagawa, H., & Hamada, H. (2003). 5-Fluorouracil efficiently enhanced apoptosis induced by adenovirus-mediated transfer of caspase-8 in DLD-1 colon cancer cells. *The Journal of Gene Medicine, 5*, 287–299.

Vernole, P., Muzi, A., Volpi, A., Dorio, A. S., Terrinoni, A., Shah, G. M., et al. (2009). Inhibition of homologous recombination by treatment with BVDU (brivudin) or by RAD51 silencing increases chromosomal damage induced by bleomycin in mismatch repair-deficient tumour cells. *Mutation Research, 664*, 39–47.

Wakasa, T., Inoue, T., Kawai, N., Murakami, J., Kishi, K., & Fukui, K. (2002). The combination of ionizing radiation and expression of a wild type p53 gene via recombinant adenovirus induced a prominent tumour suppressing effect in human oral squamous cell carcinoma. *The British Journal of Radiology, 75*, 657–662.

Walker, J. R., McGeagh, K. G., Sundaresan, P., Jorgensen, T. J., Rabkin, S. D., & Martuza, R. L. (1999). Local and systemic therapy of human prostate adenocarcinoma with the conditionally replicating herpes simplex virus vector G207. *Human Gene Therapy, 10*, 2237–2243.

Weichselbaum, R. R., Hallahan, D. E., Beckett, M. A., Mauceri, H. J., Lee, H., Sukhatme, V. P., et al. (1994). Gene therapy targeted by radiation preferentially radiosensitizes tumor cells. *Cancer Research, 54*, 4266–4269.

Weichselbaum, R. R., & Kufe, D. (2009). Translation of the radio- and chemo-inducible TNFerade vector to the treatment of human cancers. *Cancer Gene Therapy, 16*, 609–619.

Weichselbaum, R. R., Kufe, D. W., Hellman, S., Rasmussen, H. S., King, C. R., Fischer, P. H., et al. (2002). Radiation-induced tumour necrosis factor-alpha expression: Clinical application of transcriptional and physical targeting of gene therapy. *The Lancet Oncology, 3*, 665–671.

White, C. L., Menghistu, T., Twigger, K. R., Searle, P. F., Bhide, S. A., Vile, R. G., et al. (2008). Escherichia coli nitroreductase plus CB1954 enhances the effect of radiotherapy in vitro and in vivo. *Gene Therapy, 15*, 424–433.

Winter, R. N., Rhee, J. G., & Kyprianou, N. (2004). Caspase-1 enhances the apoptotic response of prostate cancer cells to ionizing radiation. *Anticancer Research, 24*, 1377–1386.
Worthington, J., McCarthy, H. O., Barrett, E., Adams, C., Robson, T., & Hirst, D. G. (2004). Use of the radiation-inducible WAF1 promoter to drive iNOS gene therapy as a novel anti-cancer treatment. *The Journal of Gene Medicine, 6*, 673–680.
Xiang, D. B., Chen, Z. T., Wang, D., Li, M. X., Xie, J. Y., Zhang, Y. S., et al. (2008). Chimeric adenoviral vector Ad5/F35-mediated APE1 siRNA enhances sensitivity of human colorectal cancer cells to radiotherapy in vitro and in vivo. *Cancer Gene Therapy, 15*, 625–635.
Xu, Y., Liu, Z., Kong, H., Sun, W., Liao, Z., Zhou, F., et al. (2011). Co-expression of interleukin 12 enhances antitumor effects of a novel chimeric promoter-mediated suicide gene therapy in an immunocompetent mouse model. *Biochemical and Biophysical Research Communications, 412*, 763–768.
Xu, L., Pirollo, K. F., Tang, W.-H., Rait, A., & Chang, E. H. (1999). Transferrin-liposome-mediated systemic p53 gene therapy in combination with radiation results in regression of human head and neck xenografts. *Human Gene Therapy, 10*, 2941–2952.
Yacoub, A., Hamed, H., Emdad, L., Dos Santos, W., Gupta, P., Broaddus, W. C., et al. (2008). MDA-7/IL-24 plus radiation enhance survival in animals with intracranial primary human GBM tumors. *Cancer Biology & Therapy, 7*, 917–933.
Yanamandra, N., Kondraganti, S., Srinivasula, S. M., Gujrati, M., Olivero, W. C., Dinh, D. H., et al. (2004). Activation of caspase-9 with irradiation inhibits invasion and angiogenesis in SNB19 human glioma cells. *Oncogene, 23*, 2339–2346.
Yang, Y. J., Chen, D. Z., Li, L. X., Sheng, Q. S., Jin, Z. K., & Zhao, D. F. (2009). Targeted IL-24 gene therapy inhibits cancer recurrence after liver tumor resection by inducing tumor cell apoptosis in nude mice. *Hepatobiliary & Pancreatic Diseases International, 8*, 174–178.
Yang, D. C., Elliott, R. L., & Head, J. F. (2002). Gene targets of antisense therapies in breast cancer. *Expert Opinion on Therapeutic Targets, 6*, 375–385.
Yang, Y., Liu, S. Z., & Fu, S. B. (2004). Anti-tumor effects of pNEgr-mIL-12 recombinant plasmid induced by X-irradiation and its mechanisms. *Biomedical and Environmental Sciences, 17*, 135–143.
Yu, H., & Jove, R. (2004). The STATs of cancer—New molecular targets come of age. *Nature Reviews. Cancer, 4*, 97–105.
Yu, D., Sekine, E., Fujimori, A., Ochiya, T., & Okayasu, R. (2008). Down regulation of BRCA2 causes radio-sensitization of human tumor cells in vitro and in vivo. *Cancer Science, 99*, 810–815.
Yuan, S., Qiao, T., & Chen, W. (2011). CpG oligodeoxynucleotide 1826 enhances the Lewis lung cancer response to radiotherapy in murine tumor. *Cancer Biotherapy & Radiopharmaceuticals, 26*, 203–208.
Zaiss, A. K., Liu, Q., Bowen, G. P., Wong, N. C., Bartlett, J. S., & Muruve, D. A. (2002). Differential activation of innate immune responses by adenovirus and adeno-associated virus vectors. *Journal of Virology, 76*, 4580–4590.
Zhou, B. B., & Bartek, J. (2004). Targeting the checkpoint kinases: Chemosensitization versus chemoprotection. *Nature Reviews. Cancer, 4*, 216–225.
Zhou, B. B., Chaturvedi, P., Spring, K., Scott, S. P., Johanson, R. A., Mishra, R., et al. (2000). Caffeine abolishes the mammalian G(2)/M DNA damage checkpoint by inhibiting ataxia-telangiectasia-mutated kinase activity. *The Journal of Biological Chemistry, 275*, 10342–10348.
Zimmerman, R. J., Marafino, B. J., Jr., Chan, A., Landre, P., & Winkelhake, J. L. (1989). The role of oxidant injury in tumor cell sensitivity to recombinant human tumor necrosis factor in vivo. Implications for mechanisms of action. *Journal of Immunology, 142*, 1405–1409.

CHAPTER EIGHT

# Oncolytic Adenoviruses for Cancer Immunotherapy: Data from Mice, Hamsters, and Humans

**Vincenzo Cerullo*,†,1, Anniina Koski†, Markus Vähä-Koskela†, Akseli Hemminki†,1**

*Laboratory of Immunovirotherapy, Division of Biopharmaceutics and Pharmacokinetics, Faculty of Pharmacy, University of Helsinki, Helsinki, Finland

†Cancer Gene Therapy Group, Molecular Cancer Biology Program & Transplantation Laboratory & Haartman Institute, University of Helsinki, Helsinki, Finland

[1]Corresponding authors: e-mail address: vincenzo.cerullo@helsinki.fi; akseli.hemminki@helsinki.fi

## Contents



## Abstract

Adenovirus is one of the most commonly used vectors for gene therapy and two products have already been approved for treatment of cancer in China (Gendicine[R] and Oncorine[R]). An intriguing aspect of oncolytic adenoviruses is that by their very nature they potently stimulate multiple arms of the immune system. Thus, combined tumor killing via oncolysis and inherent immunostimulatory properties in fact make these viruses *in situ*

*Advances in Cancer Research*, Volume 115
ISSN 0065-230X
http://dx.doi.org/10.1016/B978-0-12-398342-8.00008-2

tumor vaccines. When further engineered to express cytokines, chemokines, tumor-associated antigens, or other immunomodulatory elements, they have been shown in various preclinical models to induce antigen-specific effector and memory responses, resulting both in full therapeutic cures and even induction of life-long tumor immunity. Here, we review the *state of the art* of oncolytic adenovirus, in the context of their capability to stimulate innate and adaptive arms of the immune system and finally how we can modify these viruses to direct the immune response toward cancer.

# 1. INTRODUCTION

## 1.1. Oncolytic adenovirus and immunotherapy

Adenovirus is one of the most commonly used vectors for gene therapy, and two drugs have already been approved in China but not elsewhere (Garber, 2006).

Spurred initially by intriguing case reports of virus infections or vaccination resulting in tumor regression, several different wild-type viruses were tested in cancer patients already in the 1950s (Vaha-Koskela, Heikkila, & Hinkkanen, 2007). From that time onward, replicating viruses have been investigated as cancer therapeutics. However, progress was slow until molecular biology had developed sufficiently to allow construction of recombinant tumor selective viruses and their rational analysis *in vitro* and in animal models. The first modern oncolytic virus trials were performed in the late 1990s (Edelman & Nemunaitis, 2003; Nemunaitis, Senzer, Cunningham, & Dubensky, 2001; Nemunaitis et al., 2000). Tumors make a good substrates for the replication of oncolytic viruses (virotherapeutics) as oncogenic transformation often manifests as increased cell proliferation, increased DNA/RNA synthesis (building blocks also for viruses) and reduced antiviral defenses (Vaha-Koskela et al., 2007). Indeed, the tumor selectivity of many classes of oncolytic viruses relates to deficiency in interferon (IFN) signaling(Maheshwari, Banerjee, Waechter, Olden, & Friedman, 1980; Maheshwari, Husain, Attallah, & Friedman, 1983; Schuster, Nechansky, & Kircheis, 2006), while others take advantage of dysregulation of central growth control pathways, a universal feature of advanced tumors (Alemany, 2007; Kirn, 2001).

Oncolytic viruses may also be administered systemically to target tumor metastases (Breitbach et al., 2011). Furthermore, an important asset of oncolytic viruses is that by their very nature they potently stimulate multiple arms of the immune system. Thus, combined tumor killing via oncolysis and immunostimulation in fact render oncolytic viruses *in situ* tumor vaccines. When further engineered to express cytokines, chemokines, tumor-associated antigens (TAAs), or other immunomodulatory elements, oncolytic viruses (OVs) have been shown in various preclinical models to induce antigen-specific effector and memory responses, resulting both in full therapeutic cures and even induction of life-long tumor immunity (Kaur, Cripe, & Chiocca, 2009; Liu et al., 2011; Prestwich, Harrington, Pandha, et al., 2008; Prestwich, Harrington, Vile, & Melcher, 2008; Tuve et al., 2009).

Recently, many new therapeutic strategies, such as monoclonal antibodies or tyrosine kinase inhibitors, have displayed promising efficacy in cancer patients (Berruti, Pia, & Terzolo, 2011; Bottini et al., 2006; Correale et al., 2012). Nevertheless, these agents target only specific pathways, and given the tremendous capacity of cancers to develop resistance to any cytostatic intervention, agents with even broader mechanisms of action—preferably lytic instead of static—are urgently needed (Huber & Wolfel, 2004). In this regard, OVs can retain efficacy even if apoptosis pathways of the cancer cells are blocked. In fact, several OVs have been shown to reverse resistance to chemotherapies, facilitating synergizing combination therapy resulting in true translational potential in cancer patients (Han et al., 2011; Jiang, Alonso, Gomez-Manzano, Piao, & Fueyo, 2006; Mantwill et al., 2006; Qi, Chang, Song, Gao, & Shen, 2011). This is in part due to the capacity of the viruses to simultaneously interfere with central cellular pathways, including DNA repair, translation, and transcription (Kalu et al., 2010).

While clinical trial data from oncolytic and other types of viruses shows they are safe, no oncolytic virus has yet reached marketable status in the Western world, although China approved an oncolytic adenovirus, H101 (Oncorine), for treatment of head and neck cancer in 2005. In this regard, four phase III trials are concurrently underway: herpes simplex type 1 expressing GMCSF (granulocyte monocyte colony stimulating factor; Oncovex-GMCSF) by Amgen Inc., human reovirus type 3 by Oncolytics Biotech and thymidine-kinase deleted vaccinia virus JX-594 expressing GMCSF by Jennerex/Transgene Inc., with positive results anticipated by the end of 2012 and Svend Freytag's with oncolytic adenovirus coding for TK and CD in pancreatic cancer. However, both preclinical and clinical tests preceding these studies have shown that while provoking a measurable antitumor effect is possible, long-lasting therapeutic benefit is not to be expected in all patients, especially in the context of advanced disease (Rowan, 2010). Even if some OVs elicit lysis-mediated immune responses even without arming, these responses are generally too unspecific or targeted primarily against the virus. Specific tumor immunotherapy utilizing predefined epitopes, however, has been limited by immunosuppressive forces as well as exclusive reliance on host factors, which has limited efficacy in patients with advanced cancers (Hierholzer, 1992; Veltrop-Duits et al., 2011). Therefore, an ideal approach would be able to kill cancer cells in multipronged fashion to overcome intrinsic cellular resistance mechanisms, to induce specific antitumor immune responses while at the same time overcoming cancer associated immunosuppression.

Oncolytic viruses have two main interactions after their administration into the body; one with tumor cells which they can infect and kill and the second with the normal cells of the host (Fig. 8.1). If appropriately maneuvered, we believe OVs can be used to achieve the goals described above, but this requires a profound understanding of the basic biology of the viruses and their replication cycle in the host, including cell and tissue tropism and immunological responses (Cerullo et al., 2010; Koski et al., 2010).

# 2. ADENOVIRUSES AS THERAPEUTIC AGENTS AGAINST CANCER

## 2.1. Adenovirus biology

### 2.1.1 General information

Adenoviruses were first discovered in 1953, when Rowe et al. identified them as a novel cytopathogenic agent in tissue cultures extracted from human adenoids (Rowe, Huebner, Gilmore, Parrott, & Ward, 1953). Thus far,

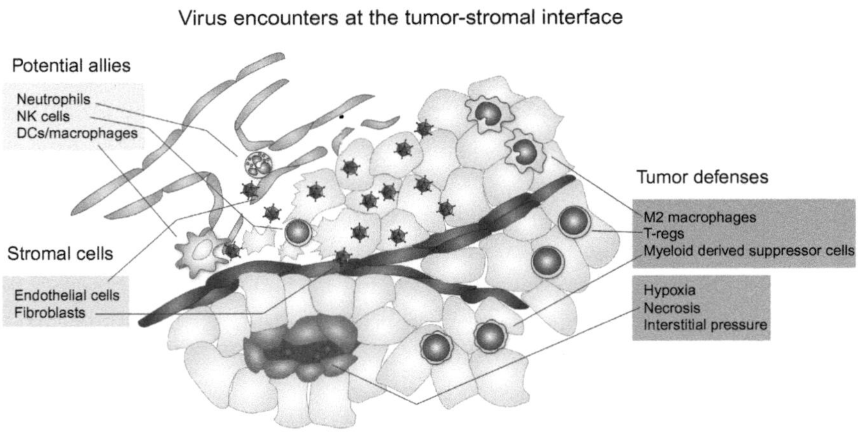


**Figure 8.1** Interactions between adenovirus and cells of the host. Oncolytic viruses have two main interactions after their administration into the body; one with tumor cells which they can infect and kill and the second with the normal cells of the host, which are less permissive but important from the immunological and safety point of view. (See Page 7 in Color Section at the back of book.)

55 different serotypes of human adenoviruses have been identified (Khare, Chen, Weaver, & Barry, 2011; Khare, May, et al., 2011). Originally different human adenoviruses were separated into subgroups based on their capacity to cross-neutralize and agglutinate erythrocytes. Currently, new viruses are assigned to appropriate subgroups by genotyping. To date, seven subgroups (A–G) of human adenoviruses are acknowledged (Khare, Chen, et al., 2011; Khare, May, et al., 2011), but the classification and designation systems for adenoviruses are under constant debate (Imperiale & Enquist, 2011). Besides humans, adenoviruses exhibit a wide range of other hosts within vertebrates. Adenoviruses, however, tend to be species specific in their replication cycle. Nevertheless, they can enter and infect other mammalian cells and express some of the early genes, but human adenoviruses do not replicate productively in these, with a few exceptions. Thus human adenoviruses are generally nonpathogenic to animals, similarly as other animal adenoviruses are nonpathogenic to humans (Wold & Horwitz, 2007). Exceptions to species specificity of human adenoviruses include reports of Ad5 replication in cotton rats (Pacini, Dubovi, & Clyde, 1984; Toth et al., 2005), New Zealand rabbits (Gordon, Romanowski, & Araullo-Cruz, 1992), and Syrian hamsters (Hjorth et al., 1988; Thomas et al., 2006). In addition, human adenoviruses can cause malignant transformation of rodent cells in culture. The ability to transform cells has been linked mainly to persistent E1A, E1B, and E4 protein expression, in the absence of subsequent lysis due to species mediated nonpermissivity (Berk, 2007). Furthermore, some adenoviruses can cause tumorigenesis in newborn rodent pups (Orend, Linkwitz, & Doerfler, 1994). Some serotypes are more oncogenic, such as Ad12, than others. Ad2 and Ad5 are examples of nontumorigenic serotypes. The differences in oncogenicity relate to differences in early genes of serotypes and interplay with the host's immune system. Despite extensive investigation, nocorrelation has been established with adenoviruses and human cancers, which supports the notion of tumorigenicity in only nonpermissive hosts (Berk, 2007). In contrast, in permissive hosts, both E1A expression and adenovirus replication have significant antitumor activity (Lee, Wen, Varnum, & Hung, 2002).

### 2.1.2 Structure of the adenovirus

Adenoviruses are non-enveloped double-stranded DNA viruses surrounded by an icosahedral protein capsid. The capsid comprises mainly of penton and hexon proteins, with knobbed fibers protruding out from the vertices of the capsid, and a number of other minor proteins such as IIIa, IVa2, VI, VIII, and IX. Hexon is the most abundant protein and the main structural component. Hexon trimers are arranged into 20 interlocking facets, and feature hypervariable regions (the most important targets for antibodies) present on surface loops facing outward. The minor capsid protein IX acts as cement between hexon molecules (Rux & Burnett, 2004). Each of the 12 vertices of the virus contains a penton complex consisting of a pentamer of penton protein and the attached trimeric fiber. The penton has flexible loops on its surface, containing an arginine–glycine–aspartic acid (RGD) motif involved in cellular binding and internalization. The fiber has an N-terminal tail attaching to the penton base, a central shaft domain comprising repeating triple β-spiral motifs, and a globular knob involved in trimerization of the fiber and cellular interactions (Campos & Barry, 2007). The length of the fiber protein varies between different serotypes, ranging from 6 repeats of the 15 residue motif in Ad3, to 22 repeats in Ad5, and up to 23 in Ad12 (Law & Davidson, 2005). The shaft of Ad2 and Ad5 includes a nonconsensus sequence of a lysine–lysine–threonine–lysine (KKTK) motif in the third repeat (Smith et al., 2003). In addition, the third repeat of the long shaft of Ad5 appears to create a flexible kink that allows for bending of the shaft. This flexibility may be essential for overcoming steric barriers in cellular interactions and internalization of Ad5 (Wu et al., 2003).

Inside the capsid, the viral DNA is intermingled with the highly basic protein VII and a small protein X, also known as *mu* (Campos & Barry, 2007). The genetic material of adenovirus is a ca. 36 kbp strand of linear double-stranded DNA (Russell, 2000). The 5′ ends of the linear DNA strand are capped by terminal proteins. The protein V helps packing the DNA-protein complex and also provides structural attachment to the capsid hexons via protein VI (Campos & Barry, 2007). The viral Ad protease also locates in the core of the virus and cleaves several capsid and core proteins to their mature functional forms (Webster, Russell, Talbot, Russell, & Kemp, 1989).

### 2.1.3 Life cycle of adenovirus

When an adenovirus particle comes into contact with a cell, *in vitro* data suggest that it binds via knob domains to its primary receptor on the cell with high affinity (Campos & Barry, 2007). Thereafter, secondary interactions take place between penton base and cellular components and trigger the dynamin-dependent clathrin-mediated endocytosis of the virus particle (Wang, Huang, Kapoor-Munshi, & Nemerow, 1998). Mouse data suggests that *in vivo* interactions between other parts of the virus through clotting factors may also play a critical role in transduction (Waddington et al., 2008). Once inside the endocytotic vesicle the virion first releases most of its fibers and then the natural process of acidification of the endosome initiates conformational changes of the capsid (Campos & Barry, 2007). This process leads to virus escape from the endosome, via mechanism mediated by the adenoviral protein VI (Wiethoff, Wodrich, Gerace, & Nemerow, 2005). Thereafter the virion moves to the nuclear pore complex and finally releases its genome into the nucleoplasm for gene expression and genome replication (Campos & Barry, 2007).

The genome can be divided into early genes that are transcribed prior to replication of DNA, and late genes transcribed after it (Russell, 2000). E1A, a subunit of E1, is referred to as an immediate early gene and it is the first viral gene that is expressed when the viral genome reaches the nucleus (Volpers & Kochanek, 2004). The main functions of E1A are to activate transcription of other early genes and to modulate cell metabolism to make the cell more susceptible to viral DNA replication by induction of S-phase. Among other things, E1A binds to Rb, thus releasing E2F, which is critical for activation of adenovirus E2 gene expression cassette and synthesis of a range of S-phase components (Russell, 2000).

Thereafter the other early genes E1B, E2, E3, and E4 are transcribed. E1B gene products are involved in prolonging cell survival by inhibiting apoptosis and necrosis and also in viral replication and transport of viral RNAs (Russell, 2000). E2 gene encodes proteins necessary for replication of viral DNA (Volpers & Kochanek, 2004). E3 genes are dispensable with regard to viral replication *in vitro* but have important roles in battling host cell defense mechanisms and inhibiting initiation of anti-adenoviral immune responses and therefore impact propagation in immune competent hosts. For example, the E3-gp19K gene product delays expression of major histocompatibility complex I (MHC I) and prevents its translocation to the cellular plasma membrane, where it would present peptides to immunological cells. Adenoviral death protein (ADP) involved in lysis of the host cell and release of virions is also coded by E3 (Lichtenstein, Toth, Doronin, Tollefson, & Wold, 2004). E4 gene products have a variety of functions in virus in replication and transcription of viral DNA and production of late proteins and progeny virions (Leppard, 1997). For example, E4orf3 and E4orf6 proteins prevent activation of the cell's DNA damage detection and correction, thus enabling viral replication (Berk, 2007).

Transcription of early genes is followed by DNA replication, initiating from both inverted terminal repeats (Russell, 2000). After onset of viral

DNA replication, the intermediate genes IVa2 and IX are expressed and—among other functions—play a role in the activation of the major late promoter, which then drives the transcription of late genes L1–L5. These late genes encode for structural proteins for assembly of progeny viruses. Thereafter new virions are encapsulated, released from nucleus to the cytoplasm (Russell, 2000). Finally, the progeny viruses are released by disintegration of the host cell membrane through a mechanism where ADP is essential (Tollefson et al., 1996). Recently, evidence is cumulating that the lysis and death of infected cells is mediated via processes of autophagy (Jiang et al., 2011; Rajecki et al., 2009; Rodriguez-Rocha et al., 2011).

## 2.2. Oncolytic adenoviruses

Oncolytic adenoviruses are engineered so that they replicate selectively in cancer tissues, leading to lysis of the cancer cells and release of progeny virus, that is, oncolysis (Fig. 8.2). Although they can enter cells of normal tissues as well, modifications in the virus genome prevent them from efficiently replicating there. To achieve cancer cell selectivity, two major classes of modifications have been employed. The first alternative has been to induce small deletions in the essential viral genes needed for replication in normal cells, thereby restricting replication to cancer cells which possess phenotypic alterations complementing these deletions. The first oncolytic virus to employ this strategy was ONYX-15 (initially described as dl1520), an adenovirus lacking a functional E1B-55k gene (Bischoff et al., 1996). This defect was initially expected to allow replication only in cells with deficiencies in tumor suppressor p53 gene. However, the same authors have subsequently proposed other hypotheses for tumor selectivity, including the ubiquitous nature of p53/p14ARF pathway defects and mechanisms related to nuclear export of viral mRNA and protein trafficking (O'Shea et al., 2004).

A 24 bp deletion of the E1A gene is another strategy to achieve tumor selectivity (Bauerschmitz et al., 2002; Fueyo et al., 2000; Heise et al., 2000). This deletion results in the inability of E1A to bind to Rb and to release eukaryotic initiation factor E2F, which in the case of wild-type adenovirus would result in S-phase induction in normal cells. Therefore, the "delta-24" virus is unable to induce S-phase in host cells and no viral replication follows. In contrast to normal cells, most if not all cancer cells have a defective Rb/p16 pathway, rendering the Rb binding property of E1A dispensable (Cody & Douglas, 2009).

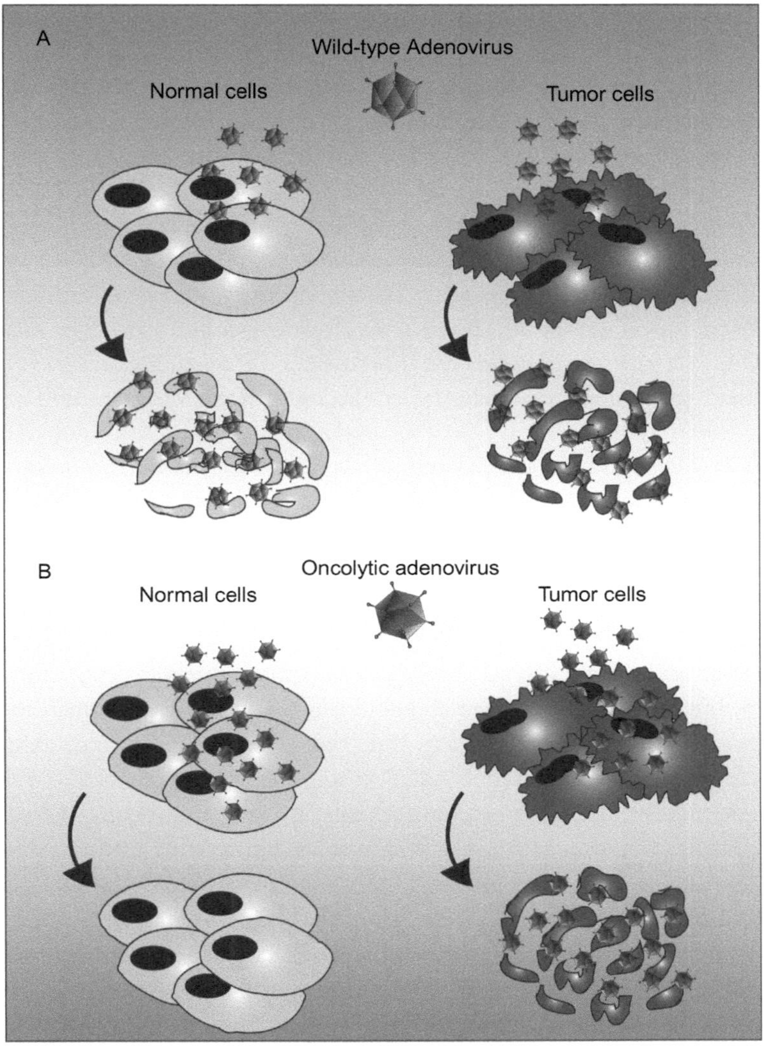


**Figure 8.2** Mechanism of action of oncolytic adenoviruses. *Wild-type* adenoviruses are able to replicate in and kill permissive cells as well as tumor cells (A). Oncolytic adenoviruses are engineered so that they replicate selectively in cancer tissue, leading to lysis of the cancer cells and release of progeny virus in a process called oncolysis (B). (See Page 8 in Color Section at the back of book.)

The other major way of engineering oncolytic viruses involves insertion of tumor- or tissue-specific promoters to control viral replication. Typically, these promoters are placed in a position to drive expression of E1A, and the promoters are chosen such that they are only active in specific cancer tissues. The first example of this type of modification was an adenovirus with prostate-specific antigen promoter driving expression of E1A (Rodriguez et al., 1997). Since then a multitude of different tissue-specific promoters have been used, such as α-fetoprotein for hepatic cancer (Kim et al., 2002), tyrosinase for melanoma (Zhang et al., 2002b), and carcino embryonic antigen for colorectal cancer (Li et al., 2003). Also, tissue-specific promoters that are activated in a variety of cancer types have been employed, including cyclooxygenase 2 promoter (Bauerschmitz et al., 2006; Kanerva et al., 2004; Ono et al., 2005; Pesonen et al., 2010a), L-plastin promoter (Akbulut, Zhang, Tang, & Deisseroth, 2003; Zhang et al., 2002b), and human telomerase reverse transcriptase promoter (Hemminki, Bauerschmitz, et al., 2011; Wirth, Kuhnel, & Kubicka, 2005).

## 2.3. Chimeric and non-Ad5 serotype oncolytic adenoviruses

Coxsackie and adenovirus receptor (CAR) is the primary receptor for serotype 5, but its expression is variable and often low in many human tumors (Kanerva & Hemminki, 2004). In fact, several reports have indicated that CAR expression is downregulated in progression of malignancy and CAR may even have a tumor suppressor role in the epithelium (Coyne & Bergelson, 2005; Kanerva et al., 2002). This can lead to undesirably high transduction of nontarget tissues expressing CAR, while transduction of tumor tissues scarce in CAR expression may remain inefficiently low. To overcome this hurdle, properties of the protein capsid have been modified to redirect virus infection through alternative receptors. These modifications include insertions of peptide sequences to the C-terminus of the fiber, to the HI-loop of the fiber knob, penton base, hexon hypervariable region, and C-terminus of protein IX. Also, chemical modifications and molecular adaptors attached on the virion surface have been investigated for redirecting tropism, but as these approaches involve noncovalent binding of the coating molecule or chemical to the virus their stability and usability *in vivo* poses a problem, especially in the context of replication competent agents (Campos & Barry, 2007).

Insertions of multiple lysine residues to the C-terminus and RGD peptides to the HI-loop are popular modifications that enhance transduction of cells by Ad5. These effects are most prominent with cells naturally expressing low levels of CAR, by redirecting the cell binding to Heparan Sulfate Proteoglycans (HSPGs) and alpha-v-beta class integrins, respectively (Dmitriev et al., 1998; Glasgow et al., 2004; Wickham, Roelvink, Brough, & Kovesdi, 1996a). As hexon is the most abundant protein on the adenovirus capsid, it is also a particularly attractive locus for tropism altering modifications. Hypervariable regions HVR2 and HVR5 seem to be the most suitable for this (Wu et al., 2005) and insertion of RGD to HVR5 has proven to yield a viable virus with ability to transduce cells independent of CAR (Vigne et al., 1999). In addition, penton base modifications have been investigated in attempts to redirect adenovirus tropism and infectivity (Wickham, Carrion, & Kovesdi, 1995; Wickham et al., 1996b). Furthermore, the minor capsid protein IX has been investigated for insertion of tropism modifying ligands and insertion of polylysine sequences has proven able to enhance adenoviral transduction and broaden the viral tropism (Dmitriev, Kashentseva, & Curiel, 2002). This site has also been established as a flexible site able to incorporate even larger molecules for targeting and vector tracking (Campos & Barry, 2007). However, it has been observed in comparative studies that cell targeting with a variety of high-affinity receptor-binding ligands is most effective when transduction is redirected through the fiber protein (Campos & Barry, 2006; Kurachi et al., 2007).

In addition to inserting new ligands also ablation of natural tropism has been investigated. Interestingly, ablation of CAR and/or integrin bindings can have a great effect on *in vitro* cell transduction, whereas these modifications do not exert as great impact on *in vivo* biodistribution (Alemany & Curiel, 2001; Bayo-Puxan et al., 2006; Mizuguchi et al., 2002; Nakamura, Sato, & Hamada, 2003). In contrast, mutation of the putative HSPG binding site on the shaft, the KKTK motif, results in significant changes in biodistribution of the virus. In particular, liver and spleen transduction is reduced with this modification (Bayo-Puxan et al., 2006). However, this mutation seems to affect more than merely cell transduction via HSPG. Specifically, it has been hypothesized that the mutation also affects the flexibility of the Ad5 shaft, as the KKTK motif locates in the proximity of the region suggested to constitute a bend in the three-dimensional structure of the fiber (Bayo-Puxan et al., 2006; Wu et al., 2003). Interestingly, the KKTK mutated adenovirus also exhibits reduced transduction of cancer cell lines *in vitro* which can be rescued only partially by inserting a transduction enhancing ligand such as RGD into the HI-loop of the fiber. Furthermore, while this

KKTK mutated, RGD-retargeted virus does exhibit efficient liver detargeting *in vivo*, unfortunately also tumor transduction is reduced compared to a vector with wild-type capsid (Bayo-Puxan et al., 2006). Thus, probably the change in the conformation of the shaft, caused by the KKTK mutation, prevents effective retargeting through the RGD/integrin interaction.

Other ways to modify cell transduction properties of adenoviruses included chimeric constructs between serotypes. Already much before the actual receptor of Ad3 was identified, it was noticed that it is highly expressed on cancer cells (Kanerva et al., 2002; Tuve et al., 2006). Placing the Ad3 fiber knob into the Ad5 backbone has resulted in an Ad5/3 chimera that displays the cell binding properties of serotype 3 (Kanerva et al., 2002; Krasnykh, Mikheeva, Douglas, & Curiel, 1996). These chimeras exhibit enhanced gene delivery and antitumor efficacy in preclinical assays with cell lines, fresh clinical specimens, and animal models featuring a multitude of tumor types (Guse et al., 2007; Kanerva et al., 2002, 2003; Kangasniemi et al., 2006; Rajecki et al., 2007; Ranki et al., 2007; Volk et al., 2003; Zheng et al., 2007) and also in cancer initiating cells (Eriksson et al., 2007). Importantly, toxicity, blood clearance or biodistribution, and gene transfer to normal tissues are not adversely affected in preclinical systems in comparison to Ad5 which has an excellent safety record in cancer trials (Kanerva et al., 2002). Interestingly, the 5/3 chimerism approach works best with the long Ad5 shaft. If the short, bendless Ad3 shaft is employed, transduction is impacted adversely. Also, other serotype chimeras have been used successfully (DiPaolo et al., 2006).

Recently, we developed a modified oncolytic adenovirus based fully on serotype 3 Ad and driven by human telomerase reverse transcriptase (hTERT) promoter (Ad3hTERT) (Alba, Bosch, & Chillon, 2005; Danthinne & Imperiale, 2000; Hemminki, Bauerschmitz, et al., 2011; Hemminki, Diaconu, et al., 2011). One of the most prominent advantages of such virus is the possibility to fully overcome preexisting antibodies to serotype 5. Moreover, by utilizing the tumor-associated desmoglein 2 receptor, the problem of CAR being downregulated in advanced cancers is circumvented. Moreover, this virus has shown safety and efficacy in mouse model and human patients (Hemminki, Diaconu, et al., 2011). A similar approach was also investigated with other different serotypes of adenovirus, such as Ad11 (Sandberg, Papareddy, Silver, Bergh, & Mei, 2009). They showed that the oncolytic capacity of such virus was 100 times higher in prostate cancer cell line. The oncolysis was independent of the level of expression of p53 in the cells or on the absence of E1B55k expression in the vector.

## 2.4. *Armed* oncolytic adenoviruses

A popular method for placing transgenes under endogenous control elements involves replacing the E3 region with the transgene (Zhang et al., 1996). Also replacing only some of the E3 genes such as E3B (Kim et al., 2002), ADP, or 6.7K/gp19K or merely gp19K by the transgene is feasible (Cody & Douglas, 2009). These insertions couple the expression of the transgene with viral replication without abolishing the oncolytic potential. Transgenes may also be inserted in L3 region, resulting in even more stringent coupling to viral replication (Robinson et al., 2008). Additional possible transgene insertion loci have also been identified using a transposon-based mechanism (Kretschmer, Jin, Chartier, & Hermiston, 2005).

An alternative method for using endogenous viral gene control elements involves linking the transgene to a viral gene by an internal ribosome entry site (IRES). This method allows for expression of the transgene along with the viral gene in the same transcript. Often used is the fiber gene that is expressed late in the viral cycle (Cody & Douglas, 2009). This type of expression not only allows for high levels but also allows for using possibly cytotoxic transgenes that might interfere with viral replication if expressed earlier. In addition to the fiber region, IRES linked transgenes in other loci have been investigated (Rivera et al., 2004). Moreover, Alemany's group has explored the possibility to express transgene downstream of fiber-RGD using an Ad5 IIIa protein splice acceptor (Rojas et al., 2010).

Another option for transgene insertion is to use an exogenous promoter, a tissue-/tumor-specific promoter or a constituently active one, for driving transgenes. These promoter-transgene sequences may then be linked with an IRES linker to viral genes or used to directly replace genes. If IRES linking is not utilized, then the promoter-transgene sequence is usually used to replace the E1B-55k or E3 region, in whole or in part (Freytag, Rogulski, Paielli, Gilbert, & Kim, 1998; van Beusechem, van den Doel, Grill, Pinedo, & Gerritsen, 2002). If transgenes are linked with the E1A gene, expression early in the replication cycle results (Akbulut et al., 2003).

Promoter driven transgene systems for oncolytic viruses not only allow for efficient transgene expression but can also be used to target tumor cells at multiple levels: for example, the 24 bp deletion of E1A or the E1B-55k deletion can be first used to restrict viral replication to tumor cells, and transgene expression can be further restricted by a tumor-specific promoter. These kinds of approaches are appealing due to their safety aspects and can be used to deliver even potentially highly cytotoxic transgenes.

Regarding transgenes used to arm oncolytic viruses, a multitude of different approaches have been investigated. Tumor-suppressor genes such as p53 have been used to enhance oncolytic cell killing regardless of the p53 status of the cancer cell line (van Beusechem et al., 2002). Prodrug-converting enzyme-based systems commonly employ either cytosine deaminase for 5-fluorocytosine conversion to 5-fluorouacil (Akbulut et al., 2003; Zhan et al., 2005) or HSV-tk for ganciclovir conversion to its active metabolite (Zhang et al., 2010a) or both (Freytag et al., 1998). Antiangiogenic molecules have also been used for arming (Guse et al., 2009), in addition to various other molecules such as human sodium iodide symporter used to concentrate radioiodine to target cells (Hakkarainen et al., 2009). Furthermore, immunostimulatory cytokines such as GMCSF (Cerullo et al., 2010; Chang et al., 2009; Lei et al., 2009; Ramesh et al., 2006) aimed to boost antitumoral immunity have been under active investigation as transgenes. Along this line, very promising results have been obtained with the cancer terminator oncolytic virus (CTV) (Sarkar, Su, & Fisher, 2006; Sarkar et al., 2005a,b, 2007, 2008). This virus bears unique properties of tumor-specificity due to the insertion of the tumor-specific promoter *PEG* in combination with production of a cancer-selective cytotoxic cytokine, melanoma differentiation associated gene-7/interleukin-24 (mda-7/IL-24) (Ad.PEG.E1A-*md-7*), which embodies potent bystander antitumor activity. Different version of this CTV has been made expressing IFN-γ (Ad.PEG.E1A-*IFN-g*) (Sarkar et al., 2005a,b).

## 3. IMMUNE RECOGNITION OF ADENOVIRUSES

Innate immune responses are triggered when pattern-recognition receptors (PRRs) recognize specific conserved molecular patterns on pathogens. Several classes of PRRs, including toll-like receptors (TLRs), NOD-like receptors (NLRs), and various cytoplasmic receptors recognize distinct microbial components and can thereafter directly activate immune cells. Exposure of immune cells to the ligands of these receptors activates intracellular signaling cascades that rapidly induce the expression of a variety of both overlapping and unique genes involved in the ensuing immune responses. Thus, the activation of PRRs results in the production of large amounts of type I IFNs and several other proinflammatory cytokines (Akira, Uematsu, & Takeuchi, 2006; Kanneganti, Lamkanfi, & Nunez, 2007; Kawai & Akira, 2006; Shaw, Reimer, Kim, & Nunez, 2008). These responses are important in controlling pathogen replication and they also provide a critical initiation signal, which modulates and controls the adaptive immune response (Akira et al., 2006; Kanneganti et al., 2007; Kawai & Akira, 2006; Shaw et al., 2008).

Adenovirus is the most commonly used gene therapy vector. It is mostly used in the context of genetic diseases and cancer, or for vaccination purposes (Edelman & Nemunaitis, 2003; Galanis et al., 2005; McLoughlin et al., 2005; Mundt et al., 2004; Nemunaitis, Khuri, et al., 2001; Nemunaitis, Senzer, et al., 2007; Nemunaitis, Vorhies, Pappen, & Senzer, 2007; Nemunaitis et al., 2003, 2006, 2009; Reid et al., 2002; Tong et al., 2005; Yu & Fang, 2007). Upon entry, adenoviruses interact with multiple PRRs (Fig. 8.3), quickly eliciting a robust cytokine

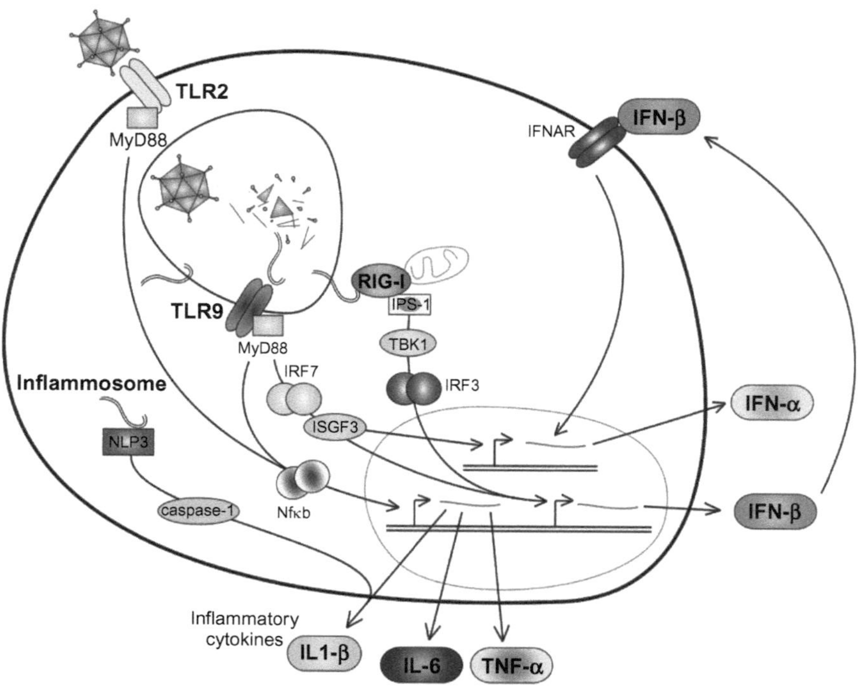


**Figure 8.3** Interactions between adenovirus and pattern recognition receptors (PRRs). At the cellular level, adenovirus interacts with several receptors of the innate immune system inside and outside of the cell.

response (Brunetti-Pierri et al., 2004; Raper et al., 2003). In context of conventional gene delivery, virus-induced inflammation can lead to premature vector elimination. Therefore, high doses of vector may be needed to obtain sufficient level of gene delivery which may in turn lead to adenovirus associated toxicity (Brunetti-Pierri et al., 2004; Raper et al., 2003).

To prolong gene expression by the vector and avoid strong immune reactions, helper-dependent adenoviruses have been developed. Also called "high capacity" or "gutless," they are devoid of all viral genes and only posses 5′ and 3′ inverted terminal region sequences and the packaging signal, thus allowing a large cloning capacity (about 36 kb) for transgenes. Advantages of these viruses include complete lack of adaptive immune response and consequently a long-term sustained gene expression, rendering them appealing for gene replacement approaches where sustained transgene expression is needed (Brunetti-Pierri & Ng, 2009, 2011; Seiler, Cerullo, & Lee, 2007a). The high levels of liver transduction required for the long-term sustained expression of transgene, necessary for the rescue of a monogenic disease, necessitate a dose of vector which is not compatible with the safety of the procedure (Brunetti-Pierri et al., 2004). This problem has been overcome by delivering the vector directly into a surgically isolated liver (Brunetti-Pierri et al., 2006).

In general, a critical aspect of adenoviral virotherapy relates to neutralizing antibodies. Approaches that have been tested for their evasion include switching the serotype (Kanerva et al., 2002, 2003). This allows overcoming preexisting antibodies remaining due to past natural adenovirus infections or previous virus treatments, but does not prevent *de novo* antibody formation. Another approach employs viral shielding by chemical methods such as polyethylene glycol (Croyle et al., 2005). Further, *ex vivo* infected cells can be used as carriers. In addition, in order to suppress immune responses and allow for better replication of the virus, pharmacological agents have been used to induce generalized immunosuppression of the host (Alemany & Cascallo, 2009). A popular preclinical approach employs high-dose cyclophosphamide for dampening of both cellular and antibody responses (Cerullo et al., 2011).

In an interesting contrast, others and we have hypothesized that in context of cancer gene therapy we can exploit the natural propensity of adenovirus to activate the innate and adaptive arms of the immune system to enhance antitumor effects. In this scenario, oncolysis by the virus would provide initial cell killing with an associated release of a variety of TAAs, whereas immunological recognition of the virus would provide the required "danger signal" to mature and stimulate antigen presenting cells (APCs; e.g., dendritic cells, DCs) and thus boost antitumor immune reactions. Along this line, investigators have recently demonstrated that oncolytic adenoviruses indeed have the capability to stimulate a specific antitumor immune response (Cerullo et al., 2010; Koski et al., 2010). Epitope spreading may be a particularly useful phenomenon in this regard, allowing antiviral reactions to influence also antitumor immunity.

We hypothesized that oncolysis delivers a double punch on behalf of antitumor immunity. On the one hand, lysis of tumor cells results in an abundant source of TAAs. On the other hand, oncolysis induces danger signals to the immune system to enhance cross-priming of APCs rather than cross-tolerization (Fig. 8.4). Moreover, oncolytic adenovirus can be armed with immunomodulating molecules to specifically stimulate pathways of the immune system. Importantly, cross-presentation and epitope spreading have been demonstrated, suggesting that cellular antiviral responses can result in immunity against TAA as well. In particular, memory responses against TAA may be important for long-term responses and survival of patients. Even

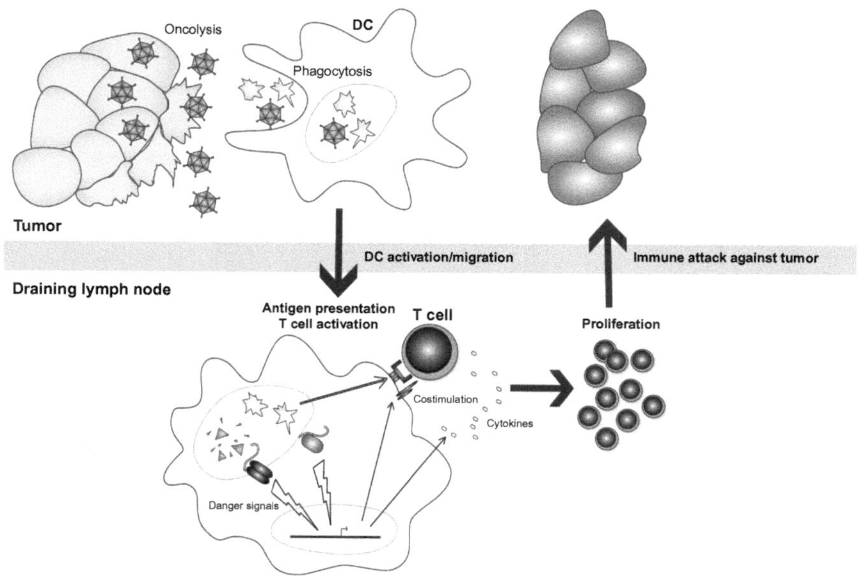


**Figure 8.4** The dual immunological effect of oncolytic adenovirus. Lysis of tumor cells results is an abundant source of tumor-associated antigens (TAAs). Also, oncolysis induces danger signals facilitating cross-priming of antigen presenting cells (APCs) rather than cross-tolerization.

though the main goal of arming is typically enhancing efficacy, also safety could be increased, if lower doses of the vector may be used or if the arming device results in less systemic dissemination or faster clearance of the virus.

## 3.1. The innate immune system

### *3.1.1 Adenovirus and TLRs*

The innate immune system recognizes intruding pathogens through PRRs that detect conserved microbial components called pathogen-associated molecular patterns (PAMPs). PAMPs represent molecules vital for microbial survival such as flagellin, nucleic acid structures unique to bacteria and viruses (CpG DNA, dsRNA), and bacterial cell-wall components such as lipopolysaccharide, lipoteichoic acid, and peptidoglycan (Akira et al., 2006). The protein Toll was identified as a key regulator of innate immune signaling in *Drosophila melanogaster* already more than a decade ago (Hoffmann, 2003). Thereafter, mammalian TLRs have been recognized for their ability to sense a wide array of microbial and self-ligands at the cell surface and within endosomes (Rajecki et al., 2009). TLRs comprise of 11 different receptors that recognize motifs found on a wide range of pathogens, and activation of TLRs results in the production of large amounts of type I IFNs and several proinflammatory cytokines. These cytokine responses are important in controlling pathogen replication and they also provide an initiation signal for the adaptive immune response.

Adenovirus capsids activate the innate immune system through mechanisms independent of viral replication and gene expression (Bowen et al., 2002; Brunetti-Pierri et al., 2004; Cerullo et al., 2007; Liu et al., 2003; Muruve, Barnes, Stillman, & Libermann, 1999; Zaiss et al., 2002; Zhang et al., 2001). For example, following exposure to UV-inactivated adenovirus, human peripheral blood mononuclear cells (PBMCs) rapidly produce many cytokines, including IL-6, IL-1β, GMCSF, IL-8, and TNF-α (Higginbotham, Seth, Blaese, & Ramsey, 2002). A similar cytokine profile is also found in the serum of mice and nonhuman primates following intravenous administration (Brunetti-Pierri et al., 2004; Cerullo et al., 2007; Zhang et al., 2001). Also, DCs produce large amounts of cytokines and type I IFNs immediately after infection with adenovirus (Andoniou et al., 2005; Edukulla et al., 2009).

Recent literature has started to link these phenomena with specific stimulation of TLRs (Fig. 8.3). We have showed that adenovirus DNA triggers innate responses in part via TLR9 (Cerullo et al., 2007), one of the PRRs located in the endosome and responsible for detecting double-stranded DNA. Myeloid differentiation primary response gene 88 (MyD88) is a universal adapter protein used by all TLRs (except TLR3) to activate the transcription factor NF-κB and trigger the immune response. Intriguingly, it has been observed that mice lacking the MyD88 showed significantly reduced cytokine secretion when challenged with high dose of adenovirus (Suzuki et al., 2010). These data were confirmed by Suzuki et al. (2010) who showed that TLR2 and TLR9 are responsible for induction of cytokines and gene silencing in mice following adenoviral vector administration. Interestingly, they showed that when *LacZ*-expressing adenoviral vectors were administered intravenously in MyD88 knockout mice (thus lacking nearly all TLRs-mediated responses) not only was cytokine secretion significantly lower but also transgene expression was significantly prolonged due to absence of a normal immune response (Suzuki et al., 2010). These results highlighted the importance of the TLRs not only in initiation of the innate immune response but also in modulation and shaping of the adaptive response (Bachmann & Jennings, 2010)

### 3.1.2 Adenovirus and cytosolic sensors

Nucleotide-binding and oligomerization domain NLRs comprise a large family of intracellular PRRs that are characterized by the presence of a conserved NOD (Inohara, McDonald, & Nunez, 2005). Together with retinoid acid-inducible gene I (RIG-I)-like receptors, NODs detect microbial components in the cytosol (Lyons et al., 2008). NLRs containing NOD1 sense the dipeptide γ-D-glutamyl-*meso*-diaminopimelic acid (iE-DAP) and NOD2 containing NLRs sense muramyl dipeptide (O'Neill, 2008). Both of these molecules are breakdown products of the bacterial peptidoglycan cell wall. NOD1 is ubiquitously expressed and occurs in most NLRs, whereas NOD2 are restricted to monocytes, macrophages, DCs, and intestinal Paneth cells (Shaw et al., 2008).

The correlation between adenovirus and NLRs recognition is to date poorly understood. Recently, it was reported that challenging of NOD2 KO mice with intravenously delivered nonreplicating adenoviral vectors led to reduced proinflammatory cytokine secretion and significantly higher transgene expression compared to wild-type mice. Moreover, experiments in NOD2/MyD88 double KO mice showed further reduced innate responses to adenoviral vectors compared to responses in singly deleted mice, indicating that NOD2 signaling contributes independently of MyD88 (Suzuki et al., 2011).

The NALP proteins are cytoplasmic NLRs (Fritz, Ferrero, Philpott, & Girardin, 2006), of which NALP3 is best characterized and also known as cryopyrin or NLRP3. NALP3 senses exogenous and host ligands such as bacterial peptidoglycan, ATP or uric acid (Petrilli, Dostert, Muruve, & Tschopp, 2007). After recognition, NALP3 recruits the inflammatory caspase-1 into a molecular complex termed the NALP-inflammasome via the action of adaptor protein ASC (Petrilli et al., 2007). Once activated, the caspase-1 processes pro-IL-1b and pro-IL-18 to their active and secreted forms. Other NLRs that are known to form IL-1b-processing inflammosomes include NALP1 and IPAF, the latter of which directly activates caspase-1 in response to bacterial flagellin (Franchi et al., 2008).

In an elegant study, Petrilli and colleagues showed that internalized adenovirus DNA triggers an innate immune response dependent on the activation of the NALP-inflammosome complex (NALP3 and ASC) (Muruve et al., 2008). Already earlier, Nociari, Ocheretina, Schoggins, and Falck-Pedersen (2007) demonstrated that TLR-independent adenovirus DNA recognition led to IRF3 activation and type I IFN and proinflammatory cytokine expression. Also, cytosolic DNA recognized by AIM2 was shown to induce IL-1b secretion through a caspase-1-dependent inflammosome pathway (Hornung et al., 2009).

### 3.1.3 Adenovirus, complement, and endothelial cells

When adenoviruses are introduced to the vascular system, they come into contact not only with preexisting and natural antibodies but also complement proteins, blood cells, and endothelial cells. Antibodies and complement play an important role in vector opsonization and clearance. Adenovirus has been shown to activate complement via antibodies in individuals having preexisting immunity (Appledorn et al., 2008). Interestingly, binding by complement protein C1q has also been shown to increase adenoviral cell transduction (Tsai, Varghese, Ravindran, Ralston, & Vellekamp, 2008). In addition, binding of adenovirus to human complement receptor 1 bridges Ad5 interaction to erythrocytes that can also bind adenovirus particles directly with CAR (Carlisle et al., 2009). Indeed, human blood cells may bind the majority of blood-borne virus, whereas in mice most intravascular human adenovirus remains free in plasma (Lyons et al., 2006). Also, direct binding to C3-derived fragments has been reported (Jiang, Wang, Serra, Frank, & Amalfitano, 2004). Such interactions can probably contribute to inflammatory responses associated with virotherapy (Seregin et al., 2010). In fact, steps have been taken to mask complement-mediated recognition of adenovirus particles with the goal of increasing vector efficacy (Diaz et al., 2007; Thomas & Fraser, 2003). Thrombocytopenia is caused by interactions between adenoviral particles and the coagulation system, resulting in platelet activation, binding to endothelial cell surfaces, and formation of platelet-leukocyte aggregates (Othman, Labelle, Mazzetti, Elbatarny, & Lillicrap, 2007). It has been proposed that these virus-platelet aggregates are then trapped in the liver sinusoids and engulfed by the liver Kupffer cells (KCs) for degradation (Stone et al., 2007).

Finally, Ad particles that escape entrapment by soluble blood factors and blood cells directly and indirectly activate endothelial cells primarily via recognition of virus capsid (Liu et al., 2003). Whereas general endothelial cells lining the vascular system could represent a large putative sink for virus accumulation, infection of these cells does seem to be particularly productive (Khare, May, et al., 2011). In contrast, the specialized endothelial cells lining liver sinusoids appear to play a major role in the fate of intravascular adenovirus particles (Ganesan et al., 2011). These liver sinusoidal endothelial cells (LSECs) belong to the reticuloendothelial system and play a role in clearing materials from the blood stream, along with KCs of the liver and similar macrophages of the spleen. All of these cell types are involved in clearing adenovirus particles from blood (Khare, Chen, et al., 2011), and the mechanism of viral uptake is suggested to occur via scavenger receptors expressed on the surface of the cells (Xu, Tian, Smith, & Byrnes, 2008). KCs are known to exert a major influence on adenovirus sequestration and uptake results in rapid destruction of engulfed virus as well as KC degradation (Khare, Chen, et al., 2011; Manickan et al., 2006). Although LSECs and KC take up large amounts of virus, neither cell type is productively infected with adenoviruses nor does cell entry generally lead to gene expression. Nonetheless, with very high dose of virus, a low level of transduction may be observed (Hegenbarth et al., 2000; Wheeler, Yamashina, Froh, Rusyn, & Thurman, 2001). Thus, LSEC and KC uptake limit the bioavailability of the viruses to target tissues. Due to their immunological capacity, they are also important with regard to treatment related toxicity through inflammatory cytokine responses (Lieber et al., 1997; Nunes, Furth, Wilson, & Raper, 1999; Shayakhmetov, Gaggar, Ni, Li, & Lieber, 2005).

## 3.2. The adaptive immune system

### *3.2.1 Adenovirus and B cells*

As consequence of the activation of the innate recognition receptors, a rapid increase in several cytokines particularly IL-6, IFN-$\alpha/\beta$, RANTES, IL-12 (p40), IL-5, G-CSF, and GMCSF is observed (Seiler et al., 2007a). Furthermore, a complex set of interactions between the innate and the adaptive immune system results in activation of CD4+ and CD8+ T cells, and B cells (Seiler et al., 2007a,b). Type I IFN signaling is important for T help-dependent antibody formation by B cells. IFNs also induce DC maturation by upregulating costimulatory molecules such as CD80, CD86, and CD40. Neutralizing antibodies against IFN-α and IFN-β have been found to be effective in blocking both innate as well as adaptive immune responses to viral vectors (Zhu, Huang, & Yang, 2007).

Understanding the humoral immune response to adenovirus is of importance for gene delivery for at least two reasons. First, presence and prevalence of neutralizing antibodies (NABs) against adenovirus might hinder the efficiency of transduction. Second, the presence of NAB might influence the outcome of the therapy at posttranductional steps. In fact, epidemiological studies on NAB in different populations have shown that most people globally carry some levels of antibodies in their serum, although some geographical variation does occur (Mast et al., 2010). Nevertheless, it should be emphasized that the specificity and immunogenicity of adenovirus type 5 NAB elicited by natural infection is quantitatively and qualitatively different than NABs induced by immunization with an adenoviral vector (Serafini et al., 2004). In this study, 1904 participants were enrolled in a cross-sectional serological survey at seven sites in Africa, Brazil, and Thailand to assess NAB for adenovirus types Ad5, Ad6, Ad26, and Ad36. Samples from a clinical trial of a T cell-based AIDS vaccine delivered with recombinant adenovirus type 5 were used to assess NAB titers from the United States and Europe. The proportions of participants that were completely negative were 14.8% (Ad5), 31.5% (Ad6); 41.2% (Ad26) and 53.6% (Ad36). The study was conducted to correlate high Ad5 titers and the inefficiency of Ad5-based vaccine for HIV and concluded that natural Ad5 infection compromises Ad5 vaccine-induced immunity to weak immunogens, such as HIV-1 Gag, used in the original clinical trial.

It would be very interesting to conduct a similar study correlating the efficacy of oncolytic adenovirus treatment in terms of overall survival or progression free survival with the preexisting NAB presence. Thus far, in our studies, we have not seen correlation between oncolytic virus treated patients and efficacy (Cerullo et al., 2010, 2011; Koski et al., 2010; Nokisalmi et al., 2010). Interestingly, we saw lower NAB levels than have been reported in previous, studies. Differences between different reports could be due to geographical factors, the patient cohorts studied (cancer patients vs. healthy volunteers), methodological issues (sensitivity and specificity of the test used or the definition of a significant titer) or a combination thereof.

### *3.2.2 Adenovirus and T cells*

CD4+ and CD8+ T cells cross-reactive against different adenovirus serotypes have been found among human PBMCs(Nayak & Herzog, 2010). Adenovirus-specific CD4+ T cells often recognize epitopes conserved among different serotypes, with the majority of people developing a long-lived memory response (Nayak & Herzog, 2010).

Further, adenovirus-specific secretion of IFN-$\gamma$ from PBMCs has been reported to occur within 12 h of exposure, suggesting prior activation of adenovirus-specific CD8+ cells. Transduction of APCs by adenoviruses also contributes to CD8+ responses, which can be directed against both viral gene and transgene products and are dependent on CD4+ help (Nayak & Herzog, 2010).

It has in fact been demonstrated that already adenovirus *per se*, given its ability to interact with a variety of receptors of the innate immune system, is able to trigger a T cell immune response in context of cancer therapy. Tuve and colleagues have shown using a mouse model of *neu*-positive syngeneic mammary-cancer (MMC) syngeneic MMC that intratumoral injection of replication-deficient, transgene-devoid adenovirus induced immunological responses at two different anatomical sites: the tumor-draining lymph nodes and the tumor microenvironment. Inside the tumor microenvironment only adenovirus-specific T cells expanded, whereas the lymph nodes supported the generation of both *neu*- and virus-specific T effector cells. Importantly, Ad-specific T cells were antitumor-reactive despite the presence of active regulatory T cell-mediated immune tolerance inside tumors. Moreover, efficacy was increased by preimmunization regardless of NAB (Tuve et al., 2009). We have treated more than 200 cancer patients with armed and unarmed oncolytic adenoviruses and have observed activation of CD8+ T cells against both virus and tumor epitopes (Cerullo et al., 2010; Koski et al., 2010; Nokisalmi et al., 2010; Pesonen et al., 2010a,b, 2011a). Thus, our data suggest that adenovirus can be a useful platform for combining immunotherapy with oncolytic gene therapy as "immunovirotherapy."

## 4. ONCOLYTIC ADENOVIRUSES AS IMMUNOTHERAPEUTIC AGENTS

### 4.1. Immunotherapy

Antitumor immunotherapy refers to an approach in which scientists explore ways to awaken or engage the immune system to recognize and kill tumor cells in a more or less specific fashion. The first observations on the role of the immune system in inducing tumor regression date back to 1700s, with a description of a sporadic tumor regressions after episodes of infection (Wiemann & Starnes, 1994). In the late 1800s, William Coley injected bacterial products (known as Coley's vaccine) directly into the tumor achieving high degree of response (Coley, 1891). These studies are considered the first empiric evidence sets suggesting that the immune system can result in tumor regression. Almost a century after Coley's studies experimental and clinical data showed that the involvement of the immune system with exogenous cytokines such as IL-2 and IFN-$\alpha$ clearly contributed to tumor regression (Atkins et al., 1999; Kirkwood et al., 1985; Mazumder & Rosenberg, 1984).

A revolution in the field has been represented by the identification of tumor-specific "signatures" or antigens that are specifically recognized by the T cell receptor (Traversari et al., 1992; van der Bruggen et al., 1991). van der Bruggen et al. (1991) conducted a landmark study by identifying MAGE-1 as the first human TAA. This study was almost immediately followed by the first human tumor-specific peptide restricted by HLA-A1 (Traversari et al., 1992). These observations initiated the vaccine based anticancer therapy field (also known as immunotherapy) by suggesting that CD8+ T cells specifically recognize and kill autologous cancer cells expressing (or overexpressing) specific tumor antigens. These important achievements, together with the discovery of various TAAs, gave scientists the necessary tools to investigate with molecular precision new strategies to enhance immune-mediated tumor rejection and develop cancer vaccines.

A logical way to apply this knowledge was the *ex vivo* generation of tumor-specific CD8+ T cells (Jena, Dotti, & Cooper, 2010). Other approaches have involved DNA-based vaccines expressing specific TAAs (Shimamura & Morishita, 2011). Both naked DNA as well as different viral platforms have given promising results in generation of TAA-specific immune responses (Lladser et al., 2011a). Nevertheless, although from an immunological point of view TAA-specific immunization reached its purpose, clinical results have always been so far disappointing and, at the present, no anticancer TAA-specific vaccine can be recommended outside of clinical trials (Wang, Panelli, & Marincola, 2006; Wang et al., 2008). A critical discovery in this regard has been that induction of an antitumor immune response is not therapeutically sufficient against the backdrop of tumor-induced immune suppression. Strikingly, a seemingly naivistic approach aiming solely at reducing immunosuppression—without any attempt at immunological induction—has resulted in promising human data in the context of melanoma (Hodi et al., 2010). In an attempt to combine immune stimulation with reduction of immunosuppression, we generated an oncolytic adenovirus coding for anti-CTLA4 monoclonal antibody. Preclinical data were promising but human data are not yet available (Dias et al., 2011).

Of note, the promise of antigen-specific immunotherapy has been realized by monoclonal antibodies such as trastuzumab and rituximab (Hodi et al., 2010). Intriguingly, it is not known how much Fc mediated immunological activity contributes vis-a-vis Fab binding mediated signaling inhibition. Nevertheless, it can be speculated that since the biggest impact of trastuzumab is in adjuvant therapy, the immunological aspects of the therapy are relevant in mediating the survival benefit (Hodi et al., 2010).

## 4.2. Strategies to exploit the immune system using oncolytic adenoviruses

### *4.2.1 Adenovirus modification for enhanced innate immunity*

Considering how articulated and complex the innate sensing of adenovirus is, and how various cellular sensors can influence and shape the long-lasting adaptive response, we believe that the innate arm of the immune system can be utilized in the treatment of cancer. These notions seem to be supported by emerging discoveries suggesting that we can exploit oncolytic viruses not only for their killing capacity but also for their ability to activate the relevant receptors. Nonetheless, human data suggest that oncolytic adenoviruses *per se* are not usually able to elicit immune responses capable of fully eradicating metastatic tumors. This could be due to the highly immune suppressive nature of advanced cancers.

Hence, researchers have now entered a new era where they are genetically manipulating adenoviruses to enhance activation of specific innate receptors. In this respect, an interesting approach is generation of an oncolytic adenovirus expressing the pan-TLR adaptor protein MyD88. Tantalizingly, intratumoral injection of Ad-MyD88 into established tumor masses enhanced adaptive immune responses and inhibited local tumor immunosuppression, resulting in significantly inhibited local and systemic growth of multiple tumor types. Further, Ad-MyD88 infection of primary human DCs, tumor-associated fibroblasts, and colorectal carcinoma cells elicited significant Th1-type cytokine responses, resulting in enhanced tumor cell lysis and expansion of human tumor antigen–specific T cells (Hartman et al., 2010).

### 4.2.2 Oncolytic Adenoviruses Armed with Immunomodulatory Transgenes

Despite emerging data showing induction of antitumor-specific immune response elicited by oncolytic adenoviruses *per se*, antitumor immune responses capable of complete eradication of advanced tumors is rarely seen. Tumors initiate from normal tissues and thus most TAA resemble self-antigens which results in lower immunogenicity in comparison to heterologous epitopes such as PAMP. Further, since tumors typically grow over a decade of more in the presence of an intact immune system, a tremendous amount of immunoediting, by means of ablation of immunogenic clones, has usually occurred. Also, as mutations accumulate, and tumor cells resemble normal cell less and less, an increasing amount of immune suppression is required for tumor sustenance. Thus, the biggest challenge for cancer immunotherapy in general and immunovirotherapy in particular could be manipulation of the tumor microenvironment in favor of immune responses rather than tolerance (Table 8.1). In this regard, encouraging results have been achieved using oncolytic adenoviruses armed with transgenes for modulation of both the innate and adaptive immune systems as reviewed in the next chapter.

### 4.2.3 Cytokine-expressing adenoviruses

Cytokines are used by the immune system for cross talk between different cell types and are thus easily harnessed as immunotherapeutic arming devices. A major advantage over autocrine or gap-junction mediated activities such as delivered by HSV-TK/ganciclovir is their physiological paracrine manner. Especially attractive with regard to gene therapy is their low systemic tolerability contrasted with high local efficacy. One widely used cytokine in this respect is IL-12 (Chang et al., 2007; Chen, Lin et al., 2008; Gabaglia et al., 2004; Gao et al., 2005, 2008; Hall et al., 2002; Hwang et al., 2005; Jin et al., 2005b; Liu et al., 2002a,b; Nasu, Ebara, & Kumon, 2004; Park et al., 2008; Raja Gabaglia et al., 2007; Satoh et al., 2003; Wen et al., 2001; Zhang & DeGroot, 2003). It is an IL naturally produced by APCs in response to antigenic stimulation. As a consequence of interaction with its receptors (IL-12R-β1 and IL-12R-β2), it activates natural killer (NK) cells and T lymphocytes (T cells) enhancing their cytotoxic activity. T and NK cells produce IFN-γ in response to IL-12 activation. Oncolytic adenoviruses expressing IL-12 have demonstrated to enhance T cell and NK activation several tumor models in mice and human (Chang et al., 2007; Chen, Wang, et al., 2008; Gabaglia et al.,

**Table 8.1** Oncolytic adenoviruses armed with immunostimulatory transgenes used in humans

| Virus | Capsid | Tumor selectivity | Arming device | Number of patients treated | Safety | Efficacy | Immunological activity | Reference |
|---|---|---|---|---|---|---|---|---|
| CG0070 | Ad5 | E2F promoter | GMCSF | 45 | No grade 3 and 4 adverse event have been reported | Response rates of 48–77% in phase I/II bladder cancer trial | Increased levels of GM-CSF in 94% of patients | Cold Genesys (f. Cell Genesys), http://coldgenesys.net/, Ramesh et al. (2006) |
| H103 | Ad5 | E1B55k deletion | HSP70 | 27 | 1 patient experienced grade 3 fever | 11.3% response, 48.4% disease control | Increase of CD4+, CD8+, and NK cells | Li et al. (2009) |
| KH901 | Ad5 | Modified hTERT promoter | GMCSF | 23 | No grade 3 and 4 adverse event have been reported | 12/19 disease control | GMCSF expression in tumors | Chang et al. (2009) |
| Ad5D24-GMCSF | Ad5 | E1A CR2 deletion | GMCSF | 93* | Good, few grade 3 and 4 adverse events, no grade 5 events | 44% disease control overall | Tumor- and virus-specific CD8+ T cells by ELISPOT and pentamer analysis | Cerullo et al. (2010) |
| Ad5/3-D24-GMCSF | Ad5/3 chimera | E1A CR2 deletion | GMCSF | 115* | Good, few grade 3 and 4 adverse events, no grade 5 events | 48% disease control overall | Tumor- and virus-specific CD8+ T cells by ELISPOT | Koski et al. (2010) |

*Continued*

**Table 8.1** Oncolytic adenoviruses armed with immunostimulatory transgenes used in humans—cont'd

| Virus | Capsid | Tumor selectivity | Arming device | Number of patients treated | Safety | Efficacy | Immunological activity | Reference |
|---|---|---|---|---|---|---|---|---|
| Ad5-D24-RGD-GMCSF | RGD-4C in HI-loop | E1A CR2 deletion | GMCSF | 64* | Good, few grade 3 and 4 adverse events, no grade 5 events | 41% disease control overall | Tumor- and virus-specific CD8+ T cells by ELISPOT | Pesonen et al. (2011a) |
| Ad5/3-hTERT-CD40L | Ad5/3 chimera | hTERT promoter | CD40L | 46* | Good, few grade 3 and 4 adverse events, no grade 5 events | 49% disease control overall | Tumor- and virus-specific CD8+ T cells by ELISPOT | Pesonen et al. (2011b), Sari Pesonen et al. (2011) |

*Overall N includes patients not yet published by the authors.

2004; Gao et al., 2005, 2008; Hall et al., 2002; Hwang et al., 2005; Jin et al., 2005a,b; Kanagawa et al., 2008; Liu et al., 2002a,b; Nasu et al., 2004; Park et al., 2008; Raja Gabaglia et al., 2007; Sangro et al., 2004; Satoh et al., 2003; Wen et al., 2001; Zhang & DeGroot, 2003).

One interesting report focused on Syrian hamsters, which is one of the few animal models considered semi-permissive for human adenovirus (Diaconu et al., 2010). An oncolytic adenovirus expressing IL-12 driven by the viral E3 promoter was capable of curing syngeneic pancreatic tumors in conjunction with an antitumor immune response measurable by T cell proliferation (Bortolanza et al., 2009). This work also suggested partial cross-reactivity between mouse and Syrian hamster cytokines since they used murine IL-12. In another study, Gabaglia et al. reported that the treatment of human PC3 prostate xenografts or TRAMP-C1 tumors with the combination Ad5-IL-12 and mifepristone produced significantly better therapeutic efficacy in comparison to controls (Gabaglia et al., 2010). In particular, they found that combination therapy increased the capacity of tumor sentinel lymph node lymphocytes to produce granzyme B in response to tumor cells. Finally, combination therapy groups had fewer CD4+/FoxP3+ T regulatory cells in local nodes.

A clinical trial using an IL-12 expressing adenovirus reported 21 patients (nine with primary liver, five with colorectal, and seven with pancreatic cancers) treated with a total of 44 injections. Ad.IL-12 was well tolerated, and dose-limiting toxicity was not reached, nor were adverse events cumulative. Frequent but transient adverse reactions, including fever, malaise, sweating, and lymphopenia, seemed to be related to vector injection rather than to transgene expression. In four of ten assessable patients, a significant increase in tumor infiltration by effector immune cells was apparent. A partial objective remission of an injected tumor mass was observed in one patient with hepatocellular carcinoma. Stable disease was observed in 29% of patients, mainly those with primary liver cancer (Prieto, Qian, Sangro, Melero, & Mazzolini, 2004).

GMCSF is among the most potent inducers of antitumor immunity (Dranoff, 2002). It acts through several mechanisms, including direct recruitment of NKs and APCs such as DCs (Andrews et al., 2005; Degli-Esposti & Smyth, 2005). GMCSF can also specifically activate DCs at the tumor site to increase their expression of costimulatory molecules to enhance cross-priming and T cell activation rather than cross-tolerance. We showed that Syrian hamsters challenged with a syngeneic pancreatic tumor developed, after treatment with the GMCSF-expressing virus, a specific antitumor response capable of protecting animals from successive challenge by the same tumor but not by different tumors (Cerullo et al., 2010). Since a virus armed with human GMSCF was more immunogenic than an unarmed virus, these results also suggested that human GMCSF is active in Syrian hamsters. Importantly, preclinical results were followed up by treatment of humans, resulting in data suggesting induction of a tumor-specific immune response also in cancer patients, as measured by INF-γ ELISPOT and pentamer staining (Cerullo et al., 2010; Koski et al., 2010; Pesonen et al., 2011a).

IL-23 is a cytokine similar to IL-12, and in fact they share their p40 subunit. However, IL-23 has a preference for CD4+ memory T cells (Oppmann et al., 2000). Recently, IL-23 together with IL-6 and TGF-b1 have been implicated in the mechanism that stimulates naive CD4+ T cells to differentiate into Th17 cells, which are distinct from classical Th1 and Th2 cells (Cua et al., 2003). Th17 cells produce IL-17, a proinflammatory cytokine that enhances T cell priming and stimulates the production of proinflammatory molecules (Boniface, Blom, Liu, & de Waal Malefyt, 2008). Reay and colleagues showed that three intratumoral injections of adenovirus expressing IL-23 significantly increased animal survival and resulted in complete rejection of 40% of tumors, with subsequent generation of protective immunity and tumor-specific cytotoxic T lymphocytes. In addition, they showed that the antitumor activity of IL-23 was independent of IL-17, perforin and Fas-ligand, but dependent on IFN-γ, CD4, and CD8 T cells (Reay, Kim, Lockhart, Kolls, & Robbins, 2009).

#### *4.2.4 Interferon-expressing adenovirus*

IFNs are small proteins made and released by the host cell to counteract the effect of pathogens, in particular viruses. IFNs are roughly divided into two subclasses: type I (*alpha*, α and *beta*, β) (Liu, 2005) and type II (also called late IFNs) such as IFN-γ (Mond & Brunswick, 1987) There is extensive data showing that IFNs are not only important in protection of normal tissues against pathogens, but they have also been shown to have antitumor activity directly on tumor cells and through activation of the immune system (Beatty & Paterson, 2001; Ikeda, Old, & Schreiber, 2002; Lukacher, 2002; Shenoy et al., 2007).

Although generating an adenovirus that expresses IFNs might seem counterintuitive due to the antiviral activity of the latter, rationale is provided by the near-universal deficiency of tumors to IFN signaling (Critchley-Thorne et al., 2009). Thus, arming with IFN can increase the therapeutic window of viruses through opposite activities in normal versus tumor tissues, which has been successfully used in several cancer types (Iqbal Ahmed et al., 2001). Interestingly, Santodonato and colleagues showed also antitumor activity in IFN-resistant tumors in mice when treated with an IFN-α expressing adenovirus (Santodonato et al., 2001). Furthermore, treatment efficacy and induction of antitumor immunity has also been reported with an IFN-α expressing adenovirus in a murine xenograft model of pancreatic cancer (Ohashi et al., 2005). Later, similar results were also obtained in a Syrian hamsters model, where injection of subcutaneous tumors with IFN-α expressing adenovirus resulted in not only regression of injected tumors but also in regression of untreated tumors both in the peritoneal cavity and at distant sites. More recently, it has been proposed that antitumor immunity can depend on the route of administration (Narumi et al., 2010) and that IFN-α resistant tumors can be killed through tumor immunity, oncolysis, and autophagy (Zhang, Dunner, & Benedict, 2010b).

An alternative but equally attractive approach is represented by expression of IFN-β. Already in 2001 Odaka et al. showed that a nonreplicating adenovirus expressing murine IFN-β was able to eradicate intraperitoneal and distant syngeneic mesothelioma tumors. In this study, the treatment effects were shown to be attributable to induction of antitumor immunity, as reactive CD8+ T cells were generated and treatment activity was lost in immunodeficient mice as well as mice specifically depleted of CD8+ T cells (Odaka et al., 2001, 2002). An interesting comparison between human and mouse IFN-β expressing adenovirus in different tumor models was reported by Qin, Beckham, Brown, Lukashev, and Barsoum (2001). Although many actions of IFN are through the adaptive arm of the immune system, an interesting role has also been proposed for macrophages (Zhang, Lu, & Dong, 2002a). A promising proof-of-concept study was performed by Park et al. (2010) who reported that a combination of oncolytic adenovirus Ad5D24RGD and a nonreplicating adenovirus coding for IFN-β resulted in a high local concentration of IFN-β. Importantly, local release of tumor antigens by oncolysis induced a strong antitumor immune response. These preclinical reports have been followed up in several clinical trials in human patients (Chiocca et al., 2008; Sterman et al., 2006, 2007, 2010).

#### *4.2.5 Surface protein expressing adenoviruses*

The basic theories of immunology predict that when an APC such as a DC is presenting an antigen to a T cell, it has the ability to determine between immune response and anergy. Normally, peptides derived from endogenously expressed proteins are presented by APC in the context of MHC class I (MHC I) to CD8+ T cells, whereas peptides obtained from exogenously derived proteins are normally loaded onto MHC class II (MHC II) for presentation to CD4+ T cells. However, exogenous antigens can be also loaded onto MHC I for "cross-presentation" to CD8+ T cells (Trinchieri, Aden, & Knowles, 1976).

In tumor-draining lymph nodes, both cross-priming and cross-tolerization have been reported, tumor antigen-specific T cell proliferation has been detected, but the numbers of T cells proliferating are typically too low, and therefore the overall effect of CD8+ T cell activation does not always result in inhibition of tumor growth (Nowak et al., 2003).

High expression of costimulatory factors that act directly on T cells has been proposed to enhance T cell activation. CD154, better known as CD40L, is one popular example. Normally, it binds to CD40 on APC, which can lead to a variation of effects depending on the target cell. In general, CD40L plays the role of a costimulatory molecule and induces activation in APC in association with T cell receptor stimulation by MHC molecules on the APC. In our laboratory, we have generated an oncolytic adenovirus expressing CD40L. This approach has shown remarkable efficacy in animal models as well as good safety and evidence of activity in human patients (Sari Pesonen et al., 2011).

A similar approach is the expression of CD80, which is also called B7-1. This is a membrane protein especially expressed by B cells, monocyte, and APCs and provides a powerful costimulatory factor for T cell activation and survival. B7-1 is the ligand for CD28 and for CTLA-4. Along this line, it has been shown that an adenovirus expressing IL-7/B7.1 induces rejection of transplanted tumors in mouse model (Willimsky & Blankenstein, 2000). The authors suggest that cancer vaccines can be effective against "minimal residual disease," but additional experimental procedures must be found against established nontransplanted tumors. A similar approach has been also described in the same year by a different group that generated an adenovirus expressing IL-12/B7.1 (Lohr et al., 2000). They showed that the efficacy of this virus was dependent on NK cells as well as T cells, and loss of efficacy was in fact observed in NK- or T cell-depleted animals. Moreover, they showed that the efficacy of this virus was further enhanced by combination with radiotherapy (Lohr et al., 2000).

More recently, a virus expressing a newly discovered member of B7 family, B7-H3, was reported. The mouse protein shares about 88% amino acid identity with the human. Unlike B7-H1 and B7-H2, its mRNA is broadly expressed in lymphoid and nonlymphoid organs (Greenwald, Freeman, & Sharpe, 2005). B7-H3 has been shown to costimulate the proliferation of CD4+ and CD8+ T cells and to stimulate IFN-γ production and cytolytic T cell activity (Sun et al., 2002). A study also demonstrates that adenoviral B7-H3 transfer is able to induce a specific cellular antitumor immune response leading to primary tumor regression and reduction of secondary metastasis *in vivo* (Lupu et al., 2007).

## 4.3. Immune-mediated antitumor activity of oncolytic adenoviruses in preclinical animal models

Adenoviruses have a strong adjuvant effect, due to their interaction with a variety of cellular receptors of the innate immune system such as TLRs (Cerullo et al., 2007; Suzuki et al., 2010), NODs (Suzuki et al., 2009), the inflammasome (Hornung et al., 2009; Muruve et al., 2008), etc. Therefore, it appears that its ability to kill cancer cells is not only due to oncolysis but it is also helped by the immune system. However, these mechanisms are still poorly understood. An important reason for this is the lack of an immunocompetent and replication-competent animal model. The "best available" model is Syrian hamsters, but they are only semi-permissive and the similarity of their immune system with the human counterpart remains unknown. Further, few immunological reagents are available which complicates analyses.

A potentially useful syngeneic (albeit not replication permissive) model for study of the immunogenicity of human adenovirus treatment is the murine melanoma OVA engineered to express the chicken ovalbumin (B16-OVA) (Linardakis et al., 2002). B16-OVA cells were derived from B16 cells by transduction with a cDNA encoding the ovalbumin gene. Importantly, C57BL/6 mice express MHC I molecules, which can present the SIINFEKL epitope derived from processing of the OVA protein. This has allowed scientists to assess specifically the amount of OVA-specific, that is, tumor-specific, reactive T cells. With a similar approach, B16-F10, expressing LCMV GP33-41 (B16.F10-gp), have been generated and currently used for the same purpose. Complete tumor regression has been showed by combination of specific peptide-expressing adenovirus with CD40 stimulation and CTLA4 blockage in syngeneic melanoma B16-F10 model (Sorensen, Holst, Steffensen, Christensen, & Thomsen, 2010).

In 2003, Hallden and colleagues screened nine murine carcinoma lines for adenovirus (Ad5) uptake, gene expression, replication, and cytopathic effects. They found that in seven of these murine cell lines the infectability and cytopathic effects were similar to those seen with human carcinoma lines, confirming that high doses of adenovirus can kill also nonpermissive cells. Surprisingly, evidence or productive viral replication was suggested for several lines; replication varied from levels similar to those for some human carcinoma lines (e.g., CMT-64) to very low levels. Seven of these lines were grown as subcutaneous xenografts in immunocompetent mice and were subsequently injected directly with Ad5, saline, or a replication-deficient control adenovirus with subsequent assessment of intratumoral viral gene expression, replication, and antitumoral effects. E1A, coat protein expression, and cytopathic effects were documented in five xenografts. Some evidence of productive Ad5 replication was demonstrated in CMT-64 and JC xenografts. With regard to efficacy, Ad5 injections were potent in both semi-replication-permissive xenografts (CMT-64, JC) and poorly permissive CMT-93 tumors underlining the immunological capacity of the virus *per se* (Tuve et al., 2009). Noteworthily, efficacy against CMT-93 tumors was significantly greater in immunocompetent mice compared to athymic mice (Hallden et al., 2003).

Edukulla et al. (2009) generated two transgenic murine tumor cell lines expressing a protein for which CD8-restricted tetramer or pentamer are available, CMT-64-OVA and KLN-205-HA. They demonstrated that oncolytic adenovirus increases cross-presentation of tumor antigens by triggering of DC and T cell infiltration resulting in enhanced antitumoral immune responses which facilitates effective viroimmunotherapy of primary tumors and established metastases (Edukulla et al., 2009).

Concurrently, the group of Dr. Lieber showed that even without replication the administration of adenovirus can elicit antitumor immune responses that result in tumor regression. The model used in their work represents syngeneic MMC. They showed that intratumoral injection with replication-deficient adenovirus induces immune responses at two different anatomical sites: the tumor-draining lymph nodes and the tumor microenvironment. Interestingly, Ad-specific T cells were antitumor-reactive despite the presence of active regulatory T cell-mediated immune tolerance inside MMC tumors and antitumor efficacy of Ad was increased by pre-immunization against Ad despite the production of Ad-neutralizing antibodies (Tuve et al., 2009).

These landmark studies identify adenovirus injections *per se* as a possible way to stimulate tumor-specific immunity. Supporting this notion, adenovirus has been used as platform to boost the immune system in a specific way by expressing tumor antigens or in a more un-specific way by expressing cytokines and other immunostimulatory molecules. Logically, also the combination of the two strategies was tested.

An interesting approach has been the combination of adenovirus with an adjuvant already known to have a strong T cell mediated antitumor effect. The first work to evaluate this approach was by VanOosten *et al.* who showed that combination of a TRAIL-expressing adenovirus and TLR9-specific stimulating oligonucleotides enhanced antitumor efficacy by triggering a potent T cell response. The effect was significantly reduced by depleting CD8+ T cells, but on the other hand it was significantly increased after depletion of CD4 or CD25 cell subsets which contain the regulatory T cells (VanOosten & Griffith, 2007). This work draws attention to the importance of suppressive regulatory mechanisms that counteract the ability of adenoviruses as well as other immunotherapeutics to stimulate tumor-specific response.

There is no optimal model for studying the immune-mediated antitumor activity of oncolytic adenovirus given the differences in the immune systems of humans and rodents, and the species specificity of human adenovirus. However, among the best available systems are Syrian hamsters (Dhar, Toth, & Wold, 2012; Thomas et al., 2006) which are semi-permissive to human serotype 5 adenovirus. We reported that Syrian hamsters bearing syngeneic tumors treated with oncolytic adenovirus were able to reject the subsequent challenge of the same tumors, but they succumbed to the challenge of different tumors which demonstrated a degree of specific antitumor response following adenoviral oncolysis (Cerullo et al., 2010). Interestingly, human GMCSF seems to be active in hamsters and complete protection against rechallenge was seen following treatment with a GMCSF coding oncolytic adenovirus (Cerullo et al., 2010). To date, the most adenovirus permissive Syrian hamsters cell lines are pancreatic carcinoma (SHPC6) (Spencer, Sagartz, Wold, & Toth, 2009), PC1 (Thomas et al., 2006) and HaP-T1 (Diaconu et al., 2010), and renal carcinoma SHRC (Spencer et al., 2009).

## 4.4. Immunotherapeutic potential of oncolytic adenoviruses in humans

Only recently, scientists have started to investigate the degree of involvement of the immune system in antitumor response observed following adenovirus treatment in human patients (Cerullo et al., 2010). We treated patients with a GMCSF coding serotype 5 virus (Ad5D24-GMCSF) bearing a 24 bp deletion in E1A gene to restrict replication to tumor cells defective in the p16/Rb pathway. We assessed tumor-specific immune response with two different techniques, by ELISPOT and by flow cytometry. ELISPOT was performed on fresh PBMCs pulsed for 12 h with tumor-specific and adenovirus-specific pools of peptides. Tumor specificity was assessed using survivin as an example of a pancarcinoma antigen commonly expressed by most tumors (Lladser, Sanhueza, Kiessling, & Quest, 2011b). Survivin was chosen since we did not have information which epitopes the patient's tumors expressed. Although survivin is widely expressed in a variety of tumors, it is not a very immunogenic antigen and hence the detection of small antisurvivin response might imply an even stronger immune response toward more immunogenic epitopes. The ELISPOT data were confirmed with tetramer staining in FACS.

Similar immunological data were observed with a serotype chimera 5/3 (Ad5/3D24-GMCSF) (Koski et al., 2010) and with an integrin targeted virus (Pesonen et al., 2011a). In an interesting contrast, when an unarmed oncolytic adenovirus (Ad3hTERT) was used in humans, antiviral responses were equally emphatic but less evidence of antitumor response was seen (Hemminki, Bauerschmitz, et al., 2011). It remains to be studied how important the immunostimulatory transgene is or if the serotype also plays a role.

Interesting results have also been reported by Li et al. (2009). They present the data of a phase I dose-escalating trial with an oncolytic adenovirus expressing the heat shock protein 70 (HSP-70) emphasizing some aspects of the antitumor immune-mediated response. Specifically, they observed elevation of the number of CD4+ and CD8+ T cells as well as NK cells in the blood of the patients after the administration of the virus (Li et al., 2009).

Similar results were also reported in another Phase I trial with an oncolytic adenovirus expressing GMCSF (Chang et al., 2009).

## 5. FINAL REMARKS

It is no longer easy to remain dogmatic on the role of the immune system in the efficacy of oncolytic virotherapy. Whether we like it or not, the immune system exerts multiple effects on the outcome of therapy. The nature and extent of the antiviral immune response to oncolytic virus infection mediate an intricate balance between safety, systemic toxicity, oncolysis, and, potentially, significant immune-mediated antitumor therapy. The challenge for the future is to understand how to accentuate the positive and how to minimize the negative. This formidable problem can be approached only by utilizing models that come as close as possible to the immune environment that will be encountered in the tumors of patients. It will require cross-interaction between the disciplines of virology and immunology with rapid human translation to understand the relevance of animal data in the context of patients. We will need to appreciate how pleiotropic agents, which either negatively or positively impact therapy in preclinical models, may be having effects on the host immune system that we have not fully appreciated. We suggest that virotherapy may act, at least in some circumstances, as much as an immunotherapy as pure oncolytic virotherapy. To date, the field has concentrated on developing viruses optimized for selective replication in tumors. In retrospect, given the tremendous complexity of the intratumoral environment, including stromal barriers, hypoxia, necrosis, etc., expecting tumor eradication by oncolysis alone, may be asking a great deal, even if antiviral responses are attenuated by the immune suppressive tumor environment. By viewing at least certain components of the immune system as partners, rather than the enemy, it should now be possible to explore additional avenues of oncolytic virus design in which immune activation becomes as much a part of the solution as it has previously been viewed as the problem.

## ACKNOWLEDGMENTS

This study was supported by the European Research Council, Marie Curie FP7-IRG-PEOPLE-2008, ASCO Foundation, HUCH Research Funds (EVO), Sigrid Juselius Foundation, Academy of Finland, Biocentrum Helsinki, Biocenter Finland, Cancer Organizations, University of Helsinki, Helsinki Biomedical Graduate School. A. H. is K. Albin Johansson Research Professor of the Foundation for the Finnish Cancer Institute.

*Conflict of Interest*: A. H. is shareholder in Oncos Therapeutics, Ltd.

## REFERENCES

Akbulut, H., Zhang, L., Tang, Y., & Deisseroth, A. (2003). Cytotoxic effect of replication-competent adenoviral vectors carrying L-plastin promoter regulated E1A and cytosine deaminase genes in cancers of the breast, ovary and colon. *Cancer Gene Therapy*, *10*, 388–395.

Akira, S., Uematsu, S., & Takeuchi, O. (2006). Pathogen recognition and innate immunity. *Cell*, *124*, 783–801.

Alba, R., Bosch, A., & Chillon, M. (2005). Gutless adenovirus: Last-generation adenovirus for gene therapy. *Gene Therapy*, *12*(Suppl. 1), S18–S27.

Alemany, R. (2007). Cancer selective adenoviruses. *Molecular Aspects of Medicine*, *28*, 42–58.

Alemany, R., & Cascallo, M. (2009). Oncolytic viruses from the perspective of the immune system. *Future Microbiology*, *4*, 527–536.

Alemany, R., & Curiel, D. T. (2001). CAR-binding ablation does not change biodistribution and toxicity of adenoviral vectors. *Gene Therapy*, *8*, 1347–1353.

Andoniou, C. E., van Dommelen, S. L., Voigt, V., Andrews, D. M., Brizard, G., Asselin-Paturel, C., et al. (2005). Interaction between conventional dendritic cells and

natural killer cells is integral to the activation of effective antiviral immunity. *Nature Immunology, 6*, 1011–1019.
Andrews, D. M., Andoniou, C. E., Scalzo, A. A., van Dommelen, S. L., Wallace, M. E., Smyth, M. J., et al. (2005). Cross-talk between dendritic cells and natural killer cells in viral infection. *Molecular Immunology, 42*, 547–555.
Appledorn, D. M., Kiang, A., McBride, A., Jiang, H., Seregin, S., Scott, J. M., et al. (2008). Wild-type adenoviruses from groups A-F evoke unique innate immune responses, of which HAd3 and SAd23 are partially complement dependent. *Gene Therapy, 15*, 885–901.
Atkins, M. B., Lotze, M. T., Dutcher, J. P., Fisher, R. I., Weiss, G., Margolin, K., et al. (1999). High-dose recombinant interleukin 2 therapy for patients with metastatic melanoma: analysis of 270 patients treated between 1985 and 1993. *Journal of Clinical Oncology, 17*, 2105–2116.
Bachmann, M. F., & Jennings, G. T. (2010). Vaccine delivery: A matter of size, geometry, kinetics and molecular patterns. *Nature Reviews. Immunology, 10*, 787–796.
Bauerschmitz, G. J., Guse, K., Kanerva, A., Menzel, A., Herrmann, I., Desmond, R. A., et al. (2006). Triple-targeted oncolytic adenoviruses featuring the cox2 promoter, E1A transcomplementation, and serotype chimerism for enhanced selectivity for ovarian cancer cells. *Molecular Therapy, 14*, 164–174.
Bauerschmitz, G. J., Nettelbeck, D. M., Kanerva, A., Baker, A. H., Hemminki, A., Reynolds, P. N., et al. (2002). The flt-1 promoter for transcriptional targeting of teratocarcinoma. *Cancer Research, 62*, 1271–1274.
Bayo-Puxan, N., Cascallo, M., Gros, A., Huch, M., Fillat, C., & Alemany, R. (2006). Role of the putative heparan sulfate glycosaminoglycan-binding site of the adenovirus type 5 fiber shaft on liver detargeting and knob-mediated retargeting. *The Journal of General Virology, 87*, 2487–2495.
Beatty, G. L., & Paterson, Y. (2001). Regulation of tumor growth by IFN-gamma in cancer immunotherapy. *Immunologic Research, 24*, 201–210.
Berk A.J. (2007). Adenoviridae: the viruses and their replication. In: B. N. Fields, D. M. Knipe, P. M. Howley (Eds.), *Fields Virology*. fifth ed. (pp. 2355–2394). Wolters Kluwer Health/Lippincott Williams & Wilkins: Philadelphia.
Berruti, A., Pia, A., & Terzolo, M. (2011). Advances in pancreatic neuroendocrine tumor treatment. *The New England Journal of Medicine, 364*, 1871–1872 author reply 1873–1874.
Bischoff, J. R., Kirn, D. H., Williams, A., Heise, C., Horn, S., Muna, M., et al. (1996). An adenovirus mutant that replicates selectively in p53-deficient human tumor cells. *Science, 274*, 373–376.
Boniface, K., Blom, B., Liu, Y. J., & de Waal Malefyt, R. (2008). From interleukin-23 to T-helper 17 cells: Human T-helper cell differentiation revisited. *Immunological Reviews, 226*, 132–146.
Bortolanza, S., Bunuales, M., Otano, I., Gonzalez-Aseguinolaza, G., Ortiz-de-Solorzano, C., Perez, D., et al. (2009). Treatment of pancreatic cancer with an oncolytic adenovirus expressing interleukin-12 in Syrian hamsters. *Molecular Therapy, 17*, 614–622.
Bottini, A., Generali, D., Brizzi, M. P., Fox, S. B., Bersiga, A., Bonardi, S., et al. (2006). Randomized phase II trial of letrozole and letrozole plus low-dose metronomic oral cyclophosphamide as primary systemic treatment in elderly breast cancer patients. *Journal of Clinical Oncology, 24*, 3623–3628.
Bowen, G. P., Borgland, S. L., Lam, M., Libermann, T. A., Wong, N. C., & Muruve, D. A. (2002). Adenovirus vector-induced inflammation: Capsid-dependent induction of the C-C chemokine RANTES requires NF-kappa B. *Human Gene Therapy, 13*, 367–379.
Breitbach, C. J., Burke, J., Jonker, D., Stephenson, J., Haas, A. R., Chow, L. Q., et al. (2011). Intravenous delivery of a multi-mechanistic cancer-targeted oncolytic poxvirus in humans. *Nature, 477*, 99–102.

Brunetti-Pierri, N., & Ng, P. (2009). Progress towards liver and lung-directed gene therapy with helper-dependent adenoviral vectors. *Current Gene Therapy*, *9*, 329–340.
Brunetti-Pierri, N., & Ng, P. (2011). Helper-dependent adenoviral vectors for liver-directed gene therapy. *Human Molecular Genetics*, *20*, R7–R13.
Brunetti-Pierri, N., Ng, T., Iannitti, D. A., Palmer, D. J., Beaudet, A. L., Finegold, M. J., et al. (2006). Improved hepatic transduction, reduced systemic vector dissemination, and long-term transgene expression by delivering helper-dependent adenoviral vectors into the surgically isolated liver of nonhuman primates. *Human Gene Therapy*, *17*, 391–404.
Brunetti-Pierri, N., Palmer, D. J., Beaudet, A. L., Carey, K. D., Finegold, M., & Ng, P. (2004). Acute toxicity after high-dose systemic injection of helper-dependent adenoviral vectors into nonhuman primates. *Human Gene Therapy*, *15*, 35–46.
Campos, S. K., & Barry, M. A. (2006). Comparison of adenovirus fiber, protein IX, and hexon capsomeres as scaffolds for vector purification and cell targeting. *Virology*, *349*, 453–462.
Campos, S. K., & Barry, M. A. (2007). Current advances and future challenges in adenoviral vector biology and targeting. *Current Gene Therapy*, *7*, 189–204.
Carlisle, R. C., Di, Y., Cerny, A. M., Sonnen, A. F., Sim, R. B., Green, N. K., et al. (2009). Human erythrocytes bind and inactivate type 5 adenovirus by presenting Coxsackie virus-adenovirus receptor and complement receptor 1. *Blood*, *113*, 1909–1918.
Cerullo, V., Diaconu, I., Kangasniemi, L., Rajecki, M., Escutenaire, S., Koski, A., et al. (2011). Immunological effects of low-dose cyclophosphamide in cancer patients treated with oncolytic adenovirus. *Molecular Therapy*, *19*, 1737–1746.
Cerullo, V., Pesonen, S., Diaconu, I., Escutenaire, S., Arstila, P. T., Ugolini, M., et al. (2010). Oncolytic adenovirus coding for granulocyte macrophage colony-stimulating factor induces antitumoral immunity in cancer patients. *Cancer Research*, *70*, 4297–4309.
Cerullo, V., Seiler, M. P., Mane, V., Brunetti-Pierri, N., Clarke, C., Bertin, T. K., et al. (2007). Toll-like receptor 9 triggers an innate immune response to helper-dependent adenoviral vectors. *Molecular Therapy*, *15*, 378–385.
Chang, C. J., Chen, Y. H., Huang, K. W., Cheng, H. W., Chan, S. F., Tai, K. F., et al. (2007). Combined GM-CSF and IL-12 gene therapy synergistically suppresses the growth of orthotopic liver tumors. *Hepatology*, *45*, 746–754.
Chang, J., Zhao, X., Wu, X., Guo, Y., Guo, H., Cao, J., et al. (2009). A Phase I study of KH901, a conditionally replicating granulocyte-macrophage colony-stimulating factor: Armed oncolytic adenovirus for the treatment of head and neck cancers. *Cancer Biology & Therapy*, *8*, 676–682.
Chen, X., Lin, X., Zhao, J., Shi, W., Zhang, H., Wang, Y., et al. (2008). A tumor-selective biotherapy with prolonged impact on established metastases based on cytokine gene-engineered MSCs. *Molecular Therapy*, *16*, 749–756.
Chen, J., Wang, J., Li, J., Wu, Q., Chu Lim, F., Yang, P., et al. (2008). Enhancement of cytotoxic T-lymphocyte response in aged mice by a novel treatment with recombinant AdIL-12 and wild-type adenovirus in rapid succession. *Molecular Therapy*, *16*, 1500–1506.
Chiocca, E. A., Smith, K. M., McKinney, B., Palmer, C. A., Rosenfeld, S., Lillehei, K., et al. (2008). A phase I trial of Ad.hIFN-beta gene therapy for glioma. *Molecular Therapy*, *16*, 618–626.
Cody, J. J., & Douglas, J. T. (2009). Armed replicating adenoviruses for cancer virotherapy. *Cancer Gene Therapy*, *16*, 473–488.
Coley, W. B. (1891). II. Contribution to the knowledge of sarcoma. *Annals of Surgery*, *14*, 199–220.
Correale, P., Botta, C., Cusi, M., Del Vecchio, M., De Santi, M., Gori Savellini, G., et al. (2012). Cetuximab +/- chemotherapy enhances dendritic cell-mediated phagocytosis of colon cancer cells and ignites a highly efficient colon cancer antigen-specific cytotoxic T-cell response in vitro. *International Journal of Cancer*, *130*, 1577–1589.

Coyne, C. B., & Bergelson, J. M. (2005). CAR: a virus receptor within the tight junction. *Advanced Drug Delivery Reviews*, *57*, 869–882.

Critchley-Thorne, R. J., Simons, D. L., Yan, N., Miyahira, A. K., Dirbas, F. M., Johnson, D. L., et al. (2009). Impaired interferon signaling is a common immune defect in human cancer. *Proceedings of the National Academy of Sciences of the United States of America*, *106*, 9010–9015.

Croyle, M. A., Le, H. T., Linse, K. D., Cerullo, V., Toietta, G., Beaudet, A. L., et al. (2005). PEGylated helper-dependent adenoviral vectors: Highly efficient vectors with an enhanced safety profile. *Gene Therapy*, *12*, 579–587.

Cua, D. J., Sherlock, J., Chen, Y., Murphy, C. A., Joyce, B., Seymour, B., et al. (2003). Interleukin-23 rather than interleukin-12 is the critical cytokine for autoimmune inflammation of the brain. *Nature*, *421*, 744–748.

Danthinne, X., & Imperiale, M. J. (2000). Production of first generation adenovirus vectors: A review. *Gene Therapy*, *7*, 1707–1714.

Degli-Esposti, M. A., & Smyth, M. J. (2005). Close encounters of different kinds: Dendritic cells and NK cells take centre stage. *Nature Reviews. Immunology*, *5*, 112–124.

Dhar, D., Toth, K., & Wold, W. S. (2012). Syrian hamster tumor model to study oncolytic Ad5-based vectors. *Methods in Molecular Biology*, *797*, 53–63.

Diaconu, I., Cerullo, V., Escutenaire, S., Kanerva, A., Bauerschmitz, G. J., Hernandez-Alcoceba, R., et al. (2010). Human adenovirus replication in immunocompetent Syrian hamsters can be attenuated with chlorpromazine or cidofovir. *The Journal of Gene Medicine*, *12*, 435–445.

Dias, J. D., Hemminki, O., Diaconu, I., Hirvinen, M., Bonetti, A., Guse, K., et al. (2011). Targeted cancer immunotherapy with oncolytic adenovirus coding for a fully human monoclonal antibody specific for CTLA-4. *Gene Therapy*, Nov 10, 1–11. doi: 10.1038/gt.2011.176.

Diaz, R. M., Galivo, F., Kottke, T., Wongthida, P., Qiao, J., Thompson, J., et al. (2007). Oncolytic immunovirotherapy for melanoma using vesicular stomatitis virus. *Cancer Research*, *67*, 2840–2848.

DiPaolo, N., Ni, S., Gaggar, A., Strauss, R., Tuve, S., Li, Z. Y., et al. (2006). Evaluation of adenovirus vectors containing serotype 35 fibers for vaccination. *Molecular Therapy*, *13*, 756–765.

Dmitriev, I. P., Kashentseva, E. A., & Curiel, D. T. (2002). Engineering of adenovirus vectors containing heterologous peptide sequences in the C terminus of capsid protein IX. *Journal of Virology*, *76*, 6893–6899.

Dmitriev, I., Krasnykh, V., Miller, C. R., Wang, M., Kashentseva, E., Mikheeva, G., et al. (1998). An adenovirus vector with genetically modified fibers demonstrates expanded tropism via utilization of a coxsackievirus and adenovirus receptor-independent cell entry mechanism. *Journal of Virology*, *72*, 9706–9713.

Dranoff, G. (2002). GM-CSF-based cancer vaccines. *Immunological Reviews*, *188*, 147–154.

Edelman, J., & Nemunaitis, J. (2003). Adenoviral p53 gene therapy in squamous cell cancer of the head and neck region. *Current Opinion in Molecular Therapeutics*, *5*, 611–617.

Edukulla, R., Woller, N., Mundt, B., Knocke, S., Gurlevik, E., Saborowski, M., et al. (2009). Antitumoral immune response by recruitment and expansion of dendritic cells in tumors infected with telomerase-dependent oncolytic viruses. *Cancer Research*, *69*, 1448–1458.

Eriksson, M., Guse, K., Bauerschmitz, G., Virkkunen, P., Tarkkanen, M., Tanner, M., et al. (2007). Oncolytic adenoviruses kill breast cancer initiating CD44+CD24-/low cells. *Molecular Therapy*, *15*, 2088–2093.

Franchi, L., Park, J. H., Shaw, M. H., Marina-Garcia, N., Chen, G., Kim, Y. G., et al. (2008). Intracellular NOD-like receptors in innate immunity, infection and disease. *Cellular Microbiology*, *10*, 1–8.

Freytag, S. O., Rogulski, K. R., Paielli, D. L., Gilbert, J. D., & Kim, J. H. (1998). A novel three-pronged approach to kill cancer cells selectively: concomitant viral, double suicide gene, and radiotherapy. *Human Gene Therapy, 9*, 1323–1333.

Fritz, J. H., Ferrero, R. L., Philpott, D. J., & Girardin, S. E. (2006). Nod-like proteins in immunity, inflammation and disease. *Nature Immunology*, 7, 1250–1257.

Fueyo, J., Gomez-Manzano, C., Alemany, R., Lee, P. S., McDonnell, T. J., Mitlianga, P., et al. (2000). A mutant oncolytic adenovirus targeting the Rb pathway produces anti-glioma effect in vivo. *Oncogene, 19*, 2–12.

Gabaglia, C. R., DeLaney, A., Gee, J., Halder, R., Graham, F. L., Gauldie, J., et al. (2010). Treatment combining RU486 and Ad5IL-12 vector attenuates the growth of experimentally formed prostate tumors and induces changes in the sentinel lymph nodes of mice. *Journal of Translational Medicine, 8*, 98.

Gabaglia, C. R., Sercarz, E. E., Diaz-De-Durana, Y., Hitt, M., Graham, F. L., Gauldie, J., et al. (2004). Life-long systemic protection in mice vaccinated with L. major and adenovirus IL-12 vector requires active infection, macrophages and intact lymph nodes. *Vaccine, 23*, 247–257.

Galanis, E., Okuno, S., Nascimento, A., Lewis, B., Lee, R., Oliveira, A., et al. (2005). Phase I-II trial of ONYX-015 in combination with MAP chemotherapy in patients with advanced sarcomas. *Gene Therapy, 12*, 437–445.

Ganesan, L. P., Mohanty, S., Kim, J., Clark, K. R., Robinson, J. M., & Anderson, C. L. (2011). Rapid and efficient clearance of blood-borne virus by liver sinusoidal endothelium. *PLoS Pathogens*, 7, e1002281.

Gao, J. Q., Kanagawa, N., Xu, D. H., Han, M., Sugita, T., Hatanaka, Y., et al. (2008). Combination of two fiber-mutant adenovirus vectors, one encoding the chemokine FKN and another encoding cytokine interleukin 12, elicits notably enhanced anti-tumor responses. *Cancer Immunology, Immunotherapy, 57*, 1657–1664.

Gao, J. Q., Sugita, T., Kanagawa, N., Iida, K., Eto, Y., Motomura, Y., et al. (2005). A single intratumoral injection of a fiber-mutant adenoviral vector encoding interleukin 12 induces remarkable anti-tumor and anti-metastatic activity in mice with Meth-A fibrosarcoma. *Biochemical and Biophysical Research Communications, 328*, 1043–1050.

Garber, K. (2006). China approves world's first oncolytic virus therapy for cancer treatment. *Journal of the National Cancer Institute, 98*, 298–300.

Glasgow, J. N., Kremer, E. J., Hemminki, A., Siegal, G. P., Douglas, J. T., & Curiel, D. T. (2004). An adenovirus vector with a chimeric fiber derived from canine adenovirus type 2 displays novel tropism. *Virology, 324*, 103–116.

Gordon, Y. J., Romanowski, E., & Araullo-Cruz, T. (1992). An ocular model of adenovirus type 5 infection in the NZ rabbit. *Investigative Ophthalmology & Visual Science, 33*, 574–580.

Greenwald, R. J., Freeman, G. J., & Sharpe, A. H. (2005). The B7 family revisited. *Annual Review of Immunology, 23*, 515–548.

Guse, K., Diaconu, I., Rajecki, M., Sloniecka, M., Hakkarainen, T., Ristimaki, A., et al. (2009). Ad5/3-9HIF-Delta24-VEGFR-1-Ig, an infectivity enhanced, dual-targeted and antiangiogenic oncolytic adenovirus for kidney cancer treatment. *Gene Therapy, 16*, 1009–1020.

Guse, K., Ranki, T., Ala-Opas, M., Bono, P., Sarkioja, M., Rajecki, M., et al. (2007). Treatment of metastatic renal cancer with capsid-modified oncolytic adenoviruses. *Molecular Cancer Therapeutics, 6*, 2728–2736.

Hakkarainen, T., Rajecki, M., Sarparanta, M., Tenhunen, M., Airaksinen, A. J., Desmond, R. A., et al. (2009). Targeted radiotherapy for prostate cancer with an oncolytic adenovirus coding for human sodium iodide symporter. *Clinical Cancer Research, 15*, 5396–5403.

Hall, S. J., Canfield, S. E., Yan, Y., Hassen, W., Selleck, W. A., & Chen, S. H. (2002). A novel bystander effect involving tumor cell-derived Fas and FasL interactions following Ad.HSV-tk and Ad.mIL-12 gene therapies in experimental prostate cancer. *Gene Therapy*, *9*, 511–517.
Hallden, G., Hill, R., Wang, Y., Anand, A., Liu, T. C., Lemoine, N. R., et al. (2003). Novel immunocompetent murine tumor models for the assessment of replication-competent oncolytic adenovirus efficacy. *Molecular Therapy*, *8*, 412–424.
Han, Z., Hong, Z., Gao, Q., Chen, C., Hao, Z., Ji, T., et al. (2011). A potent oncolytic adenovirus selectively blocks the STAT3 signaling pathway and potentiates cisplatin antitumor activity in ovarian cancer. *Human Gene Therapy*, *23*, 32–45.
Hartman, Z. C., Osada, T., Glass, O., Yang, X. Y., Lei, G. J., Lyerly, H. K., et al. (2010). Ligand-independent toll-like receptor signals generated by ectopic overexpression of MyD88 generate local and systemic antitumor immunity. *Cancer Research*, *70*, 7209–7220.
Hegenbarth, S., Gerolami, R., Protzer, U., Tran, P. L., Brechot, C., Gerken, G., et al. (2000). Liver sinusoidal endothelial cells are not permissive for adenovirus type 5. *Human Gene Therapy*, *11*, 481–486.
Heise, C., Hermiston, T., Johnson, L., Brooks, G., Sampson-Johannes, A., Williams, A., et al. (2000). An adenovirus E1A mutant that demonstrates potent and selective systemic anti-tumoral efficacy. *Nature Medicine*, *6*, 1134–1139.
Hemminki, O., Bauerschmitz, G., Hemmi, S., Lavilla-Alonso, S., Diaconu, I., Guse, K., et al. (2011). Oncolytic adenovirus based on serotype 3. *Cancer Gene Therapy*, *18*, 288–296.
Hemminki, O., Diaconu, I., Hemmi, S., Kanerva, A., Cerullo, V., Escutenaire, S., et al. (2011). Human data with a desmoglein 2 binding oncolytic adenovirus Ad3-hTERT-E1A. *Human Gene Therapy*, *22*, A28–A29.
Hierholzer, J. C. (1992). Adenoviruses in the immunocompromised host. *Clinical Microbiology Reviews*, *5*, 262–274.
Higginbotham, J. N., Seth, P., Blaese, R. M., & Ramsey, W. J. (2002). The release of inflammatory cytokines from human peripheral blood mononuclear cells in vitro following exposure to adenovirus variants and capsid. *Human Gene Therapy*, *13*, 129–141.
Hjorth, R. N., Bonde, G. M., Pierzchala, W. A., Vernon, S. K., Wiener, F. P., Levner, M. H., et al. (1988). A new hamster model for adenoviral vaccination. *Archives of Virology*, *100*, 279–283.
Hodi, F. S., O'Day, S. J., McDermott, D. F., Weber, R. W., Sosman, J. A., Haanen, J. B., et al. (2010). Improved survival with ipilimumab in patients with metastatic melanoma. *The New England Journal of Medicine*, *363*, 711–723.
Hoffmann, J. A. (2003). The immune response of Drosophila. *Nature*, *426*, 33–38.
Hornung, V., Ablasser, A., Charrel-Dennis, M., Bauernfeind, F., Horvath, G., Caffrey, D. R., et al. (2009). AIM2 recognizes cytosolic dsDNA and forms a caspase-1-activating inflammasome with ASC. *Nature*, *458*, 514–518.
Huber, C. H., & Wolfel, T. (2004). Immunotherapy of cancer: From vision to standard clinical practice. *Journal of Cancer Research and Clinical Oncology*, *130*, 367–374.
Hwang, K. S., Cho, W. K., Yoo, J., Yun, H. J., Kim, S., & Im, D. S. (2005). Adenovirus-mediated interleukin-12 gene transfer combined with cytosine deaminase followed by 5-fluorocytosine treatment exerts potent antitumor activity in Renca tumor-bearing mice. *BMC Cancer*, *5*, 51.
Ikeda, H., Old, L. J., & Schreiber, R. D. (2002). The roles of IFN gamma in protection against tumor development and cancer immunoediting. *Cytokine & Growth Factor Reviews*, *13*, 95–109.
Imperiale, M. J., & Enquist, L. W. (2011). What's in a Name? *Journal of Virology*, *85*, 5245.

Inohara, C., McDonald, C., & Nunez, G. (2005). NOD-LRR proteins: Role in host-microbial interactions and inflammatory disease. *Annual Review of Biochemistry*, *74*, 355–383.

Iqbal Ahmed, C. M., Johnson, D. E., Demers, G. W., Engler, H., Howe, J. A., Wills, K. N., et al. (2001). Interferon alpha2b gene delivery using adenoviral vector causes inhibition of tumor growth in xenograft models from a variety of cancers. *Cancer Gene Therapy*, *8*, 788–795.

Jena, B., Dotti, G., & Cooper, L. J. (2010). Redirecting T-cell specificity by introducing a tumor-specific chimeric antigen receptor. *Blood*, *116*, 1035–1044.

Jiang, H., Alonso, M. M., Gomez-Manzano, C., Piao, Y., & Fueyo, J. (2006). Oncolytic viruses and DNA-repair machinery: Overcoming chemoresistance of gliomas. *Expert Review of Anticancer Therapy*, *6*, 1585–1592.

Jiang, H., Wang, Z., Serra, D., Frank, M. M., & Amalfitano, A. (2004). Recombinant adenovirus vectors activate the alternative complement pathway, leading to the binding of human complement protein C3 independent of anti-ad antibodies. *Molecular Therapy*, *10*, 1140–1142.

Jiang, H., White, E. J., Rios-Vicil, C. I., Xu, J., Gomez-Manzano, C., & Fueyo, J. (2011). Human adenovirus type 5 induces cell lysis through autophagy and autophagy-triggered caspase activity. *Journal of Virology*, *85*, 4720–4729.

Jin, H. S., Park, E. K., Lee, J. M., NamKoong, S. E., Kim, D. G., Lee, Y. J., et al. (2005). Immunization with adenoviral vectors carrying recombinant IL-12 and E7 enhanced the antitumor immunity to human papillomavirus 16-associated tumor. *Gynecologic Oncology*, *97*, 559–567.

Jin, H. T., Youn, J. I., Kim, H. J., Lee, J. B., Ha, S. J., Koh, J. S., et al. (2005). Enhancement of interleukin-12 gene-based tumor immunotherapy by the reduced secretion of p40 subunit and the combination with farnesyltransferase inhibitor. *Human Gene Therapy*, *16*, 328–338.

Kalu, S. U., Loeffelholz, M., Beck, E., Patel, J. A., Revai, K., Fan, J., et al. (2010). Persistence of adenovirus nucleic acids in nasopharyngeal secretions: A diagnostic conundrum. *The Pediatric Infectious Disease Journal*, *29*, 746–750.

Kanagawa, N., Gao, J. Q., Motomura, Y., Yanagawa, T., Mukai, Y., Yoshioka, Y., et al. (2008). Antitumor mechanism of intratumoral injection with IL-12-expressing adenoviral vector against IL-12-unresponsive tumor. *Biochemical and Biophysical Research Communications*, *372*, 821–825.

Kanerva, A., Bauerschmitz, G. J., Yamamoto, M., Lam, J. T., Alvarez, R. D., Siegal, G. P., et al. (2004). A cyclooxygenase-2 promoter-based conditionally replicating adenovirus with enhanced infectivity for treatment of ovarian adenocarcinoma. *Gene Therapy*, *11*, 552–559.

Kanerva, A., & Hemminki, A. (2004). Modified adenoviruses for cancer gene therapy. *International Journal of Cancer*, *110*, 475–480.

Kanerva, A., Mikheeva, G. V., Krasnykh, V., Coolidge, C. J., Lam, J. T., Mahasreshti, P. J., et al. (2002). Targeting adenovirus to the serotype 3 receptor increases gene transfer efficiency to ovarian cancer cells. *Clinical Cancer Research*, *8*, 275–280.

Kanerva, A., Zinn, K. R., Chaudhuri, T. R., Lam, J. T., Suzuki, K., Uil, T. G., et al. (2003). Enhanced therapeutic efficacy for ovarian cancer with a serotype 3 receptor-targeted oncolytic adenovirus. *Molecular Therapy*, *8*, 449–458.

Kangasniemi, L., Kiviluoto, T., Kanerva, A., Raki, M., Ranki, T., Sarkioja, M., et al. (2006). Infectivity-enhanced adenoviruses deliver efficacy in clinical samples and orthotopic models of disseminated gastric cancer. *Clinical Cancer Research*, *12*, 3137–3144.

Kanneganti, T. D., Lamkanfi, M., & Nunez, G. (2007). Intracellular NOD-like receptors in host defense and disease. *Immunity*, *27*, 549–559.

Kaur, B., Cripe, T. P., & Chiocca, E. A. (2009). "Buy one get one free": armed viruses for the treatment of cancer cells and their microenvironment. *Current Gene Therapy, 9*, 341–355.

Kawai, T., & Akira, S. (2006). TLR signaling. *Cell Death and Differentiation, 13*, 816–825.

Khare, R., Chen, C. Y., Weaver, E. A., & Barry, M. A. (2011). Advances and future challenges in adenoviral vector pharmacology and targeting. *Current Gene Therapy, 11*, 241–258.

Khare, R., May, S. M., Vetrini, F., Weaver, E. A., Palmer, D., Rosewell, A., et al. (2011). Generation of a Kupffer cell-evading adenovirus for systemic and liver-directed gene transfer. *Molecular Therapy: The Journal of the American Society of Gene Therapy, 19*, 1254–1262.

Kim, J., Lee, B., Kim, J. S., Yun, C. O., Kim, J. H., Lee, Y. J., et al. (2002). Antitumoral effects of recombinant adenovirus YKL-1001, conditionally replicating in alpha-fetoprotein-producing human liver cancer cells. *Cancer Letters, 180*, 23–32.

Kirkwood, J. M., Ernstoff, M. S., Davis, C. A., Reiss, M., Ferraresi, R., & Rudnick, S. A. (1985). Comparison of intramuscular and intravenous recombinant alpha-2 interferon in melanoma and other cancers. *Annals of Internal Medicine, 103*, 32–36.

Kirn, D. (2001). Clinical research results with dl1520 (Onyx-015), a replication-selective adenovirus for the treatment of cancer: What have we learned? *Gene Therapy, 8*, 89–98.

Koski, A., Kangasniemi, L., Escutenaire, S., Pesonen, S., Cerullo, V., Diaconu, I., et al. (2010). Treatment of cancer patients with a serotype 5/3 chimeric oncolytic adenovirus expressing GMCSF. *Molecular Therapy, 18*, 1874–1884.

Krasnykh, V. N., Mikheeva, G. V., Douglas, J. T., & Curiel, D. T. (1996). Generation of recombinant adenovirus vectors with modified fibers for altering viral tropism. *Journal of Virology, 70*, 6839–6846.

Kretschmer, P. J., Jin, F., Chartier, C., & Hermiston, T. W. (2005). Development of a transposon-based approach for identifying novel transgene insertion sites within the replicating adenovirus. *Molecular Therapy, 12*, 118–127.

Kurachi, S., Koizumi, N., Sakurai, F., Kawabata, K., Sakurai, H., Nakagawa, S., et al. (2007). Characterization of capsid-modified adenovirus vectors containing heterologous peptides in the fiber knob, protein IX, or hexon. *Gene Therapy, 14*, 266–274.

Law, L. K., & Davidson, B. L. (2005). What does it take to bind CAR? *Molecular Therapy, 12*, 599–609.

Lee, W. P., Wen, Y., Varnum, B., & Hung, M. C. (2002). Akt is required for Axl-Gas6 signaling to protect cells from E1A-mediated apoptosis. *Oncogene, 21*, 329–336.

Lei, N., Shen, F. B., Chang, J. H., Wang, L., Li, H., Yang, C., et al. (2009). An oncolytic adenovirus expressing granulocyte macrophage colony-stimulating factor shows improved specificity and efficacy for treating human solid tumors. *Cancer Gene Therapy, 16*, 33–43.

Leppard, K. N. (1997). E4 gene function in adenovirus, adenovirus vector and adeno-associated virus infections. *The Journal of General Virology, 78*(Pt 9), 2131–2138.

Li, Y., Chen, Y., Dilley, J., Arroyo, T., Ko, D., Working, P., et al. (2003). Carcinoembryonic antigen-producing cell-specific oncolytic adenovirus, OV798, for colorectal cancer therapy. *Molecular Cancer Therapeutics, 2*, 1003–1009.

Li, J.-L., Liu, H.-L., Zhang, X.-R., Xu, J.-P., Hu, W.-K., Liang, M., et al. (2009). A phase I trial of intratumoral administration of recombinant oncolytic adenovirus overexpressing HSP70 in advanced solid tumor patients. *Gene Therapy, 16*, 376–382.

Lichtenstein, D. L., Toth, K., Doronin, K., Tollefson, A. E., & Wold, W. S. (2004). Functions and mechanisms of action of the adenovirus E3 proteins. *International Reviews of Immunology, 23*, 75–111.

Lieber, A., He, C. Y., Meuse, L., Schowalter, D., Kirillova, I., Winther, B., et al. (1997). The role of Kupffer cell activation and viral gene expression in early liver toxicity after infusion of recombinant adenovirus vectors. *Journal of Virology, 71*, 8798–8807.

Linardakis, E., Bateman, A., Phan, V., Ahmed, A., Gough, M., Olivier, K., et al. (2002). Enhancing the efficacy of a weak allogeneic melanoma vaccine by viral fusogenic membrane glycoprotein-mediated tumor cell-tumor cell fusion. *Cancer Research*, *62*, 5495–5504.

Liu, Y. J. (2005). IPC: professional type 1 interferon-producing cells and plasmacytoid dendritic cell precursors. *Annual Review of Immunology*, *23*, 275–306.

Liu, Y., Ehtesham, M., Samoto, K., Wheeler, C. J., Thompson, R. C., Villarreal, L. P., et al. (2002). In situ adenoviral interleukin 12 gene transfer confers potent and long-lasting cytotoxic immunity in glioma. *Cancer Gene Therapy*, *9*, 9–15.

Liu, Y., Tuve, S., Persson, J., Beyer, I., Yumul, R., Li, Z. Y., et al. (2011). Adenovirus-mediated intratumoral expression of immunostimulatory proteins in combination with systemic Treg inactivation induces tumor-destructive immune responses in mouse models. *Cancer Gene Therapy*, *18*, 407–418.

Liu, Q., Zaiss, A. K., Colarusso, P., Patel, K., Haljan, G., Wickham, T. J., et al. (2003). The role of capsid-endothelial interactions in the innate immune response to adenovirus vectors. *Human Gene Therapy*, *14*, 627–643.

Liu, Y., Zhang, X., Zhang, W., Chen, Z., Chan, T., Ali, K., et al. (2002). Adenovirus-mediated CD40 ligand gene-engineered dendritic cells elicit enhanced CD8(+) cytotoxic T-cell activation and antitumor immunity. *Cancer Gene Therapy*, *9*, 202–208.

Lladser, A., Mougiakakos, D., Tufvesson, H., Ligtenberg, M. A., Quest, A. F., Kiessling, R., et al. (2011). DAI (DLM-1/ZBP1) as a genetic adjuvant for DNA vaccines that promotes effective antitumor CTL immunity. *Molecular Therapy*, *19*, 594–601.

Lladser, A., Sanhueza, C., Kiessling, R., & Quest, A. F. (2011). Is survivin the potential Achilles' heel of cancer? *Advances in Cancer Research*, *111*, 1–37.

Lohr, F., Hu, K., Haroon, Z., Samulski, T. V., Huang, Q., Beaty, J., et al. (2000). Combination treatment of murine tumors by adenovirus-mediated local B7/IL12 immunotherapy and radiotherapy. *Molecular Therapy*, *2*, 195–203.

Lukacher, A. E. (2002). IFN-gamma suspends the killing license of anti-tumor CTLs. *The Journal of Clinical Investigation*, *110*, 1407–1409.

Lupu, C. M., Eisenbach, C., Lupu, A. D., Kuefner, M. A., Hoyler, B., Stremmel, W., et al. (2007). Adenoviral B7-H3 therapy induces tumor specific immune responses and reduces secondary metastasis in a murine model of colon cancer. *Oncology Reports*, *18*, 745–748.

Lyons, A., Longfield, J., Kuschner, R., Straight, T., Binn, L., Seriwatana, J., et al. (2008). A double-blind, placebo-controlled study of the safety and immunogenicity of live, oral type 4 and type 7 adenovirus vaccines in adults. *Vaccine*, *26*, 2890–2898.

Lyons, M., Onion, D., Green, N. K., Aslan, K., Rajaratnam, R., Bazan-Peregrino, M., et al. (2006). Adenovirus type 5 interactions with human blood cells may compromise systemic delivery. *Molecular Therapy*, *14*, 118–128.

Maheshwari, R. K., Banerjee, D. K., Waechter, C. J., Olden, K., & Friedman, R. M. (1980). Interferon treatment inhibits glycosylation of a viral protein. *Nature*, *287*, 454–456.

Maheshwari, R. K., Husain, M. M., Attallah, A. M., & Friedman, R. M. (1983). Tunicamycin treatment inhibits the antiviral activity of interferon in mice. *Infection and Immunity*, *41*, 61–66.

Manickan, E., Smith, J. S., Tian, J., Eggerman, T. L., Lozier, J. N., Muller, J., et al. (2006). Rapid Kupffer cell death after intravenous injection of adenovirus vectors. *Molecular Therapy*, *13*, 108–117.

Mantwill, K., Kohler-Vargas, N., Bernshausen, A., Bieler, A., Lage, H., Kaszubiak, A., et al. (2006). Inhibition of the multidrug-resistant phenotype by targeting YB-1 with a conditionally oncolytic adenovirus: implications for combinatorial treatment regimen with chemotherapeutic agents. *Cancer Research*, *66*, 7195–7202.

Mast, T. C., Kierstead, L., Gupta, S. B., Nikas, A. A., Kallas, E. G., Novitsky, V., et al. (2010). International epidemiology of human pre-existing adenovirus (Ad) type-5, type-6, type-26 and type-36 neutralizing antibodies: Correlates of high Ad5 titers and implications for potential HIV vaccine trials. *Vaccine*, *28*, 950–957.

Mazumder, A., & Rosenberg, S. A. (1984). Successful immunotherapy of natural killer-resistant established pulmonary melanoma metastases by the intravenous adoptive transfer of syngeneic lymphocytes activated in vitro by interleukin 2. *The Journal of Experimental Medicine*, *159*, 495–507.

McLoughlin, J. M., McCarty, T. M., Cunningham, C., Clark, V., Senzer, N., Nemunaitis, J., et al. (2005). TNFerade, an adenovector carrying the transgene for human tumor necrosis factor alpha, for patients with advanced solid tumors: surgical experience and long-term follow-up. *Annals of Surgical Oncology*, *12*, 825–830.

Mizuguchi, H., Koizumi, N., Hosono, T., Ishii-Watabe, A., Uchida, E., Utoguchi, N., et al. (2002). CAR- or alphav integrin-binding ablated adenovirus vectors, but not fiber-modified vectors containing RGD peptide, do not change the systemic gene transfer properties in mice. *Gene Therapy*, *9*, 769–776.

Mond, J. J., & Brunswick, M. (1987). A role for IFN-gamma and NK cells in immune responses to T cell-regulated antigens types 1 and 2. *Immunological Reviews*, *99*, 105–118.

Mundt, A. J., Vijayakumar, S., Nemunaitis, J., Sandler, A., Schwartz, H., Hanna, N., et al. (2004). A Phase I trial of TNFerade biologic in patients with soft tissue sarcoma in the extremities. *Clinical Cancer Research*, *10*, 5747–5753.

Muruve, D. A., Barnes, M. J., Stillman, I. E., & Libermann, T. A. (1999). Adenoviral gene therapy leads to rapid induction of multiple chemokines and acute neutrophil-dependent hepatic injury in vivo. *Human Gene Therapy*, *10*, 965–976.

Muruve, D. A., Pétrilli, V., Zaiss, A. K., White, L. R., Clark, S. A., Ross, P. J., et al. (2008). The inflammasome recognizes cytosolic microbial and host DNA and triggers an innate immune response. *Nature*, *452*, 103–107.

Nakamura, T., Sato, K., & Hamada, H. (2003). Reduction of natural adenovirus tropism to the liver by both ablation of fiber-coxsackievirus and adenovirus receptor interaction and use of replaceable short fiber. *Journal of Virology*, *77*, 2512–2521.

Narumi, K., Kondoh, A., Udagawa, T., Hara, H., Goto, N., Ikarashi, Y., et al. (2010). Administration route-dependent induction of antitumor immunity by interferon-alpha gene transfer. *Cancer Science*, *101*, 1686–1694.

Nasu, Y., Ebara, S., & Kumon, H. (2004). Adenovirus-mediated interleukin-12 gene therapy for prostate cancer. *Nippon Rinsho*, *62*, 1181–1191.

Nayak, S., & Herzog, R. W. (2010). Progress and prospects: Immune responses to viral vectors. *Gene Therapy*, *17*, 295–304.

Nemunaitis, J., Cunningham, C., Tong, A. W., Post, L., Netto, G., Paulson, A. S., et al. (2003). Pilot trial of intravenous infusion of a replication-selective adenovirus (ONYX-015) in combination with chemotherapy or IL-2 treatment in refractory cancer patients. *Cancer Gene Therapy*, *10*, 341–352.

Nemunaitis, J., Ganly, I., Khuri, F., Arseneau, J., Kuhn, J., McCarty, T., et al. (2000). Selective replication and oncolysis in p53 mutant tumors with ONYX-015, an E1B-55kD gene-deleted adenovirus, in patients with advanced head and neck cancer: a phase II trial. *Cancer Research*, *60*, 6359–6366.

Nemunaitis, J., Khuri, F., Ganly, I., Arseneau, J., Posner, M., Vokes, E., et al. (2001). Phase II trial of intratumoral administration of ONYX-015, a replication-selective adenovirus, in patients with refractory head and neck cancer. *Journal of Clinical Oncology*, *19*, 289–298.

Nemunaitis, J., Meyers, T., Senzer, N., Cunningham, C., West, H., Vallieres, E., et al. (2006). Phase I trial of sequential administration of recombinant DNA and adenovirus expressing L523S protein in early stage non-small-cell lung cancer. *Molecular Therapy*, *13*, 1185–1191.

Nemunaitis, J., Senzer, N., Cunningham, C., & Dubensky, T. W. (2001). Virus-mediated killing of cells that lack p53 activity. *Drug Resistance Updates*, *4*, 289–291.
Nemunaitis, J., Senzer, N., Sarmiento, S., Zhang, Y. A., Arzaga, R., Sands, B., et al. (2007). A phase I trial of intravenous infusion of ONYX-015 and enbrel in solid tumor patients. *Cancer Gene Therapy*, *14*, 885–893.
Nemunaitis, J., Tong, A., Nemunaitis, M., Senzer, N., Phadke, A. P., Chen, S., et al. (2009). Phase I study of intratumoral injection with telomerase specific replication competent oncolytic adenovirus, telomelysin (OBP-301) in advanced cancer. *Molecular Therapy*, *17*, S283.
Nemunaitis, J., Vorhies, J. S., Pappen, B., & Senzer, N. (2007). 10-year follow-up of gene-modified adenoviral-based therapy in 146 non-small-cell lung cancer patients. *Cancer Gene Therapy*, *14*, 762–763.
Nociari, M., Ocheretina, O., Schoggins, J. W., & Falck-Pedersen, E. (2007). Sensing infection by adenovirus: Toll-like receptor-independent viral DNA recognition signals activation of the interferon regulatory factor 3 master regulator. *Journal of Virology*, *81*, 4145–4157.
Nokisalmi, P., Pesonen, S., Escutenaire, S., Sarkioja, M., Raki, M., Cerullo, V., et al. (2010). Oncolytic adenovirus ICOVIR-7 in patients with advanced and refractory solid tumors. *Clinical Cancer Research*, *16*, 3035–3043.
Nowak, A. K., Lake, R. A., Marzo, A. L., Scott, B., Heath, W. R., Collins, E. J., et al. (2003). Induction of tumor cell apoptosis in vivo increases tumor antigen cross-presentation, cross-priming rather than cross-tolerizing host tumor-specific CD8 T cells. *Journal of Immunology*, *170*, 4905–4913.
Nunes, F. A., Furth, E. E., Wilson, J. M., & Raper, S. E. (1999). Gene transfer into the liver of nonhuman primates with E1-deleted recombinant adenoviral vectors: safety of readministration. *Human Gene Therapy*, *10*, 2515–2526.
Odaka, M., Sterman, D. H., Wiewrodt, R., Zhang, Y., Kiefer, M., Amin, K. M., et al. (2001). Eradication of intraperitoneal and distant tumor by adenovirus-mediated interferon-beta gene therapy is attributable to induction of systemic immunity. *Cancer Research*, *61*, 6201–6212.
Odaka, M., Wiewrodt, R., DeLong, P., Tanaka, T., Zhang, Y., Kaiser, L., et al. (2002). Analysis of the immunologic response generated by Ad.IFN-beta during successful intraperitoneal tumor gene therapy. *Molecular Therapy*, *6*, 210–218.
Ohashi, M., Yoshida, K., Kushida, M., Miura, Y., Ohnami, S., Ikarashi, Y., et al. (2005). Adenovirus-mediated interferon alpha gene transfer induces regional direct cytotoxicity and possible systemic immunity against pancreatic cancer. *British Journal of Cancer*, *93*, 441–449.
O'Neill, L. A. (2008). When signaling pathways collide: Positive and negative regulation of toll-like receptor signal transduction. *Immunity*, *29*, 12–20.
Ono, H. A., Davydova, J. G., Adachi, Y., Takayama, K., Barker, S. D., Reynolds, P. N., et al. (2005). Promoter-controlled infectivity-enhanced conditionally replicative adenoviral vectors for the treatment of gastric cancer. *Journal of Gastroenterology*, *40*, 31–42.
Oppmann, B., Lesley, R., Blom, B., Timans, J. C., Xu, Y., Hunte, B., et al. (2000). Novel p19 protein engages IL-12p40 to form a cytokine, IL-23, with biological activities similar as well as distinct from IL-12. *Immunity*, *13*, 715–725.
Orend, G., Linkwitz, A., & Doerfler, W. (1994). Selective sites of adenovirus (foreign) DNA integration into the hamster genome: Changes in integration patterns. *Journal of Virology*, *68*, 187–194.
O'Shea, C. C., Johnson, L., Bagus, B., Choi, S., Nicholas, C., Shen, A., et al. (2004). Late viral RNA export, rather than p53 inactivation, determines ONYX-015 tumor selectivity. *Cancer Cell*, *6*, 611–623.

Othman, M., Labelle, A., Mazzetti, I., Elbatarny, H. S., & Lillicrap, D. (2007). Adenovirus-induced thrombocytopenia: The role of von Willebrand factor and P-selectin in mediating accelerated platelet clearance. *Blood*, *109*, 2832–2839.

Pacini, D. L., Dubovi, E. J., & Clyde, W. A., Jr. (1984). A new animal model for human respiratory tract disease due to adenovirus. *J Infect Dis*, *150*, 92–97.

Park, E. K., Bae, S. M., Kwak, S. Y., Lee, S. J., Kim, Y. W., & Han, C. H. (2008). Photodynamic therapy with recombinant adenovirus AdmIL-12 enhances anti-tumour therapy efficacy in human papillomavirus 16 (E6/E7) infected tumour model. *Immunology*, *124*, 461–468.

Park, M. Y., Kim, D. R., Jung, H. W., Yoon, H. I., Lee, J. H., & Lee, C. T. (2010). Genetic immunotherapy of lung cancer using conditionally replicating adenovirus and adenovirus-interferon-beta. *Cancer Gene Therapy*, *17*, 356–364.

Pesonen, S., Cerullo, V., Escutenaire, S., Raki, M., Kangasniemi, L., Nokisalmi, P., et al. (2010). Treatment of patients with advanced and refractory solid tumors with oncolytic adenoviruses Ad5-D24-RGD and Ad5-RGD-D24-GMCSF. *Molecular Therapy*, *18*, 1874–1884.

Pesonen, S., Diaconu, I., Cerullo, V., Escutenaire, S., Raki, M., Kangasniemi, L., et al. (2011). Integrin targeted oncolytic adenoviruses Ad5-D24-RGD and Ad5-RGD-D24-GMCSF for treatment of patients with advanced chemotherapy refractory solid tumors. *International Journal of Cancer*, *130*, 1937–1947.

Pesonen, S., Diaconu, I., Cerullo, V., Ranki, T., Kangasniemi, L., Escutenaire, S., et al. (2011). Chimeric oncolytic adenovirus Ad5/3-hTERT-E1A-CD40L for the treatment of advanced solid tumors: Preclinical and clinical evaluation. *Human Gene Therapy*, *22*, A120–A.

Pesonen, S., Nokisalmi, P., Escutenaire, S., Sarkioja, M., Raki, M., & Cerullo, V. (2010). Prolonged systemic circulation of chimeric oncolytic adenovirus Ad5/3-Cox2L-D24 in patients with metastatic and refractory solid tumors. *Gene Therapy*, *17*, 892–904.

Petrilli, V., Dostert, C., Muruve, D. A., & Tschopp, J. (2007). The inflammasome: A danger sensing complex triggering innate immunity. *Current Opinion in Immunology*, *19*, 615–622.

Prestwich, R. J., Harrington, K. J., Pandha, H. S., Vile, R. G., Melcher, A. A., & Errington, F. (2008). Oncolytic viruses: A novel form of immunotherapy. *Expert Review of Anticancer Therapy*, *8*, 1581–1588.

Prestwich, R. J., Harrington, K. J., Vile, R. G., & Melcher, A. A. (2008). Immunotherapeutic potential of oncolytic virotherapy. *The Lancet Oncology*, *9*, 610–612.

Prieto, J., Qian, C., Sangro, B., Melero, I., & Mazzolini, G. (2004). Biologic therapy of liver tumors. *The Surgical Clinics of North America*, *84*, 673–696.

Qi, X., Chang, Z., Song, J., Gao, G., & Shen, Z. (2011). Adenovirus-mediated p53 gene therapy reverses resistance of breast cancer cells to adriamycin. *Anti-Cancer Drugs*, *22*, 556–562.

Qin, X. Q., Beckham, C., Brown, J. L., Lukashev, M., & Barsoum, J. (2001). Human and mouse IFN-beta gene therapy exhibits different anti-tumor mechanisms in mouse models. *Molecular Therapy*, *4*, 356–364.

Raja Gabaglia, C., Diaz de Durana, Y., Graham, F. L., Gauldie, J., Sercarz, E. E., & Braciak, T. A. (2007). Attenuation of the glucocorticoid response during Ad5IL-12 adenovirus vector treatment enhances natural killer cell-mediated killing of MHC class I-negative LNCaP prostate tumors. *Cancer Research*, *67*, 2290–2297.

Rajecki, M., af Hallstrom, T., Hakkarainen, T., Nokisalmi, P., Hautaniemi, S., Nieminen, A. I., et al. (2009). Mre11 inhibition by oncolytic adenovirus associates with autophagy and underlies synergy with ionizing radiation. *International Journal of Cancer*, *125*, 2441–2449.

Rajecki, M., Kanerva, A., Stenman, U. H., Tenhunen, M., Kangasniemi, L., & Sarkioja, M. (2007). Treatment of prostate cancer with Ad5/3Delta24hCG allows non-invasive detection of the magnitude and persistence of virus replication in vivo. *Molecular Cancer Therapeutics*, *6*, 742–751.

Ramesh, N., Ge, Y., Ennist, D. L., Zhu, M., Mina, M., Ganesh, S., et al. (2006). CG0070, a conditionally replicating granulocyte macrophage colony-stimulating factor-armed oncolytic adenovirus for the treatment of bladder cancer. *Clinical Cancer Research*, *12*, 305–313.

Ranki, T., Sarkioja, M., Hakkarainen, T., von Smitten, K., Kanerva, A., & Hemminki, A. (2007). Systemic efficacy of oncolytic adenoviruses in imagable orthotopic models of hormone refractory metastatic breast cancer. *International Journal of Cancer*, *121*, 165–174.

Raper, S. E., Chirmule, N., Lee, F. S., Wivel, N. A., Bagg, A., Gao, G.-P., et al. (2003). Fatal systemic inflammatory response syndrome in a ornithine transcarbamylase deficient patient following adenoviral gene transfer. *Molecular Genetics and Metabolism*, *80*, 148–158.

Reay, J., Kim, S. H., Lockhart, E., Kolls, J., & Robbins, P. D. (2009). Adenoviral-mediated, intratumor gene transfer of interleukin 23 induces a therapeutic antitumor response. *Cancer Gene Therapy*, *16*, 776–785.

Reid, T., Galanis, E., Abbruzzese, J., Sze, D., Wein, L. M., Andrews, J., et al. (2002). Hepatic arterial infusion of a replication-selective oncolytic adenovirus (dl1520): Phase II viral, immunologic, and clinical endpoints. *Cancer Research*, *62*, 6070–6079.

Rivera, A. A., Wang, M., Suzuki, K., Uil, T. G., Krasnykh, V., Curiel, D. T., et al. (2004). Mode of transgene expression after fusion to early or late viral genes of a conditionally replicating adenovirus via an optimized internal ribosome entry site in vitro and in vivo. *Virology*, *320*, 121–134.

Robinson, M., Ge, Y., Ko, D., Yendluri, S., Laflamme, G., Hawkins, L., et al. (2008). Comparison of the E3 and L3 regions for arming oncolytic adenoviruses to achieve a high level of tumor-specific transgene expression. *Cancer Gene Therapy*, *15*, 9–17.

Rodriguez, R., Schuur, E. R., Lim, H. Y., Henderson, G. A., Simons, J. W., & Henderson, D. R. (1997). Prostate attenuated replication competent adenovirus (ARCA) CN706: A selective cytotoxic for prostate-specific antigen-positive prostate cancer cells. *Cancer Research*, *57*, 2559–2563.

Rodriguez-Rocha, H., Gomez-Gutierrez, J. G., Garcia-Garcia, A., Rao, X. M., Chen, L., McMasters, K. M., et al. (2011). Adenoviruses induce autophagy to promote virus replication and oncolysis. *Virology*, *416*, 9–15.

Rojas, J. J., Guedan, S., Searle, P. F., Martinez-Quintanilla, J., Gil-Hoyos, R., Alcayaga-Miranda, F., et al. (2010). Minimal RB-responsive E1A promoter modification to attain potency, selectivity, and transgene-arming capacity in oncolytic adenoviruses. *Molecular Therapy*, *18*, 1960–1971.

Rowan, K. (2010). Oncolytic viruses move forward in clinical trials. *Journal of the National Cancer Institute*, *102*, 590–595.

Rowe, W. P., Huebner, R. J., Gilmore, L. K., Parrott, R. H., & Ward, T. G. (1953). Isolation of a cytopathogenic agent from human adenoids undergoing spontaneous degeneration in tissue culture. *Proceedings of the Society for Experimental Biology and Medicine*, *84*, 570–573.

Russell, W. C. (2000). Update on adenovirus and its vectors. *The Journal of General Virology*, *81*, 2573–2604.

Rux, J. J., & Burnett, R. M. (2004). Adenovirus structure. *Human Gene Therapy*, *15*, 1167–1176.

Sandberg, L., Papareddy, P., Silver, J., Bergh, A., & Mei, Y. F. (2009). Replication-competent Ad11p vector (RCAd11p) efficiently transduces and replicates in hormone-refractory metastatic prostate cancer cells. *Human Gene Therapy*, *20*, 361–373.

Sangro, B., Mazzolini, G., Ruiz, J., Herraiz, M., Quiroga, J., Herrero, I., et al. (2004). Phase I trial of intratumoral injection of an adenovirus encoding interleukin-12 for advanced digestive tumors. *Journal of Clinical Oncology, 22,* 1389–1397.

Santodonato, L., Ferrantini, M., Palombo, F., Aurisicchio, L., Delmastro, P., La Monica, N., et al. (2001). Antitumor activity of recombinant adenoviral vectors expressing murine IFN-alpha in mice injected with metastatic IFN-resistant tumor cells. *Cancer Gene Therapy, 8,* 63–72.

Sari Pesonen, I. D., Cerullo, V., Ranki, T., Kangasniemi, L., Escutenaire, S., Kanerva, A., et al. (2011). Chimeric oncolytic adenovirus Ad5/3-hTERT-CD40L for the treatment of advanced solid tumors: Assessment of safety and immunological responses in patients. In: *6th International Conference on Oncolytic Viruses as Cancer Therapeutics* (p. 84), Las Vegas, NV: Mayo Clinic.

Sarkar, D., Lebedeva, I. V., Su, Z. Z., Park, E. S., Chatman, L., Vozhilla, N., et al. (2007). Eradication of therapy-resistant human prostate tumors using a cancer terminator virus. *Cancer Research, 67,* 5434–5442.

Sarkar, D., Su, Z. Z., & Fisher, P. B. (2006). Unique conditionally replication competent bipartite adenoviruses-cancer terminator viruses (CTV): Efficacious reagents for cancer gene therapy. *Cell Cycle, 5,* 1531–1536.

Sarkar, D., Su, Z. Z., Park, E. S., Vozhilla, N., Dent, P., Curiel, D. T., et al. (2008). A cancer terminator virus eradicates both primary and distant human melanomas. *Cancer Gene Therapy, 15,* 293–302.

Sarkar, D., Su, Z. Z., Vozhilla, N., Park, E. S., Gupta, P., & Fisher, P. B. (2005). Dual cancer-specific targeting strategy cures primary and distant breast carcinomas in nude mice. *Proceedings of the National Academy of Sciences of the United States of America, 102,* 14034–14039.

Sarkar, D., Su, Z. Z., Vozhilla, N., Park, E. S., Randolph, A., Valerie, K., et al. (2005). Targeted virus replication plus immunotherapy eradicates primary and distant pancreatic tumors in nude mice. *Cancer Research, 65,* 9056–9063.

Satoh, T., Saika, T., Ebara, S., Kusaka, N., Timme, T. L., Yang, G., et al. (2003). Macrophages transduced with an adenoviral vector expressing interleukin 12 suppress tumor growth and metastasis in a preclinical metastatic prostate cancer model. *Cancer Research, 63,* 7853–7860.

Schuster, M., Nechansky, A., & Kircheis, R. (2006). Cancer immunotherapy. *Biotechnology Journal, 1,* 138–147.

Seiler, M. P., Cerullo, V., & Lee, B. (2007). Immune response to helper dependent adenoviral mediated liver gene therapy: Challenges and prospects. *Current Gene Therapy, 7,* 297–305.

Seiler, M. P., Gottschalk, S., Cerullo, V., Ratnayake, M., Mane, V. P., Clarke, C., et al. (2007). Dendritic cell function after gene transfer with adenovirus-calcium phosphate co-precipitates. *Molecular Therapy, 15,* 386–392.

Serafini, P., Carbley, R., Noonan, K. A., Tan, G., Bronte, V., & Borrello, I. (2004). High-dose granulocyte-macrophage colony-stimulating factor-producing vaccines impair the immune response through the recruitment of myeloid suppressor cells. *Cancer Research, 64,* 6337–6343.

Seregin, S. S., Aldhamen, Y. A., Appledorn, D. M., Hartman, Z. C., Schuldt, N. J., Scott, J., et al. (2010). Adenovirus capsid-display of the retro-oriented human complement inhibitor DAF reduces Ad vector-triggered immune responses in vitro and in vivo. *Blood, 116,* 1669–1677.

Shaw, M. H., Reimer, T., Kim, Y. G., & Nunez, G. (2008). NOD-like receptors (NLRs): Bona fide intracellular microbial sensors. *Current Opinion in Immunology, 20,* 377–382.

Shayakhmetov, D. M., Gaggar, A., Ni, S., Li, Z. Y., & Lieber, A. (2005). Adenovirus binding to blood factors results in liver cell infection and hepatotoxicity. *Journal of Virology, 79,* 7478–7491.

Shenoy, A. R., Kim, B. H., Choi, H. P., Matsuzawa, T., Tiwari, S., & MacMicking, J. D. (2007). Emerging themes in IFN-gamma-induced macrophage immunity by the p47 and p65 GTPase families. *Immunobiology*, *212*, 771–784.

Shimamura, M., & Morishita, R. (2011). Naked plasmid DNA for gene therapy. *Current Gene Therapy*, *11*, 433.

Smith, T. A., Idamakanti, N., Rollence, M. L., Marshall-Neff, J., Kim, J., Mulgrew, K., et al. (2003). Adenovirus serotype 5 fiber shaft influences in vivo gene transfer in mice. *Human Gene Therapy*, *14*, 777–787.

Sorensen, M. R., Holst, P. J., Steffensen, M. A., Christensen, J. P., & Thomsen, A. R. (2010). Adenoviral vaccination combined with CD40 stimulation and CTLA-4 blockage can lead to complete tumor regression in a murine melanoma model. *Vaccine*, *28*, 6757–6764.

Spencer, J. F., Sagartz, J. E., Wold, W. S., & Toth, K. (2009). New pancreatic carcinoma model for studying oncolytic adenoviruses in the permissive Syrian hamster. *Cancer Gene Therapy*, *16*, 912–922.

Sterman, D. H., Gillespie, C. T., Carroll, R. G., Coughlin, C. M., Lord, E. M., Sun, J., et al. (2006). Interferon beta adenoviral gene therapy in a patient with ovarian cancer. *Nature Clinical Practice. Oncology*, *3*, 633–639.

Sterman, D. H., Recio, A., Carroll, R. G., Gillespie, C. T., Haas, A., Vachani, A., et al. (2007). A phase I clinical trial of single-dose intrapleural IFN-beta gene transfer for malignant pleural mesothelioma and metastatic pleural effusions: high rate of antitumor immune responses. *Clinical Cancer Research*, *13*, 4456–4466.

Sterman, D. H., Recio, A., Haas, A. R., Vachani, A., Katz, S. I., Gillespie, C. T., et al. (2010). A phase I trial of repeated intrapleural adenoviral-mediated interferon-beta gene transfer for mesothelioma and metastatic pleural effusions. *Molecular Therapy*, *18*, 852–860.

Stone, D., Liu, Y., Shayakhmetov, D., Li, Z. Y., Ni, S., & Lieber, A. (2007). Adenovirus-platelet interaction in blood causes virus sequestration to the reticuloendothelial system of the liver. *Journal of Virology*, *81*, 4866–4871.

Sun, M., Richards, S., Prasad, D. V., Mai, X. M., Rudensky, A., & Dong, C. (2002). Characterization of mouse and human B7-H3 genes. *Journal of Immunology*, *168*, 6294–6297.

Suzuki, M., Cela, R., Bertin, T. K., Sule, G., Cerullo, V., Rodgers, J. R., et al. (2011). NOD2 signaling contributes to the innate immune response against helper-dependent adenovirus vectors independently of MyD88 in vivo. *Human Gene Therapy*, *22*, 1071–1082.

Suzuki, M., Cerullo, V., Bertin, T., Cela, R., Clarke, C., Guenther, M., et al. (2009). MyD88-dependent silencing of transgene expression during the innate and adaptive immune response to Helper-dependent adenovirus. *Human Gene Therapy*, *21*, 325–336.

Suzuki, M., Cerullo, V., Bertin, T. K., Cela, R., Clarke, C., Guenther, M., et al. (2010). MyD88-dependent silencing of transgene expression during the innate and adaptive immune response to helper-dependent adenovirus. *Human Gene Therapy*, *21*, 325–336.

Thomas, D. L., & Fraser, N. W. (2003). HSV-1 therapy of primary tumors reduces the number of metastases in an immune-competent model of metastatic breast cancer. *Molecular Therapy*, *8*, 543–551.

Thomas, M. A., Spencer, J. F., La Regina, M. C., Dhar, D., Tollefson, A. E., Toth, K., et al. (2006). Syrian hamster as a permissive immunocompetent animal model for the study of oncolytic adenovirus vectors. *Cancer Research*, *66*, 1270–1276.

Tollefson, A. E., Scaria, A., Hermiston, T. W., Ryerse, J. S., Wold, L. J., & Wold, W. S. (1996). The adenovirus death protein (E3-11.6K) is required at very late stages of infection for efficient cell lysis and release of adenovirus from infected cells. *Journal of Virology*, *70*, 2296–2306.

Tong, A. W., Nemunaitis, J., Su, D., Zhang, Y., Cunningham, C., Senzer, N., et al. (2005). Intratumoral injection of INGN 241, a nonreplicating adenovector expressing the melanoma-differentiation associated gene-7 (mda-7/IL24): Biologic outcome in advanced cancer patients. *Molecular Therapy*, *11*, 160–172.

Toth, K., Spencer, J. F., Tollefson, A. E., Kuppuswamy, M., Doronin, K., Lichtenstein, D. L., et al. (2005). Cotton rat tumor model for the evaluation of oncolytic adenoviruses. *Human Gene Therapy*, *16*, 139–146.

Traversari, C., van der Bruggen, P., Luescher, I. F., Lurquin, C., Chomez, P., Van Pel, A., et al. (1992). A nonapeptide encoded by human gene MAGE-1 is recognized on HLA-A1 by cytolytic T lymphocytes directed against tumor antigen MZ2-E. *The Journal of Experimental Medicine*, *176*, 1453–1457.

Trinchieri, G., Aden, D. P., & Knowles, B. B. (1976). Cell-mediated cytotoxicity to SV40-specific tumour-associated antigens. *Nature*, *261*, 312–314.

Tsai, V., Varghese, R., Ravindran, S., Ralston, R., & Vellekamp, G. (2008). Complement component C1q and anti-hexon antibody mediate adenovirus infection of a CAR-negative cell line. *Viral Immunology*, *21*, 469–476.

Tuve, S., Liu, Y., Tragoolpua, K., Jacobs, J. D., Yumul, R. C., Li, Z. Y., et al. (2009). In situ adenovirus vaccination engages T effector cells against cancer. *Vaccine*, *27*, 4225–4239.

Tuve, S., Wang, H., Ware, C., Liu, Y., Gaggar, A., & Bernt, K. (2006). A new group B adenovirus receptor is expressed at high levels on human stem and tumor cells. *Journal of Virology*, *80*, 12109–12120.

Vaha-Koskela, M. J., Heikkila, J. E., & Hinkkanen, A. E. (2007). Oncolytic viruses in cancer therapy. *Cancer Letters*, *254*, 178–216.

van Beusechem, V. W., van den Doel, P. B., Grill, J., Pinedo, H. M., & Gerritsen, W. R. (2002). Conditionally replicative adenovirus expressing p53 exhibits enhanced oncolytic potency. *Cancer Research*, *62*, 6165–6171.

van der Bruggen, P., Traversari, C., Chomez, P., Lurquin, C., De Plaen, E., Van den Eynde, B., et al. (1991). A gene encoding an antigen recognized by cytolytic T lymphocytes on a human melanoma. *Science*, *254*, 1643–1647.

VanOosten, R. L., & Griffith, T. S. (2007). Activation of tumor-specific CD8+ T Cells after intratumoral Ad5-TRAIL/CpG oligodeoxynucleotide combination therapy. *Cancer Research*, *67*, 11980–11990.

Veltrop-Duits, L. A., van Vreeswijk, T., Heemskerk, B., Thijssen, J. C., El Seady, R., Jol-van der Zijde, E. M., et al. (2011). High titers of pre-existing adenovirus serotype-specific neutralizing antibodies in the host predict viral reactivation after allogeneic stem cell transplantation in children. *Clinical Infectious Diseases*, *52*, 1405–1413.

Vigne, E., Mahfouz, I., Dedieu, J. F., Brie, A., Perricaudet, M., & Yeh, P. (1999). RGD inclusion in the hexon monomer provides adenovirus type 5-based vectors with a fiber knob-independent pathway for infection. *Journal of Virology*, *73*, 5156–5161.

Volk, A. L., Rivera, A. A., Kanerva, A., Bauerschmitz, G., Dmitriev, I., Nettelbeck, D. M., et al. (2003). Enhanced adenovirus infection of melanoma cells by fiber-modification: Incorporation of RGD peptide or Ad5/3 chimerism. *Cancer Biology & Therapy*, *2*, 511–515.

Volpers, C., & Kochanek, S. (2004). Adenoviral vectors for gene transfer and therapy. *The Journal of Gene Medicine*, *6*(Suppl. 1), S164–S171.

Waddington, S. N., McVey, J. H., Bhella, D., Parker, A. L., Barker, K., Atoda, H., et al. (2008). Adenovirus serotype 5 hexon mediates liver gene transfer. *Cell*, *132*, 397–409.

Wang, K., Huang, S., Kapoor-Munshi, A., & Nemerow, G. (1998). Adenovirus internalization and infection require dynamin. *Journal of Virology*, *72*, 3455–3458.

Wang, E., Panelli, M., & Marincola, F. M. (2006). Autologous tumor rejection in humans: Trimming the myths. *Immunological Investigations*, *35*, 437–458.

Wang, E., Selleri, S., Sabatino, M., Monaco, A., Pos, Z., Worschech, A., et al. (2008). Spontaneous and treatment-induced cancer rejection in humans. *Expert Opinion on Biological Therapy*, *8*, 337–349.

Webster, A., Russell, S., Talbot, P., Russell, W. C., & Kemp, G. D. (1989). Characterization of the adenovirus proteinase: Substrate specificity. *The Journal of General Virology*, *70* (Pt 12), 3225–3234.

Wen, X. Y., Mandelbaum, S., Li, Z. H., Hitt, M., Graham, F. L., Hawley, T. S., et al. (2001). Tricistronic viral vectors co-expressing interleukin-12 (1L-12) and CD80 (B7-1) for the immunotherapy of cancer: Preclinical studies in myeloma. *Cancer Gene Therapy*, *8*, 361–370.

Wheeler, M. D., Yamashina, S., Froh, M., Rusyn, I., & Thurman, R. G. (2001). Adenoviral gene delivery can inactivate Kupffer cells: Role of oxidants in NF-kappaB activation and cytokine production. *Journal of Leukocyte Biology*, *69*, 622–630.

Wickham, T. J., Carrion, M. E., & Kovesdi, I. (1995). Targeting of adenovirus penton base to new receptors through replacement of its RGD motif with other receptor-specific peptide motifs. *Gene Therapy*, *2*, 750–756.

Wickham, T. J., Roelvink, P. W., Brough, D. E., & Kovesdi, I. (1996). Adenovirus targeted to heparan-containing receptors increases its gene delivery efficiency to multiple cell types. *Nature Biotechnology*, *14*, 1570–1573.

Wickham, T. J., Segal, D. M., Roelvink, P. W., Carrion, M. E., Lizonova, A., Lee, G. M., et al. (1996). Targeted adenovirus gene transfer to endothelial and smooth muscle cells by using bispecific antibodies. *Journal of Virology*, *70*, 6831–6838.

Wiemann, B., & Starnes, C. O. (1994). Coley's toxins, tumor necrosis factor and cancer research: A historical perspective. *Pharmacology & Therapeutics*, *64*, 529–564.

Wiethoff, C. M., Wodrich, H., Gerace, L., & Nemerow, G. R. (2005). Adenovirus protein VI mediates membrane disruption following capsid disassembly. *Journal of Virology*, *79*, 1992–2000.

Willimsky, G., & Blankenstein, T. (2000). Interleukin-7/B7.1-encoding adenoviruses induce rejection of transplanted but not nontransplanted tumors. *Cancer Research*, *60*, 685–692.

Wirth, T., Kuhnel, F., & Kubicka, S. (2005). Telomerase-dependent gene therapy. *Current Molecular Medicine*, *5*, 243–251.

Wold, W. S. M., & Horwitz, M. S. (2007). *Adenoviruses* (5th ed.). Philadelphia: Lippincott-Raven.

Wu, H., Han, T., Belousova, N., Krasnykh, V., Kashentseva, E., Dmitriev, I., et al. (2005). Identification of sites in adenovirus hexon for foreign peptide incorporation. *Journal of Virology*, *79*, 3382–3390.

Wu, E., Pache, L., Von Seggern, D. J., Mullen, T. M., Mikyas, Y., Stewart, P. L., et al. (2003). Flexibility of the adenovirus fiber is required for efficient receptor interaction. *Journal of Virology*, *77*, 7225–7235.

Xu, Z., Tian, J., Smith, J. S., & Byrnes, A. P. (2008). Clearance of adenovirus by Kupffer cells is mediated by scavenger receptors, natural antibodies, and complement. *Journal of Virology*, *82*, 11705–11713.

Yu, W., & Fang, H. (2007). Clinical trials with oncolytic adenovirus in China. *Current Cancer Drug Targets*, *7*, 141–148.

Zaiss, A. K., Liu, Q., Bowen, G. P., Wong, N. C., Bartlett, J. S., & Muruve, D. A. (2002). Differential activation of innate immune responses by adenovirus and adeno-associated virus vectors. *Journal of Virology*, *76*, 4580–4590.

Zhan, J., Gao, Y., Wang, W., Shen, A., Aspelund, A., Young, M., et al. (2005). Tumor-specific intravenous gene delivery using oncolytic adenoviruses. *Cancer Gene Therapy*, *12*, 19–25.

Zhang, L., Akbulut, H., Tang, Y., Peng, X., Pizzorno, G., Sapi, E., et al. (2002). Adenoviral vectors with E1A regulated by tumor-specific promoters are selectively cytolytic for breast cancer and melanoma. *Molecular Therapy*, *6*, 386–393.

Zhang, Y., Chirmule, N., Gao, G. P., Qian, R., Croyle, M., Joshi, B., et al. (2001). Acute cytokine response to systemic adenoviral vectors in mice is mediated by dendritic cells and macrophages. *Molecular Therapy*, *3*, 697–707.

Zhang, R., & DeGroot, L. J. (2003). Gene therapy of a rat follicular thyroid carcinoma model with adenoviral vectors transducing murine interleukin-12. *Endocrinology*, *144*, 1393–1398.

Zhang, X. Q., Dunner, K., Jr., & Benedict, W. F. (2010). Autophagy is induced by adenoviral-mediated interferon alpha treatment in interferon resistant bladder cancer and normal urothelial cells as a cell death protective mechanism but not by the bystander factors produced. *Cancer Gene Therapy*, *17*, 579–584.

Zhang, J. F., Hu, C., Geng, Y., Selm, J., Klein, S. B., Orazi, A., et al. (1996). Treatment of a human breast cancer xenograft with an adenovirus vector containing an interferon gene results in rapid regression due to viral oncolysis and gene therapy. *Proceedings of the National Academy of Sciences of the United States of America*, *93*, 4513–4518.

Zhang, F., Lu, W., & Dong, Z. (2002). Tumor-infiltrating macrophages are involved in suppressing growth and metastasis of human prostate cancer cells by INF-beta gene therapy in nude mice. *Clinical Cancer Research*, *8*, 2942–2951.

Zhang, J. F., Wei, F., Wang, H. P., Li, H. M., Qiu, W., Ren, P. K., et al. (2010). Potent anti-tumor activity of telomerase-dependent and HSV-TK armed oncolytic adenovirus for non-small cell lung cancer in vitro and in vivo. *Journal of Experimental & Clinical Cancer Research*, *29*, 52.

Zheng, S., Ulasov, I. V., Han, Y., Tyler, M. A., Zhu, Z. B., & Lesniak, M. S. (2007). Fiber-knob modifications enhance adenoviral tropism and gene transfer in malignant glioma. *The Journal of Gene Medicine*, *9*, 151–160.

Zhu, J., Huang, X., & Yang, Y. (2007). Innate immune response to adenoviral vectors is mediated by both Toll-like receptor-dependent and -independent pathways. *Journal of Virology*, *81*, 3170–3180.

# INDEX

Note: Page numbers followed by "*f*" indicate figures, and "*t*" indicate tables.

## C



## D



## E



## F

## G



## H



## I

## P



## R

## S



## T

## U



## V

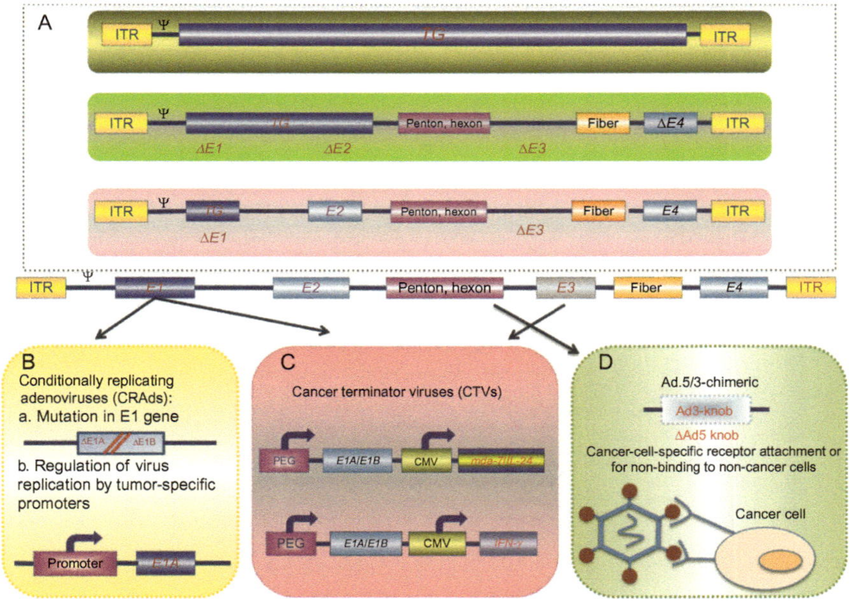


Figure 1.1, Swadesh K. Das *et al.* (See Page 6 of this volume.)

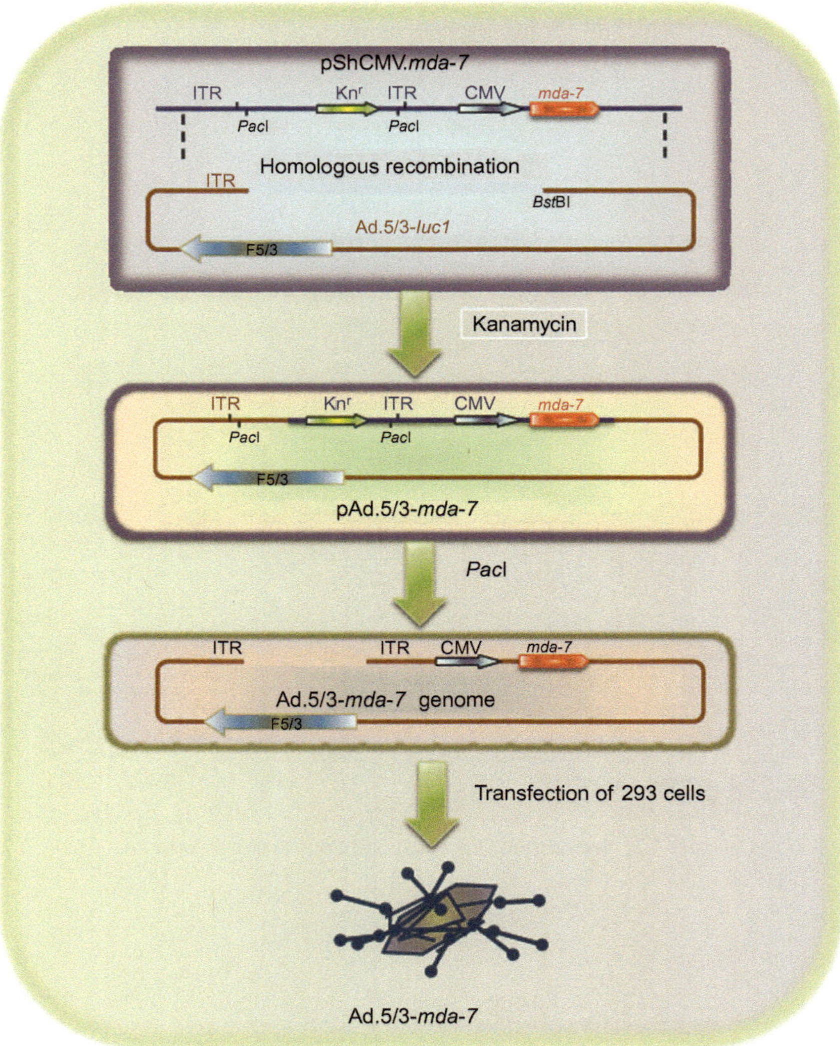


Figure 1.3, Swadesh K. Das *et al.* (See Page 23 of this volume.)

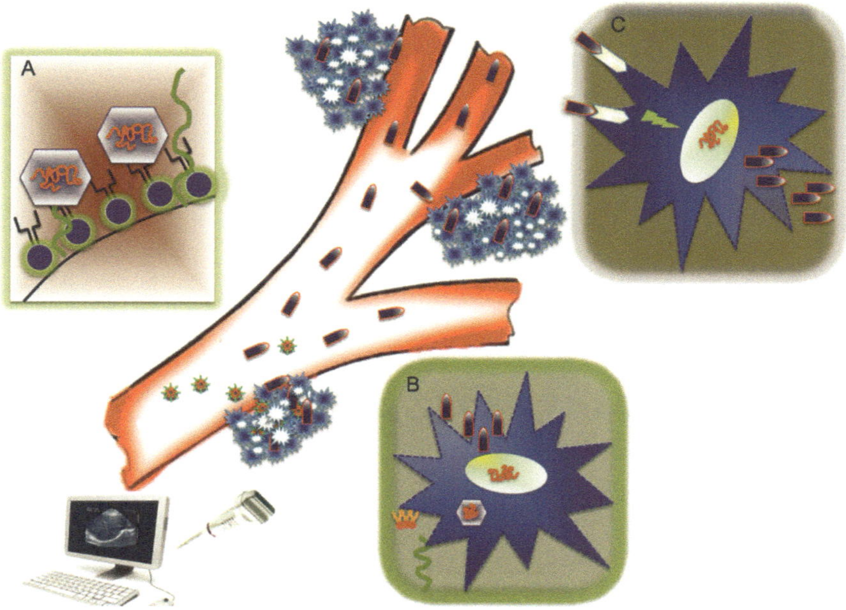


Figure 1.4, Swadesh K. Das *et al.* (See Page 27 of this volume.)

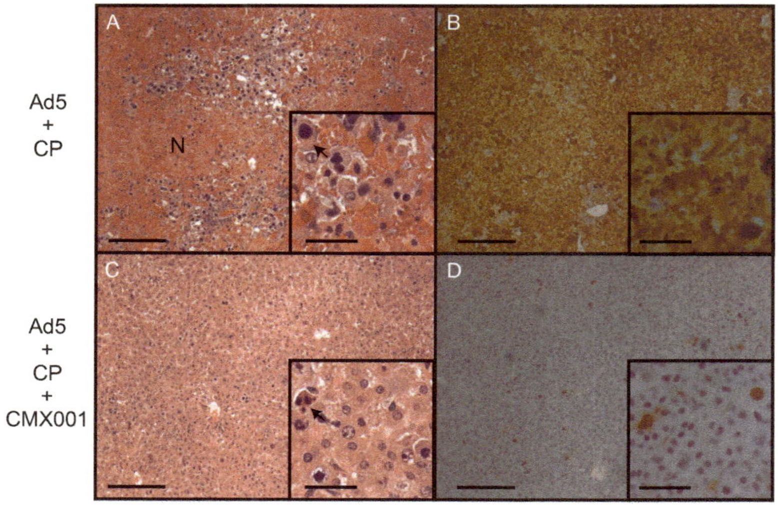


Figure 3.5, William S.M. Wold and Karoly Toth (See Page 85 of this volume.)

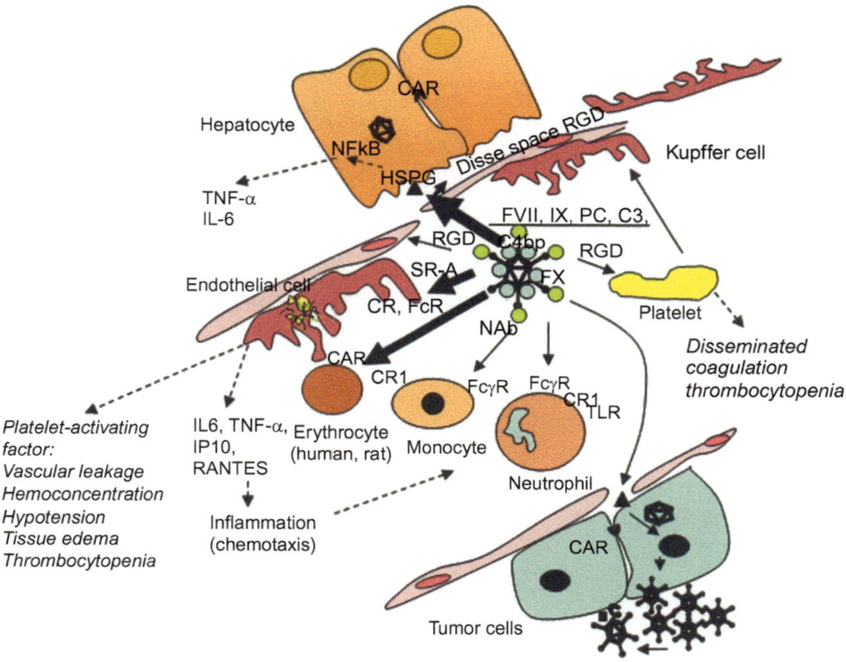


Figure 4.1, Ramon Alemany (See Page 97 of this volume.)

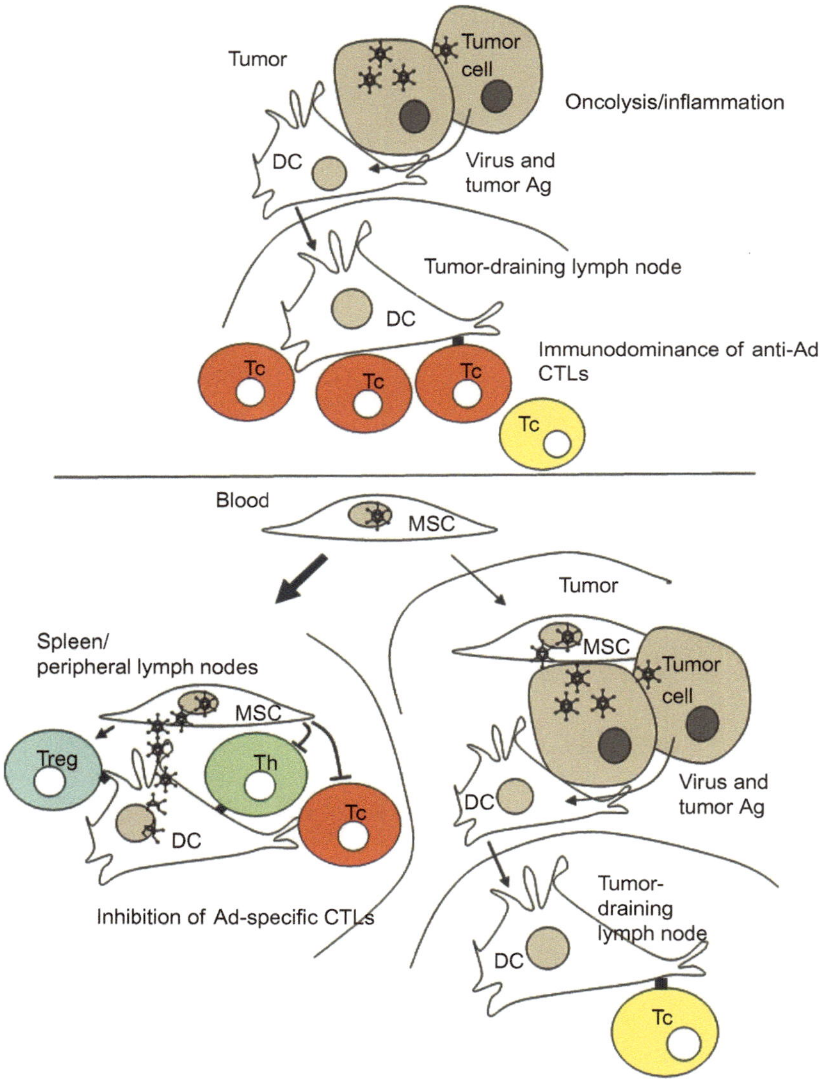


Figure 4.2, Ramon Alemany (See Page 109 of this volume.)

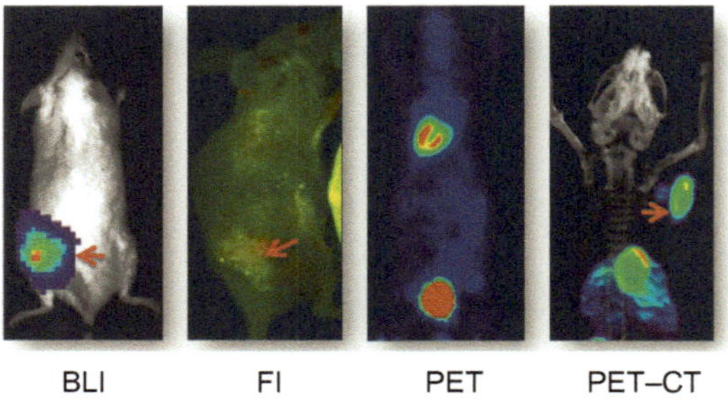


Figure 5.1, Ziyue Karen Jiang *et al.* (See Page 119 of this volume.)

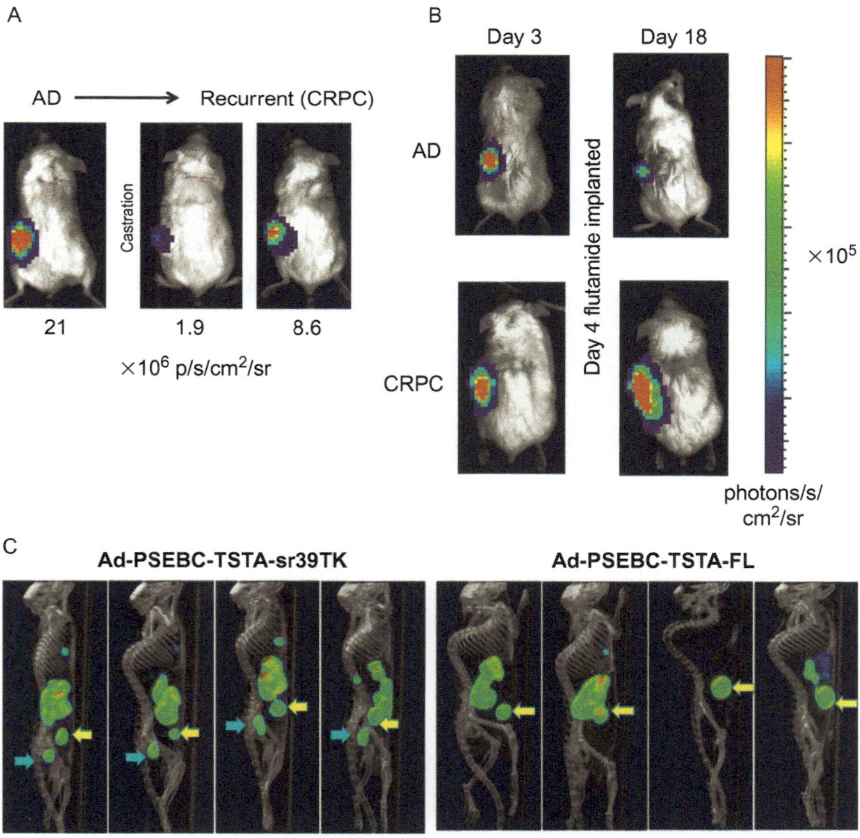


Figure 5.2, Ziyue Karen Jiang *et al.* (See Page 127 of this volume.)

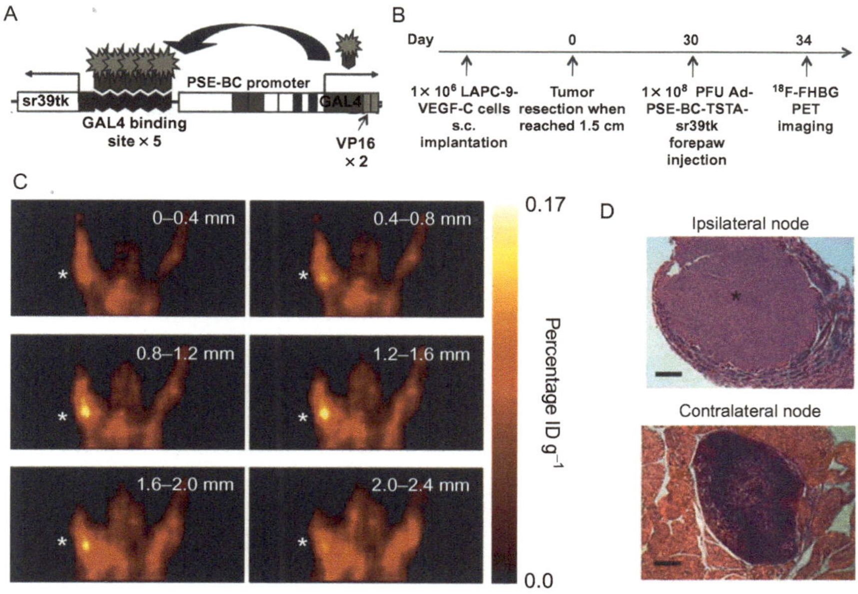


Figure 5.4, Ziyue Karen Jiang *et al.* (See Page 135 of this volume.)

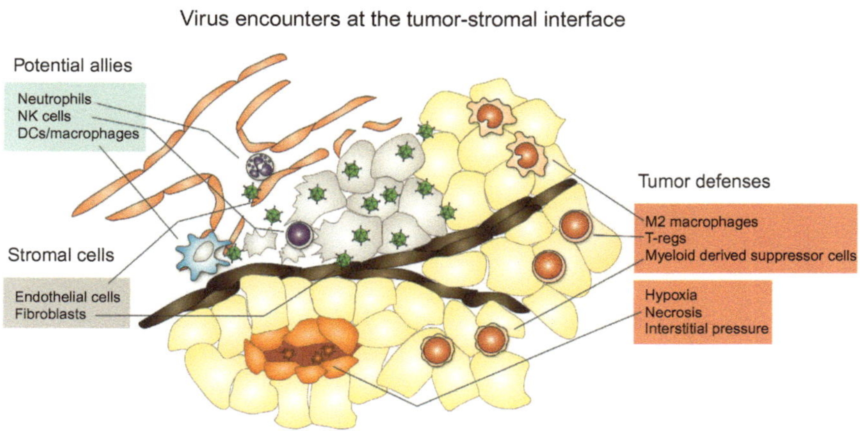


Figure 8.1, Vincenzo Cerullo *et al.* (See Page 268 of this volume.)

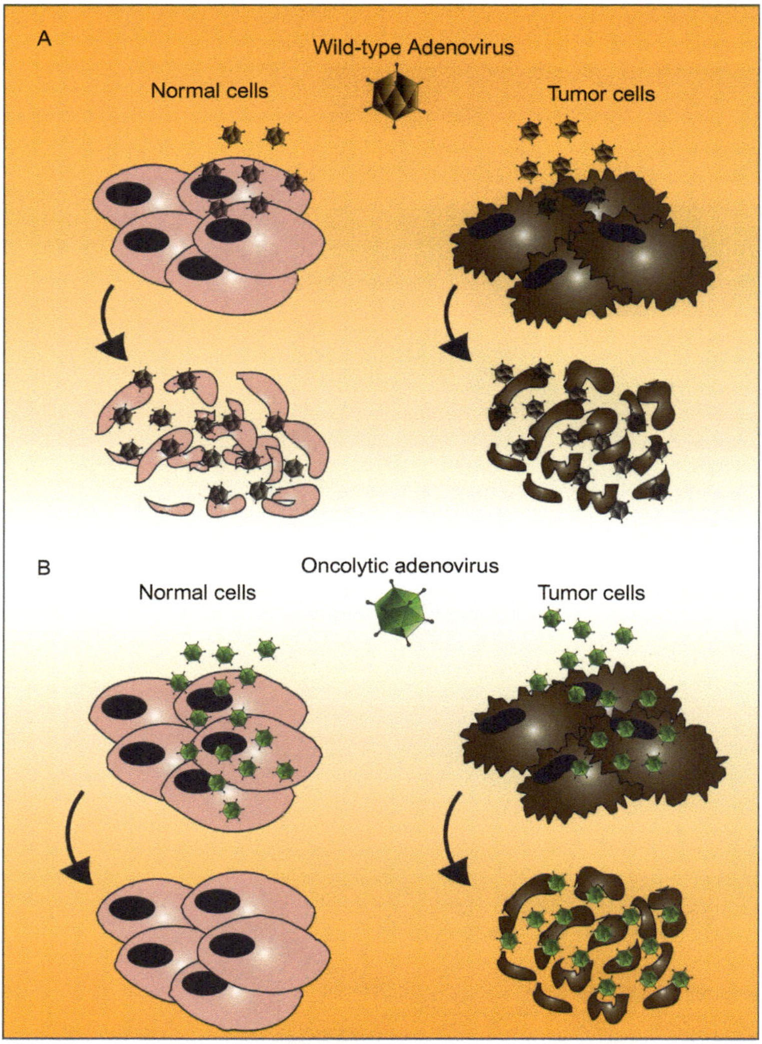


Figure 8.2, Vincenzo Cerullo *et al.* (See Page 273 of this volume.)

FSC
www.fsc.org
MIX
Papier | Fördert
gute Waldnutzung
FSC® C083411

Kolibri 360 GmbH
Ferdinand-Jühlke-Straße 7
99095 Erfurt, Deutschland
produktsicherheit@kolibri360.de